STUDIES ON WATER MANAGEMENT ISSUES

Edited by
Muthukrishnavellaisamy Kumarasamy

Studies on Water Management Issues
Edited by Muthukrishnavellaisamy Kumarasamy

Published by InTech
Janeza Trdine 9, 51000 Rijeka, Croatia

Publishing Process Manager Gorana Scerbe

Technical Editor Teodora Smiljanic

Cover Designer InTech Design Team

Additional hard copies can be obtained from orders@intechopen.com

Studies on Water Management Issues, Edited by Muthukrishnavellaisamy Kumarasamy
p. cm.
ISBN 978-953-307-961-5

For EU product safety concerns:
IN TECH d.o.o., Prolaz Marije Krucifikse Kozulić 3, 51000 Rijeka, Croatia,
info@intechopen.com or visit our website at intechopen.com.

Meet the editor

Dr. Muthukrishnavellaisamy Kumarasamy is a faculty member at School of Civil Engineering, Surveying and Construction, University of KwaZulu-Natal, Durban, South Africa. After acquiring his Bachelor's degree in civil engineering, he graduated with a Master's degree in Irrigation Water Management from Anna University, India. He received his Doctoral degree from Indian Institute of Technology Roorkee, India for his research work on pollutant transport studies and continued as a research associate to carry out sponsored research project on minimum flow requirement of river system for ecology. He has also worked in an affiliated state engineering institute. Presently, he is collaborating with a research team to carry out SANPAD sponsored ecosystem project for St. Lucia estuarine lake system in South Africa. He is actively involved in teaching and research activities at UKZN, South Africa. He has been awarded with various Indian fellowships from Ministry of HRD, Ministry of Water Resources and Council for Scientific and Industrial Research. He engages himself in professional activities as editor, editorial board member and a reviewer for various international journals.

Contents

Preface

Sustainability of ecosystem needs water which is a finite and vulnerable resource. Fortunately, water is a renewable resource. But management of water resources is always a challenge for sustainability development.

Rapid urbanization and industrialization has put more stress on water resources from increased water needs for agriculture, municipal water supply, navigation, recreation, industrial processes and ecosystem maintenance. Agriculture consumes about 70 % of world's freshwater. Industrial activities need 20 % and remaining 10 % is to be managed for water supply. Studies on human population warn that current population growth rate will lead to a demand of 6% more food and it is predicted to reach 9% in 30 years from now. In order to meet this food demand, agricultural and industrial activities will require more freshwater for agricultural production and food processes.

In the year 2007, the International Water Management Institute (IWMI), Sri Lanka conducted an assessment of agricultural water management to relate availability of freshwater, demand for food and growing population. The current availability of water for agriculture use around the world had been assessed and the assessment identified water scares sites. It found that one fifth (more than 1.2 billion) of the world's population live in very severe water shortage areas. These areas do not have adequate water to meet even all basic demands. Nearly about one and half billion people are living in areas with severe water scarcity. These areas have insufficient water resources to satisfy all demands. Although water scarcity has become an important issue in many parts of the world, studies say water scarcity may not be an urgent issue in rest of the world, but many countries realized that there are growing pressures on water availability and supplies and become conscious about water security which will be critical in future. Half of the world's population now lives in cities and towns. This number will probably increase which is expected nearly one sixth more by the year 2050. As a result there will be a huge demand for freshwater supply.

Multidirectional developmental activities, for improved quality of life, have increased the quantity of wastes. Thus there are threats due to huge amount of pollutants reaching fresh water sources and make them unfit for human consumption and ecosystem processes. Wastewater discharged from industrial and urban areas can contain a wide variety of pollutants. This means that there will be high pollution level due to huge amounts of heavy metals, salts, nutrients and pathogens. Unwisely, the fresh water resources are being exploited for assimilation of wastes beyond their self-cleaning capacities. Due to the least developments, many world countries have the lowest levels of wastewater treatment options; sometimes untreated wastes are being discharged. The farmers in those countries often use polluted water with pathogens for irrigating crops. The pathogens such as bacteria, viruses affect farmers and consumers in direct or indirect ways and kill more than a million people annually due to common illnesses, include diarrhea, cholera etc.

Too much water at the wrong time is as bad as too little water when it is required. Water availability is highly uneven in its spatial and temporal distributions both in quantity and quality. Thus, management of water resources must be viewed from local, regional and global perspectives to solve problems associated with storm water & flood management, water storage & distribution, surface water & groundwater quality, erosion & sedimentation, ecosystem protection & other natural resources etc.

People in the interdisciplinary water management field are concerned with supply of clean potable water to be available to people who need it. They also focus on water availability, supply and demand of industrial and the environmental water needs. Water management studies include a number of different topics such as sewage treatment, wetlands, soil erosion etc. Many countries have water resources departments. These departments are in charge for management of water resources and also control and direct regional offices of their own focus area of water management. One water management research group involves in researching and managing the water for the natural environment and ecosystem. This includes monitoring and analyzing the water availability for the environment, its spatial and temporal variations. Water management can also include studies on water pollution, navigation, environmental protection and flood control measures. This may include design, build and maintenance of flood control structures, restoring natural wetlands, constructing artificial wetlands and minimizing runoff from the developmental areas.

Some people in water management focus on human tendency in water use. They also focus on use of water in industry and agriculture. Many water management specialists say that it is very important to preserve water for conservation of ecosystem in many regions of the world; they also suggest recycling and reuse of treated effluent in agricultural and industrial sectors may reduce environmental impact. Thus water management is all about optimal use of water and recycle, reuse and minimal discharge of waste water which must have lowest environmental impact on the natural environment. Successful management of any resources needs accurate information of the resource available, demands for the resource, evaluation of significance and worth of the demands and good managerial policies to make decisions.

A study of water management lists out the principles of water management for people and the environment which include (i) value water; (ii) use water sustainably; (iii) develop water management institutions; (iv) maintain database on water; (v) maintain a social and cultural perspective; (vi) make sure water is accessible to everyone; (vii) use appropriate technology; (viii) act before the water related issues become severe; (ix) consider ecosystem sustainability; and (x) involve multidisciplinary teams for water related projects.

Water management project expert group may include civil engineers, water resources engineers, meteorologist, sanitation engineers, ecologists & environmentalists, biological scientists and policy/decision makers. Many studies are being carried out by researchers worldwide to gain knowledge and to develop technologies to manage water resources for sustainability. Thus, the scope of this book was influenced by the experience and knowledge gained by various experts who contributed to its preparation.

The subject of water management is too broad and has too many sub-themes to be covered to address water availability, water treatment, water distribution, waste water treatment & disposal, flood management and water quality and ecosystem protection. The objective of this book is therefore to present water management approaches under three different subsections, namely surface water & groundwater management, water quality management and water resources planning & management. The surface water & groundwater section aims to cover research on extreme rainfall, flood hazards, flow calculation, groundwater level dynamics & flow etc. The water quality section aims to cover research on water quality assessment, ecological tools & remediation and waste water treatment options. Finally, the water resources planning & management section aims to cover research works on planning and management of water resources, to protect water pollution, for sustainable drinking water and climate change impacts. It is believed that this interdisciplinary book on water management will benefit professionals involved in the field by exposing them to new methodologies & case studies and enable them to develop new approaches.

I am grateful to Dr. Barik D K, Catholic University of Daegu, Rep. of Korea, Dr. Saravanan R, Anna University, India, Dr. Suresh Babu, DHI, Singapore, Dr. Purushothaman, NIH, India, Dr. Loganathan G, Environmental Consultant, Oman, Dr. Samal K P, India for their help and support during review process of chapters. I also express my thanks to Prof. D stretch, University of KwaZulu-Natal, South Africa for his encouragement and support during the editorial process of this book. Finally, helps rendered by Ms. Gorana Scerbe, Publishing process manager, InTech Open Assess publisher and all other direct or indirect supports are much appreciated.

Muthukrishnavellaisamy Kumarasamy
School of Civil Engineering, Surveying & Construction,
University of KwaZulu-Natal,
Durban,
South Africa

Part 1

Surface Water and Groundwater Management

1

Comparing Extreme Rainfall and Large-Scale Flooding Induced Inundation Risk – Evidence from a Dutch Case-Study

E.E. Koks, H. de Moel and E. Koomen
VU University Amsterdam
The Nederlands

1. Introduction

Flood risk is an important force in shaping land use patterns. Attention for flood risk is even more important in view of climatic changes that will impact sea-level rise, river discharge and precipitation patterns. Flooding typically results from two types of events: extreme rainfall events and large-scale floods. The former can be defined as inundation due to more rainfall than the water system in a specific area can handle and the latter as a temporary covering of land by water outside its normal confines due to flooding or breaching of the primary or regional defense structures such as dikes.

In recent history, the Netherlands has seen a number of events with both extreme rainfall and large-scale flooding. For example, an extreme rainfall event occurred in 1998 that caused substantial damage in the southwestern part of the Netherlands (Smits et al., 2004) and a large-scale flood almost occurred in 1993 and 1995 (Wind et al., 1999). Extreme rainfall events generally have a high probability of occurrence but a low impact, while large-scale floods have a low probability but a high impact (Merz et al., 2009). It is, therefore, interesting to compare the flood risk for both types of events. Flood risk is interpreted here as the product of the probability of a certain flood event and the (economic) impact that the event would cause if it occurred (Sayer et al., 2002).

Recent studies have already made initial progress with comparing flood risk from large-scale floods and extreme rainfall events (van Veen, 2005; Merz et al., 2009). Other studies examined several aspects differing between extreme rainfall events and large-scale floods, including the regional differentiation in precipitation, communication, types of measures, possibilities to reduce flood risk and the possibility of insurance (Kok and Klopstra, 2009). It is still not known, however, how the flood risk (in terms of expected annual damage) of extreme rainfall exactly compares to the risk of large-scale floods. Since both inundation due to extreme rainfall and large floods from the sea or river can cause economic damage, it is interesting and valuable to calculate them in a consistent way in order to compare them.

At the moment, more policy measures are taken to mitigate and prevent large-scale floods in the Netherlands, than to mitigate extreme rainfall events (Kok and Klopstra, 2009). Due to climate change, there is the expectation that not only the occurrence of extreme weather events will increase in the future (IPCC, 2007), but also a possible increase in large-scale floods (Milly et al., 2002; Te Linde et al., 2010). This can result, in combination with various socio-economic changes, in an increase in economic damage from floods (Bouwer et al., 2010; De Moel et al., 2011). Therefore, the risk of both extreme rainfall events and large-scale floods are both likely to increase. In order to prioritize it is interesting to have both types of flood risk calculated in the same way.

The objective of this research is to assess the flood risk, in terms of 'annual expected damage' (AED) of inundation due to extreme rainfall and large floods in a consistent way in order to compare the respective types of risk. Therefore, it is interesting to know to what extent flood risk from extreme rainfall events and large-scale flooding can be compared and how they relate to each other. To do this, the main objective of this research is to make an integrated model to compare the different types of flood risk in a plausible and consistent way.

In section 2, the study area, 'Noord-Beveland', which will be used to test the model, will be briefly discussed and described. In section 3, we then discuss the different conditions that must be taken into account to be able to put the comparison of the different forms of flood risk in a proper perspective. In section 4, the methodology will be described that was used to make to integrated flood risk model. In section 5, the results will be described after using the data from the study area as input for the integrated flood risk model. In section 6, the results will be discussed to see whether or not the model matches our expectations. Finally, in section 7, conclusions will be drawn.

2. Study area

The area of focus of this chapter is the Netherlands, which is located in the western part of Europe. The country borders to the North Sea in the west. The general low altitude in the Netherlands and therefore potentially high risk of flooding has stimulated the development of an extensive network of dunes and dike-rings. Currently, the Netherlands consists of 57 dike-rings, varying from safety norms between 1/1250 in the river areas, up to 1/10000 along the coast (see section 3.2 for a further explanation about these safety norms).

The study area that is used to test the integrated flood risk model is 'Noord-Beveland', which is a municipality in the Dutch province of Zeeland. 'Noord-Beveland' is within dike ring '28' and has a safety norm of 1/4000 (TAW, 2000). Noord-Beveland is the smallest island of Zeeland and is connected with three dams and a bridge to the mainland. It is relatively safe for large-scale floods since most of its shore is located behind the Oosterschelde barrier, part of the 'Delta Works'. This barrier borders 'Noord-Beveland' in the northwestern part, whereby the northern and eastern shore of Noord-Beveland is secluded from the North Sea. The municipality has 7,408 inhabitants (CBS, 2011) and has a total area of 120 km^2 of which 34 km^2 is water. Noord-Beveland is mostly flat and is about one meter above sea level. A closer look to the land use in this area (Figure 1) reveals mainly agriculture uses, with some small villages, a few recreational areas and other rural activities. The most common agricultural land uses include: potato fields, wheat fields, pastures, corn fields, beet fields, orchards and other agriculture.

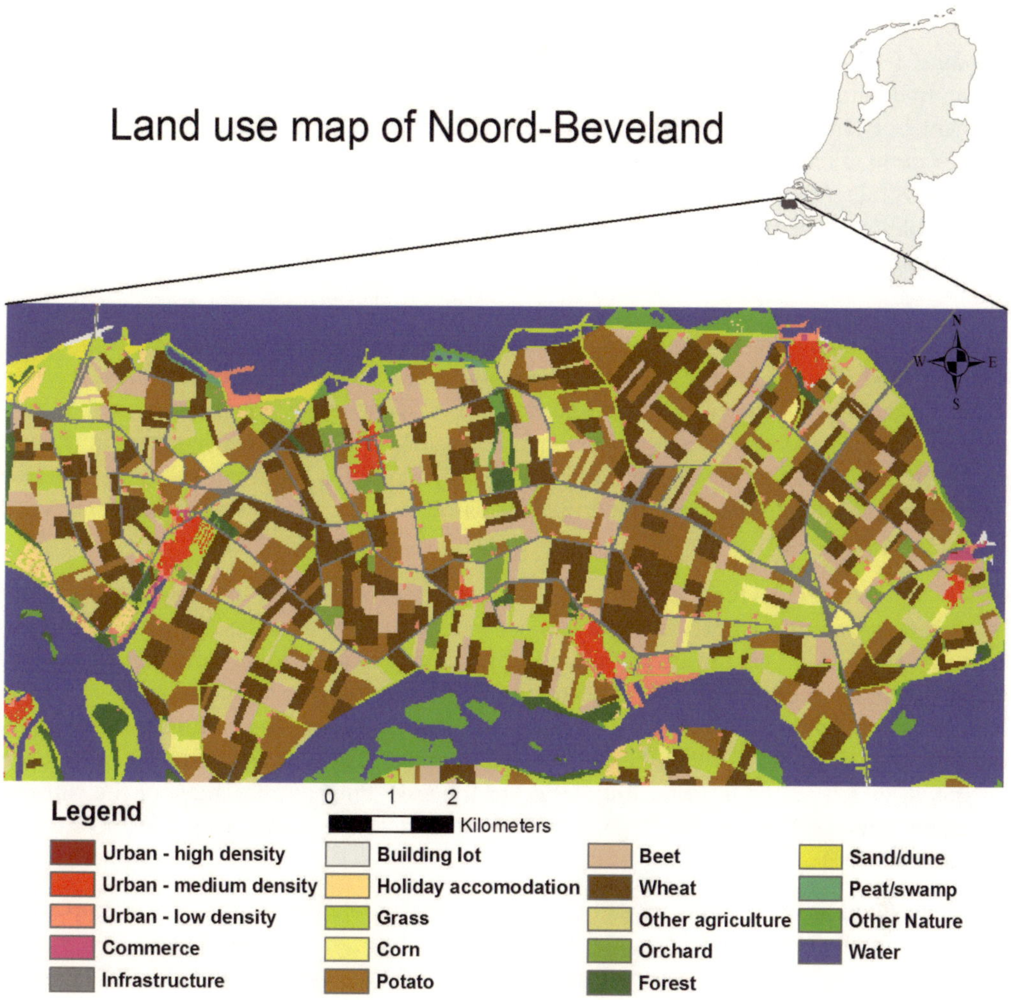

Fig. 1. Land use map of 'Noord-Beveland'. Based on the LGN4 and Land Use Scanner maps (see www.lgn.nl and Riedijk et al., 2007)

3. Differences between flood risk resulting from extreme rainfall events and large-scale floods

In this section, a number of conditions will be described that need to be taken into account to make a consistent comparison between the flood risk induced by extreme events and that resulting from large-scale floods since it is important to assess both types of risk in a comparable way. This should be done to avoid methodological biases as much as possible. Besides the issue of probability (large differences in probability for both forms of risk), there are many more fundamental differences between inundation due to heavy rainfall and large scale flooding with respect to their processes, consequences, exposure and the way they are dealt with. These differences should first be identified properly in order to allow for proper comparisons from which appropriate conclusions can be drawn.

3.1 Flood risk of extreme rainfall events

Flooding from extreme rainfall events vary from upwelling groundwater levels, which occurs frequently but with little damage to very large inundations of land which occur less frequent but with lots of damage. Inundation due to extreme rainfall events occur when there is more rainfall than the water system in a specific area can handle (Hoes, 2007). There can be several reasons for this inundation: the rainfall is not able to infiltrate into the ground or not able to properly flow away, there is insufficient pumping capacity or there is a too little storage capacity in the area to store all the water.

Besides the issue of storage capacity of a specific area, the duration of a rainfall event determines the amount of inundation. First, slow and fast reacting water systems can be distinguished in a specific area. Green houses and urban areas are for example fast reacting water systems, while pastures are an example of slow reacting systems and arable land responds usually between the two systems. For quick responsive water systems, a short period of several hours to several days with heavy rainfall is often necessary to have the land inundated, while the slow-reacting systems often need an event spread over several days (Smits et al., 2004).

Other important factors that determine the amount of inundation are the soil type, the water levels before a rainfall event and the time of occurrence in the year. The soil type determines how fast the rainfall can infiltrate in the ground. For example, if the soil is rich in sand, water can much more easily infiltrate then when the soil is rich in clay. The water levels before a rainfall event determine how much more water can be stored during the extreme rainfall event. High water levels before an extreme rainfall event means that less water can be stored, which results in faster inundation of the area. Furthermore, most of the extreme rainfall events, both short and large, occur at the end of the summer and in the autumn (Smits et al., 2004).

The consequences of extreme rainfall can sometimes be relatively large. For example, the extreme rainfall event that occurred in the autumn of 1998 caused around half a billion euro in damage. Looking at the exposure, this can be relative large but also very local. In 1998, the south-western parts of the Netherlands had problems with this extreme rainfall event (Smits et al., 2004), while in 2006, the problems occurred at a much more local scale (i.e. minor flooding in Egmond aan Zee, which is a small town located in the northwestern part of the Netherlands). One of the models that is used to predict and determine the damage of

Land-use type	Probability criteria [1/yr]
Pastures	1/10
Agriculture	1/25
High quality agriculture and horticulture	1/50
Greenhouses	1/50
Urban area	1/100

Table 1. The probability that a certain land use type may become inundated. (Nationaal Bestuursakkoord Water, 2003)

extreme rainfall events, is the model made by Hoes (2007), which determines the expected annual damage per pixel.

For extreme rainfall events, a number of probabilities have been authorized in the 'Nationaal Bestuursakkoord Water' (2003), which are necessary to take into account in the comparison between the two different types of risk. Important to note that in the Netherlands the probabilities of extreme rainfall events define the safety norms that describe how often different land uses are allowed to inundate. In Table 1 an overview of the different probabilities is given. Finally, policy on water management in the Netherlands is mostly the responsibility of Regional Water Boards.

3.2 Flood risk of large-scale flooding

Large-scale flooding can be defined as a temporary covering of land by water outside its normal confines due to flooding or breaching of primary flood defenses, which can result in large inundation depths, high damages and even human casualties (Kok and Klopstra, 2009). Important to notice for floods in the Netherlands, is that the Dutch area is divided into so-called dike rings. Dike rings are areas that are surrounded by levees, dunes or other higher areas that protect the inner area of the dike ring from flooding. Large-scale flooding happens when a dike or dune cannot stop the water from flowing into the inner part of a dike-ring. This happens when water levels exceed the height of the defense or when the water barriers breach. The results of large-scale floods can vary from only a few decimeters of inundation up to several meters of inundation.

Different safety norms apply to different dike-ring areas, which are described in the 'Water Protection Act'. These safety norms can be defined as the probability of occurrence of a certain water level and wave conditions that are higher than the dike or dune, as described in Table 2. Important to note is that these water level exceedence probabilities are different than the flood probabilities. It can happen that a dike or dune already breaches before the water level is higher than the dike or dune due to various failure mechanism (Vrijling, 2001; RWS-DWW, 2005). This means that the probabilities which are described in the 'Water Protection Act' are not always the exact flood probabilities. Also other factors, such as the probability of breaching and thus the strength of the dike at a certain place play a role (de Bruijn, 2007).

Area	Probability criteria [1/yr]
Coastal areas	Between 1/2000 to 1/10000
Areas along large lakes	Between 1/2000 to 1/4000
Areas along tidal rivers	1/2000
Other areas among main rivers	1/1250

Table 2. Flood probabilities for different areas in the Netherlands

Furthermore, as discussed in section 3.1, a number of other factors should be taken into account when describing flood risk. The consequences of large-scale floods are in general quite large. For example, the economic damage of the flood of 1953 was, in present value, around one billion euro (van Veen, 2005). With large-scale flooding, there is not only damage to crops and sewers, but also human casualties and damage to buildings and infrastructure.

Looking at exposure in the case of large-scale flooding, it is usually limited to one or maybe two dike rings or part of a dike ring, due to safety measurements before a flood will occur or during a flood. These safety measurements are for example strengthening of closely located weak parts of the dike, closing the breach or the closing of possible weirs that are in the area.

In the Netherlands, the HIS-SSM ('Hoogwater Informatie Systeem - Schade- en Slachtoffermodule') is commonly used for the determination of the flood risk of rivers and sea. With the HIS-SSM model, expected damage and the expected amount of casualties because of large-scale floods can be calculated (Kok et al., 2005). Another model, the Damage Scanner, is a simplified model of the HIS-SSM that calculates the expected damage of a large-scale flood (de Bruijn, 2006; Klijn et al., 2007). Whilst the HIS-SSM model calculates the damage per object, the Damage Scanner calculates the damage per land-use class (van der Hoeven et al., 2009). Finally, in the case of large-scale floods, policy is mostly made by 'Rijkswaterstaat', which is a governmental institution that is responsible for national water management and the roads of national importance in the Netherlands.

3.3 Comparison of the different conditions

When looking at the two sorts of flood risk described in the previous sections, a number of important differences can be determined, wherefore in Table 3 an overview is given. In this section, these dissimilarities will be further explained.

Factor	Extreme rainfall event	Large-scale flood
Occurence	Relatively frequent	Relatively unfrequent
Impact	Low	High
Exposure	Relatively unlimited	Relatively limited
Amount of inundation	Few decimeters	Few meters
Flow speed	Low	High
Human casualties	None to few	Few to many
Type of water	Fresh water	Fresh, salt or brackish water
Costs of prevention	Relatively low	Relatively high
Safety norms	Actual inundation	Possibility of overflow
Models	Hoes (2007)	HIS-SSM and Damage Scanner
Policy	Regional Water Boards	Rijkswaterstaat

Table 3. Overview of the differences between extreme rainfall events and large-scale floods

A number of observations can be made based on the table. First, there is a large difference in probability of occurrence. While the flood risk related to extreme rainfall events has a relatively high probability of occurrence, flood risk resulting from large-scale flooding has a relatively low probability of occurrence. When looking at the differences in impact (damage), we see that extreme rainfall events have a relatively low impact in comparison with large-scale floods, which have a much higher impact. With these two conditions in mind, there is now one clear difference: high probability/low damage (extreme rainfall events) versus low probability/high damage (large-scale floods) (Merz et al., 2009).

Second, while exposure for extreme rainfall events concerns almost the whole of the Netherlands (extreme precipitation can happen anywhere), exposure to large-scale flooding is relatively limited since it is confined to those areas contained within the dike rings. Also important is the amount of inundation of both forms of flood risk. Whilst the inundation of extreme rainfall events is most of the time much lower than that of large-scale floods, usually a few decimeters, the inundation for large-scale floods is much higher (up to a few meters). Not only the amount of inundation determines the damage though, but also the speed of the water flow. A high speed will usually cause much more damage, especially in terms of human casualties. With extreme rainfall events, there is usually very little or almost no flow speed, while large-scale floods can have very high flow velocities, especially near the breach. The occurrence of human casualties is an important difference between the two forms of risk. For large-scale floods the chances of human casualties are much higher than for extreme rainfall events. There can also be a difference in the 'type' of water that inundates the area. While extreme rainfall events mainly involve fresh water, large scale floods are usually salt or brackish water. The latter is especially for agriculture much more harming than fresh water inundation (Nieuwenhuizen et al., 2003). Finally, flooding from extreme rainfall events mainly occurs due to minor bottlenecks in the regional water system, while flooding from large-scale floods mainly occur due to failure of primary water defenses. Due to this difference, for extreme rainfall events minor (relative cheap) measurements are expected to prevent flooding, while for large-scale floods much larger (and more expensive) measurements are expected to be implemented. Nevertheless, Kok and Klopstra (2009) found in a simple cost-benefit analysis that the cost-effectiveness of reducing the risk of large-scale floods is in general much higher than that of reducing the risk related to extreme rainfall events.

There are also clear differences in the probability criteria. As described before, the safety norms of extreme rainfall events are not only higher than those of large-scale floods, there is also a clear difference in the interpretation. The safety norms for extreme rainfall events mean the minimum probability that there will be an actual inundation, while the safety norms for large-scale floods are defined as the levels at which the dikes could possibly overflow.

Another important difference is the determination of flood risk, since both types of flood risk are determined in different models that use different input parameters to determine the risk. For extreme rainfall events, the damage model of Hoes (2007) has been developed, while for large-scale floods, the HIS-SSM of Kok et al. (2005) is most commonly used. While looking at these two models, there are already a few differences. Not only different inundation maps are used to determine the expected inundation (e.g. starting at different depths), but also different land-use maps with different land-use classes are used. While in the model of Hoes many more agriculture classes are used, the HIS-SSM provides more variety in urban classes. Other differences are observed in the definitions of maximum damages and damage curves.

Of final importance are the differences in policy. Whilst for extreme events the Regional Water Boards are responsible for policy making, is 'Rijkswaterstaat' responsible for the policy making with large-scale floods. Due to this difference, other criteria or other processes are seen as important for flood policies.

4. Methodology of the integrated flood risk model

Since it is now clear what the conditions are that need to be taken into account and what the dissimilarities are between the flood risk of extreme events and large-scale flooding, it is possible to continue with the actual integrated flood risk model. Even though both types of risk are normally estimated using different models that differ in several aspects, both models are based on the same underlying concepts, namely: depth-damage curves and maximum damages. It should therefore be possible to integrate both approaches into a single integrated flood risk model. This is possible since the integrated flood risk model – like the models it is based on – is mainly focused on direct damage and most of the differences described in the previous section (e.g. human casualties, costs of preventing floods) do not have a direct influence on that. Several studies note that the most important factor that determines direct damage in both extreme rainfall events and large-scale floods is the flood depth (Merz et al., 2007; Penning-Rowsell et al., 1995; Wild et al., 1999). Therefore, the integrated flood risk model will be built around this parameter. In this section, a general description of the methodology will first be explained, then the input will be described and finally the damage factors and maximum damages.

4.1 General outline of the flood risk model

The integrated flood risk model uses the same approach as the Damage Scanner and the HIS-SSM model. In this approach, a land use map and inundation map are used, which are combined using damage curves and maximum damages per land use. Every land-use class has a different maximum amount of possible damage and uses a different damage function, whereby the possible amount of damage is in millions of euro per hectare. Every damage function shows a curve where the possible inundation is on the x-axis and the damage factor on the y-axis (Figure 2). To determine the amount of damage in the area, a number of steps have to be taken:

1. Inundation depth: Inundation maps determine the maximum inundation depth for each cell, which varies depending on the scenario.
2. Land-use class: Land-use maps determine the land-use for individual cells.
3. Damage factor: a damage factor is derived from the damage functions and represents the percentage of the maximum total damage. The damage function used is defined by the land-use class. Then, the inundation depth defines the damage factor, which is measured in percentage terms. These damage curves and maximum damages per land use will further be described in section 4.3.
4. Damage calculation: the final step is to determine the amount of damage for a specific cell by multiplying the damage factor with the maximum amount of damage. This quantifies the damage that occurs in each cell.

Once the calculations are done, the outcome will be a map and a table for every inundation map with the different amount of damage respectively per pixel and per land use in euro. In the table, not only the different amount of damage is described, but also the average damage, the standard deviation and the total area per land use. Once these damages are calculated, the final outcome can be determined. As described in the introduction, the flood risk is determined by multiplying the flood probability with the consequences, which can be described as the maximum amount of possible damage in a specific area that is calculated in the integrated flood risk model. The final outcome is the flood risk in terms of Expected Annual Damage (EAD).

4.2 Land-use and inundation data

The key inputs to this model come from two different maps. One is the land use map and the other is the inundation map. For the land use map, a new land use map is made which is a combination of the land use map from Land Use Scanner (described in Riedijk et al., 2007 and used in the Damage Scanner) and the 'Landgebruikskaart Nederland' (LGN4, used in the model of Hoes, 2007). The former are derived from a land use model that is applied to simulate land use changes and that is mainly focused on urban areas (see, for example, Koomen et al., 2008 and Koomen and Borsboom-van Beurden, 2011). The latter dataset is more focused on agriculture and distinguishes more classes in these categories (de Wit and Clevers 2004; de Wit 2003; van Oort et al. 2004). Since extreme rainfall events mainly damage agriculture but large-scale floods also damage urban areas and infrastructure, we combine those two to cover enough land-uses for both types of flood risk. The other map we use is the inundation map, which shows us the maximum inundation in a specific area for the different flood probabilities.

The combined land use map contains 25 different land-use classes which can be aggregated into four major land-uses: urban land-uses, agriculture, nature and infrastructure. The urban land-uses consist of five classes: Urban - high density, Urban - low-density, Urban - rural, Commerce and Building lot. Where 'Urban - high-density' are the main cities and towns (like Amsterdam or The Hague), 'Urban - low density' are suburbs and villages (like Egmond aan Zee) and 'Urban – rural' are farms and large houses between pastures and along rural roads. Commerce is all the commercial areas within the Netherlands. The agricultural land-uses consist of nine classes: Greenhouses, pastures, corn, potato, beet, wheat, orchard, bulbs and other agriculture. The nature land-uses consist of seven classes: fen meadow, forest, sand/dune, heath, peat/swamp, water and other nature. Finally, the infrastructure land-uses consist of three classes: Airport, seaport and infrastructure, where the 'infrastructure' class are all the roads, railways and other infrastructure that is not included in airport and seaport.

The inundation maps depict the inundation of extreme rainfall events or large-scale floods. These maps show the inundation in a specific area for different return periods, varying from a probability of 1/10 to a probability of 1/40000. The inundation maps used in this study for large-scale floods, which are calculated for different scenarios, are obtained from the province of Zeeland. The inundation maps can be subdivided into four scenarios: 1/4000 with RTC, 1/4000 without RTC, 1/400 with RTC and 1/40000 with RTC. "RTC (Real Time Control) is a module in the SOBEK model which allows the system to react optimally to actual water levels and weirs, sluices and pumps" (Deltares, 2010). Important to note is that for the 'North Sea-side' of Noord-Beveland all four scenarios are used, while for the 'Oosterschelde-side' only the first two scenarios are used. This is due to the fact that with high water levels the 'Delta Works' will close.

The inundation maps used in this study for extreme rainfall events are obtained from the water board. These maps, which have been calculated with the use of SOBEK RR and Channel Flow, are made for the water boards in response to the 2003 'Nationaal Bestuursakkoord Water'. For the study, the inundation maps with return periods of 1/10, 1/25, 1/50 and 1/100 are used, whereas the higher return periods have the lowest inundation depths and the lowest return periods the highest inundation depths.

Finally, two additional maps were used for a closer examination of the damage that can occur with respect to the safety norms. For large-scale floods, the Risk Map for the Netherlands (www.risicokaart.nl) has been used and for extreme rainfall, an inundation map has been made with an overall inundation of 0.165 meter, which is the average inundation level above zero of the four different inundation maps for extreme rainfall events.

To be able to use all the maps properly in the model, the land use map and the different inundation maps are modified with ArcGIS to match the same study area. Several adjustments must be made to be able to fit the different inundation maps in the same model. Since the maps for inundation from large-scale floods start at inundation above 0, all the zero values in the map mean no water. But with extreme rainfall events, a value of zero means that there is water up to the ground level. Therefore, the inundation maps of large-scale floods need to be adjusted to have no damage in areas where there is no inundation.

4.3 Maximum damage values and damage curves

Maximum damages and damage curves were created using various sources. The maximum damage for most of the land-use classes is derived from their mean damage per hectare in the damage maps of the HIS-SSM for ten meters of inundation, above which hardly any extra damage occurs. A few land use classes were new and thus not able to have their correct maximum damage derived via the HIS-SSM damage maps. These maximum damages were therefore derived by comparing the specific land use class to damages given in various other studies (Brienne et al., 2002; de Bruijn, 2006; Hoes, 2007; Klijn et al., 2007; Vanneuville et al., 2006). Urban – high density is calculated by first determining the amount of dwellings in high density residential areas (Jacobs et al., 2011) and then multiplied with the amount of damage per dwelling as described in studies of Briene et al. (2002). The maximum damage for rural area is not only derived from the maximum damage per farm, as described in studies of Briene et al. (2002), but also derived after determining the average amount of rural area in the land-use map. Once the maximum damage has been calculated, a simple calculation allows us to estimate the damage per hectare for rural areas. Finally, the maximum damages for the different types of natural land use (e.g. forest, heathland) are set to zero, since no economic valuea can be attached to these areas. This is consistent with studies of Briene et al. (2002), Hoes (2007) and Vanneuville et al. (2006). In Table 4 is an overview of the different maximum damages per hectare.

Furthermore, damage curves are developed that specify the different amount of damage for different inundations. These curves allow us to calculate the different damage factors for different possible inundations. These inundations vary from elevated groundwater levels (-0.3 meters) up to high water levels (5 meters). These curves are mainly based on results of the HIS-SSM, but also other studies (Hoes, 2007; Vanneuville et al., 2006) were used to adapt the curves to our specific land-use classes. The damage maps of the HIS-SSM were used to calculate the amount of damage for different inundation depths. By dividing the damage of a certain water depth by the total possible damage (at ten meters of inundation), the damage factor for that inundation depth can be determined. In Figure 2, the different damage curves are shown.

Land use	Million euro per hectare
1 - Urban - high density	9.9
2 - Urban - low density	5.3
3 - Rural area	1.2
4 – Commerce	7.9
5 – Seaport	5.5
6 – Airport	11
7 – Infrastructure	1.4
8 - Building lot	0.8
9 - Holiday accomodation	0.4
10 - Green houses	0.65
11 – Pastures	0.015
12 – Corn	0.025
13 – Potato	0.025
14 – Beet	0.025
15 – Wheat	0.025
16 - Other agriculture	0.025
17 – Orchard	0.140
18 – Bulbs	0.050
19 - Fen meadow	0.015
20 – Forest	0
21 - Sand/dune	0
22 – Heath	0
23 - Peat/swamp	0
24 - Other Nature	0
25 – Water	0

Table 4. Maximum damage per land use in millions of euro

A close look at Figure 2 reveals that damage curves for the agriculture classes reach the maximum amount of possible damage relatively quickly. This is consistent with the Damage

Scanner and the HIS-SSM (Klijn et al., 2007) and studies of Hoes (2007) and Vanneuville et al. (2006). This occurs because only a small amount of inundation is sufficient to harm the crops. The damage curve for airports also shows a very steep curve at the beginning, which is due to a lot of indirect damage (e.g. cancelling of flights) that will happen if there is water on the runways. Damage to the urban and other build up areas are relatively similar. A final comment is warranted on the damage curve for 'Commerce', which starts relatively flat and then rises relatively steeply above 3 meters of inundation. Limited information and large heterogeneity makes it difficult to determine the exact damage curve for commerce (Vanneuville et al., 2006).

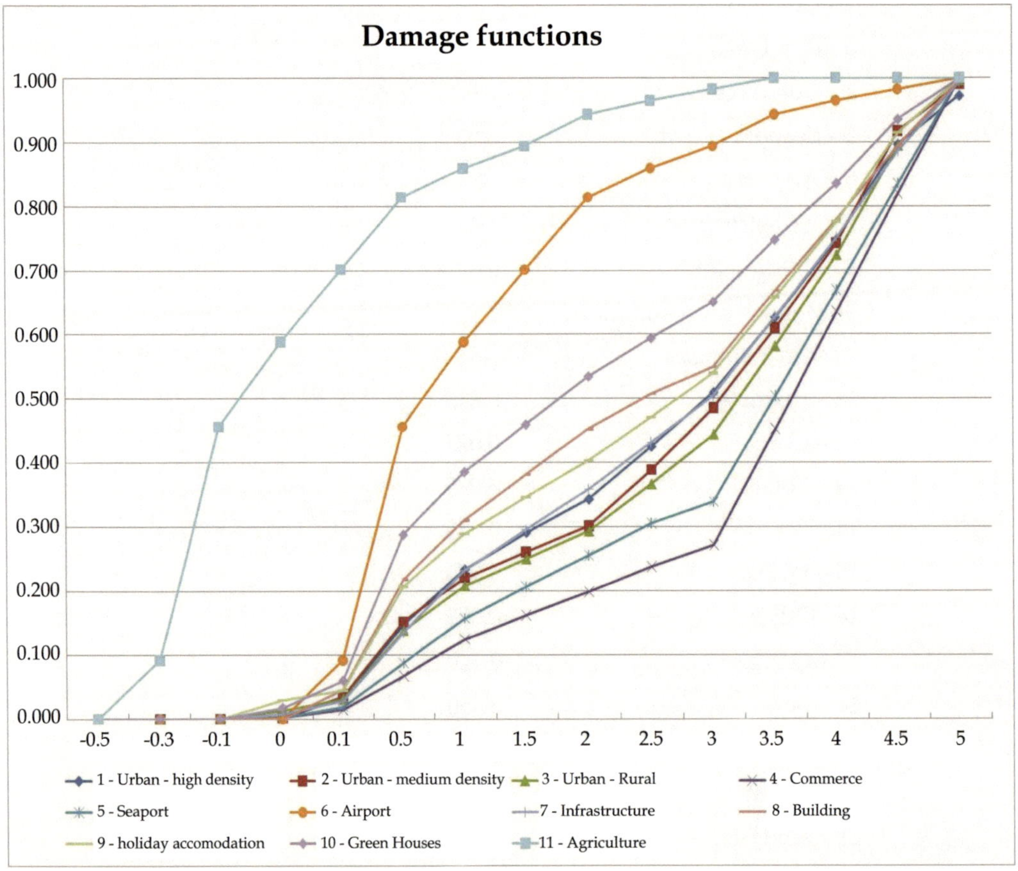

Fig. 2. Damage curves per land use type

5. Results

In this section, the outcome of the model will be described using the land use map and different inundation maps for 'Noord-Beveland'. To compare the different types of flood risk in a consistent way, we will compare them in two different ways. One of the comparisons is the 'existing situation', which describes the most plausible inundation scenarios given the characteristics of the regional water system and primary defenses. For extreme rainfall events, the return periods of 1/10, 1/25, 1/50 and 1/100 are used, while for large-scale floods the probability maps of 1/400, 1/4000 and 1/40000 are used. The other is the comparison with respect to the safety norms, which describes the amount of damage for all the land uses looking at the different safety norms (i.e. what is socially and politically acceptable). In other words, when looking to extreme rainfall events, an urban area is for example allowed to inundate once every 100 years and with a large-scale flood, the whole area is in the case of Noord-Beveland allowed to inundate once every 4000 years. This means that in the second comparison, the whole area will be inundated to see what the amount of damage will be with respect to the safety norms.

5.1 The current situation

5.1.1 Extreme rainfall events

For extreme rainfall events, four different inundation maps are used. For these different inundation maps, potential damages were calculated with the use of the model. Data showed that most of the damage occurs in agricultural area and infrastructure, and the most damage occurs in areas with wheat, potato and pastures. This is due to the fact that these simply have the largest area. The reason why mostly agricultural areas have large amount of damages reflects the fact that crops are severely damaged with only small amount of inundations.

If we take a closer look at the flood risk for the different probabilities, we will look at the annual expected damage. The annual expected damage is calculated by multiplying the probability times the total damage. In Table 5 we see an overview of total damage and the different flood risk per probability for extreme rainfall events.

Return period	Total damage (x €100,000)	Flood risk (x €100,000)
1/10	9.5	0.95
1/25	33	1.3
1/50	62	1.2
1/100	99	0.99

Table 5. Overview of the estimated total damage and flood risk (in terms of Expected Annual Damage) per probability for extreme rainfall events

In the table above, we see that higher return periods are associated with higher total damage but not higher flood risk (measured in annual expected damage). This is mainly due to the fact that when the probability of specific events becomes lower, the annual expected damage is also lower because you will multiply the total damage with a much lower factor. Interestingly, the highest total damage occurs for the return period of 1/25 and that all the return periods have almost the same flood risk in terms of EAD (about 100,000 euro per year), even though the total damage varies considerably.

It is also interesting to see where the damage exactly occurs. Figure 3 shows that even with a very low inundation probability (1/10), there is already a relative large amount of damage in the northwestern part of the area. This is mainly due to the fact that there are higher inundation levels in these areas and agricultural land uses that undergo damage at even low inundation levels. If we compare this with the land use map of the region (Figure 1), we see that these are all agricultural crops (wheat, beet, and grass).

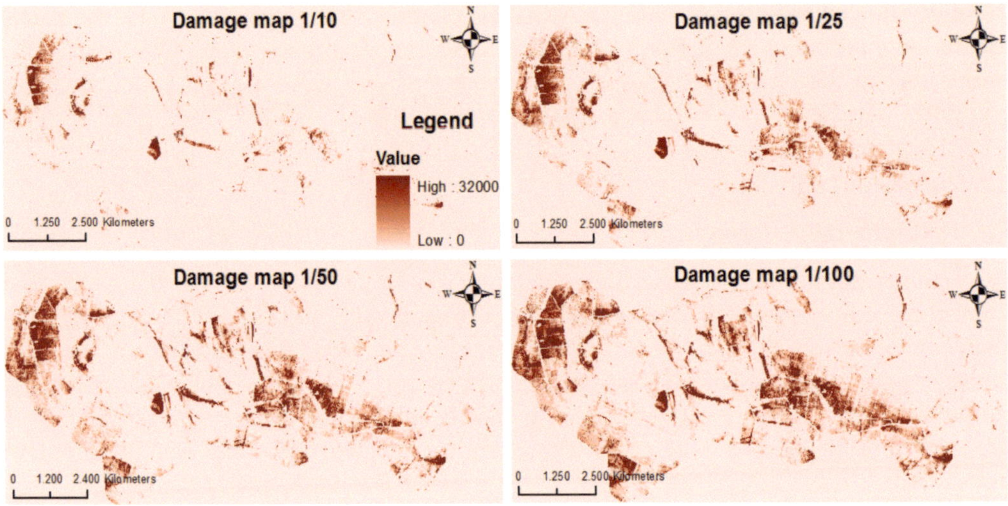

Fig. 3. Damage maps for the four return periods for extreme rainfall events

5.1.2 Large-scale floods

For large-scale floods, two scenarios are used to determine to the total damage in the area. One scenario is a flood that results from a dune breach at the 'North sea-side' of 'Noord-Beveland', the other scenario is a flood that results from a dike breach at the 'Oosterschelde-side' of 'Noord-Beveland'. For the 'North sea-side' the flood scenario is sub-divided into four more sub scenarios, which are 1/4000 with RTC, 1/4000 without RTC, 1/400 with RTC and 1/40000 with RTC. For the 'Oosterschelde-side', the flood scenario is sub-divided into two more sub scenarios, which are 1/4000 with RTC and 1/4000 without RTC (see section 4.2). The breach at the 'North Sea-side' is chosen because there is simply only one place where the dune could breach. The breach at the 'Oosterschelde-side' is chosen since this section in the dike has not been reinforced yet and has therefore at the moment a higher possibility to breach compared to other dike sections at the 'Oosterschelde-side'.

After determining the damages for the dune breach at the North Sea, results for this scenario show that the highest damages occur in the agricultural areas. In Figure 4, which shows the damage in the area with respect to the four sub scenarios, it can be seen that the flood from the North Sea mainly inundates the western part of Noord-Beveland. This area mainly consists of agricultural areas (see Figure 1). Only the inundation with the probability of 1/40000 inundates a much larger area, including a village. This can be seen in the damage map as a much darker spot in the middle of the area that inundates.

For the other breach location at the 'Oosterschelde-side', it is interesting that the results show major differences between the two sub scenarios. In the sub scenario with the RTC-module, there is much more damage. Especially the damage in the infrastructure changes from 22.3 million euro to 1.7 million euro in the sub scenario without the RTC-module. When looking at Figure 5, we see that the main reason for the higher damages is that there is a much larger area that inundates in the sub scenario with the RTC-module even though the inundation depth is lower.

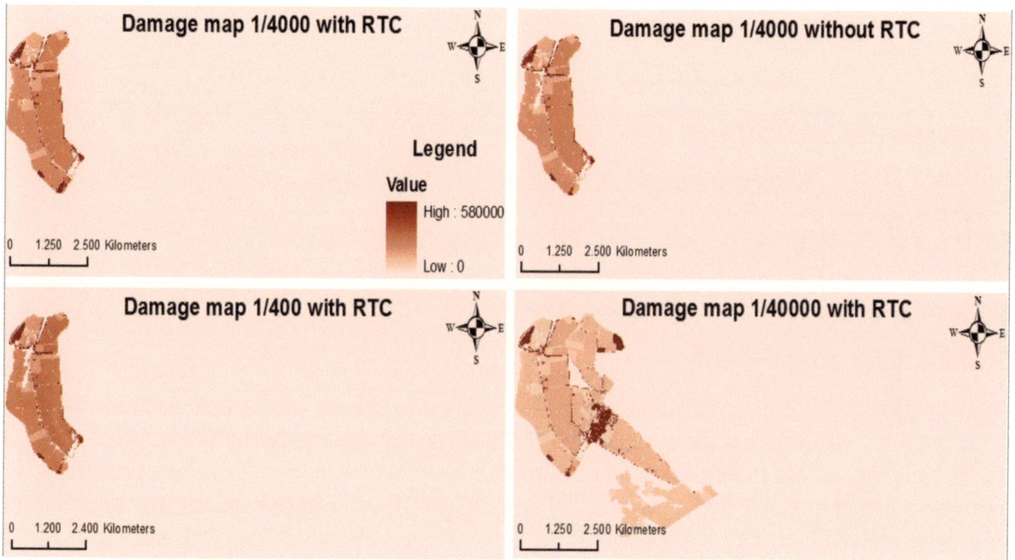

Fig. 4. Damage maps for the four different sub scenarios with the 'North Sea breach'

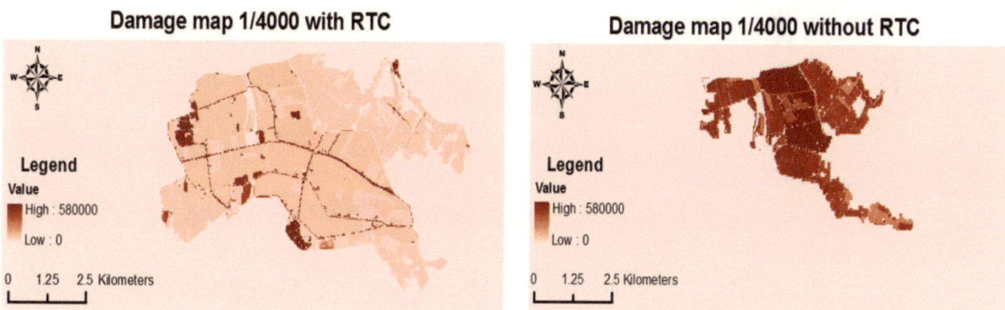

Fig. 5. Damage maps for the two different sub scenarios with the 'Oosterschelde breach'

Finally, the flood risk per sub scenario was calculated (Table 6). At first, we see the highest total damages in the 'Oosterschelde sub scenario 1/4000 with RTC' and the 'North Sea sub scenario 1/40000 with RTC'. This is mainly due to the fact that, as described above, a much larger area inundates with a lot more urban area in both these sub scenarios and a lot more infrastructural areas in the first sub scenario. If we closer examine the flood risk values, we see the highest flood risk in the 'North Sea' sub scenario 1/400 with RTC and the 'Oosterschelde' sub scenario 1/4000 with RTC. The reason why the first sub scenario has a much higher flood risk is because it has a much higher probability of occurrence. The reason why the latter has a high flood risk is simply because there are very high total damages.

	Return period	Total damage (x €100,000)	Flood risk in terms of EAD (x €100,000)
North Sea	1/4000 with RTC	162	0.04
	1/4000 without RTC	162	0.04
	1/400 with RTC	137	0.3
	1/40000 with RTC	575	0.014
Oosterschelde	1/4000 with RTC	943	0.2
	1/4000 without RTC	224	0.06

Table 6. Flood risk for all the sub scenarios with large-scale floods

5.2 Safety norms

5.2.1 Extreme rainfall events

After looking at the current situation, it is also interesting to see what the maximum damage could be if we assume that the probability of flooding equals exactly the safety standards for every cell, regardless of breach scenarios or the local water system. The safety norms for extreme rainfall events, described in Table 7, imply that different areas are allowed to inundate with different probabilities. In Table 7, we see the maximum damages and flood risk per land use if all the land is inundated with 0.165 meters of water. This inundation level is chosen because this is the average inundation above ground level for all four inundation maps.

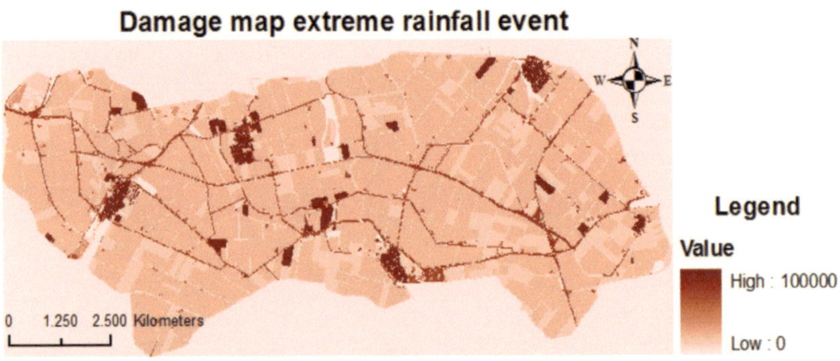

Fig. 6. Damage maps for an extreme event in Noord-Beveland (inundation of 0.165 meters)

In Table 7, the annual expected damage is calculated per land use, according to the different safety norms described in Table 1. If we look at the maximum damages, we now see that highest amount of damages are in the urban areas and infrastructure, which is in contrast with the highest damages in the 'current situation' where we saw that the highest damages were found in the agricultural land uses. Important to note is that the damage in agricultural land-uses are still much higher than in the 'current situation'. When examining the flood risk more closely, we see that the highest flood risk occurs in agricultural areas. This is mainly due to the fact that these areas have low safety norms.

Figure 6 shows the spatial distribution of the damage. In this figure, there can be seen that in urban areas the highest damages occur, which is consistent with the data for this scenario. The figure also highlights the difference in agricultural land uses. Much of lighter areas in the damage map are associated with pastures.

Land use	Maximum damage (x €100,000)	Flood risk in terms of EAD (x €100,000)	Area (ha)
1 - Urban - high density	2.9	0.03	0.56
2 - Urban - low density	340	3.4	118
3 - Rural area	62	0.6	105
4 – Commerce	6.7	0.07	3.7
7 – Infrastructure	210	2.1	330
8 - Building lot	1.3	0.01	2.1
9 - Holiday accomodation	7.4	0.07	25.6
11 – Pastures	140	14	1305
12 – Corn	28	1.1	157.4
13 – Potato	250	10	1363.4
14 – Beet	170	6.8	938
15 – Wheat	280	11.2	1538
16 - Other agriculture	220	8.8	1203
17 – Orchard	140	5.6	138.7
20 – Forest	0	0	59.7
21 - Sand/dune	0	0	1.7
23 - Peat/swamp	0	0	1.8
24 - Other Nature	0	0	57
25 – Water	0	0	4
Total	*1860*	*64*	*5598*

Table 7. Total damages and flood risk for an extreme rainfall event with an inundation of 0.165 meters

5.2.2 Large-scale floods

To determine what the maximum damage will be in 'Noord-Beveland' when looking at the safety norms for large-scale floods, the Risk Map for the Netherlands. For the creation of this map it was assumed that the complete dike ring inundates in case of flooding up to a level where flood water would spill out of the dike ring (RWS-DWW, 2005).

Calculating the damages showed that the highest damages occurred in the urban - low density areas and to infrastructure. The total damages in the agricultural land uses are almost the same as the total damages seen in Table 7 with extreme rainfall events. In Figure 7, we see that even though the dike or dunes breach at all the possible locations, not all areas are inundated. For example, a few areas in the middle are not inundated. Furthermore, the damage map clearly shows the location of villages and infrastructure, because these are the areas that incur the highest damages.

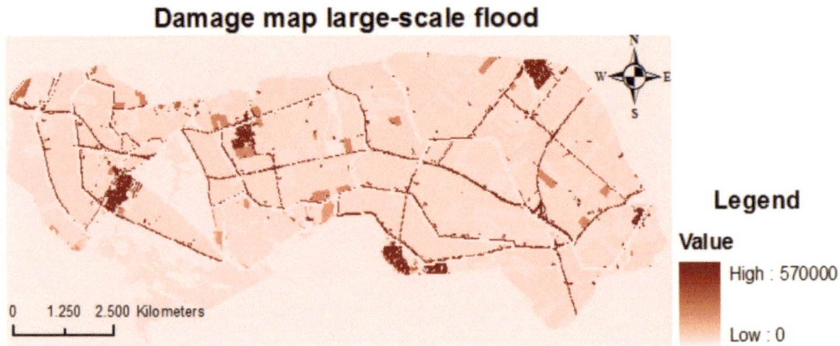

Fig. 7. Damage map for a large flood in Noord-Beveland (dike ring fills up completely)

For large-scale floods in this scenario, the total damage and flood risk are described in Table 8, where a total damage can be seen of approximately 388 million euro and an expected annual damage of approximately 97,000 euro per year.

Return Period	Total damage (x €100,000)	Flood risk in terms of EAD (x €100,000)	Total area (ha)
1/4000	3880	0.97	5598

Table 8. Total damage and flood risk for a large flood in Noord-Beveland

6. Discussion

In this section we provide some critical discussion on the structure of our model and our results. First, we consider the results from the study area according to the 'existing situation' and the safety norms. Second, we identify possible methodological issues with the integrated flood risk model and our analysis.

6.1 Total damage and flood risk estimates

In this section, a few results will be examined more closely. First, the differences between the total damage and flood risk for extreme rainfall events and large-scale floods will be examined, for respectively the 'current situation' and the safety norms. Second, the differences between the 'current situation' and the safety norms will be discussed.

If we examine the results of the 'current situation' carefully, we see two important differences. One is the difference in exposure. While for extreme rainfall events, the area of exposure is almost the whole dike-ring area, for large-scale floods the inundation area is limited to a much smaller area. Reasons for this are that there are higher areas (roads or inner dikes) within Noord-Beveland that act as secondary defenses and that simply not much water flows into the dike ring after a breach near the dunes (where the difference in elevation between the water level and the land surface is limited). The other difference is in the damage distribution. While for extreme rainfall events the highest damages are found in agricultural land uses, the highest damages for large-scale floods mainly are found in urban areas and infrastructure.

	Return period	Total damage (x €100,000)	Flood risk (in terms of EAD)
Extreme rainfall events			
	1/10	9.5	95000
	1/25	30	133000
	1/50	60	125000
	1/100	100	99000
Large-scale floods			
North Sea	1/4000 with RTC	162	4000
	1/4000 without RTC	162	4000
	1/400 with RTC	137	34000
	1/40000 without RTC	575	1400
Oosterschelde	1/4000 with RTC	943	24000
	1/4000 without RTC	22.4	6000

Table 9. Total damage and flood risk for all the different scenarios

In Table 9, all the total flood risk values are listed for the 'current situation'. In the table can be seen that flood risk for extreme rainfall events is much higher than the flood risk of large-scale floods, which is remarkable since the total damages of extreme rainfall events are in general much lower than that of large-scale floods. The main reason for this is that even though the total damages are much lower for extreme rainfall events, the probability of occurrence is much higher. This is an interesting result, since much more policy has been made to prevent or mitigate the chance of large-scale floods (Kok and Klopstra, 2009).

If we examine the results with respect to the safety norms closer, we see the same differences as in the 'current situation'. In Table 10, we see that even though the damages of large-scale floods are higher, the flood risk in terms of annual expected damage is much lower. This comparison is more interesting because of the dissimilarities between the two types of flood risk, described in section 3.3. The probabilities and safety norms for extreme rainfall events can be interpreted as 'accepted risk'. In other words, the area is allowed to inundate with these probability levels. Also important is to take into account what the effect is for both events on for example the insurances, indirect damage, human casualties and the social disturbance. These effects are not taken into account in the calculated annual expected damage but have, especially for large-scale floods, a very high effect on the total impact. Taking these unquantified effects into account would probably bring both types of risk closer together. It is, however, questionable whether the difference would completely be bridged by these additional effects given the large (64 times) difference.

	Total damage (x €100,000)	Flood risk (in terms of EAD) (x €100,000)
Extreme rainfall event	1860	64
Large-scale flood	3880	1

Table 10. Total damage and flood risk for the safety norm maps

It is important to note that the inundation maps with respect to the safety norms for both types of events are hypothetical. For extreme rainfall events, it is not plausible that the whole area inundates with the same height, since the norm is a lower limit and many areas will probably be much safer than the norm. Similarly, for large-scale floods the compartmentalization within the dike-ring area will probably prevent the whole area from inundating unless there are many dike failures at all sides. Nevertheless, by contrasting these situations we could compare the types of risk as they are 'allowed' by current policy.

It is interesting that the total damage calculated with the integrated flood risk model is much lower than the total damage calculated with the Damage Scanner and the HIS-SSM for dike-ring 28. The total damage calculated with these latter models is respectively 583 and 653 million euro (Klijn et al., 2007), while the total damage calculated in the integrated flood risk model is only 388 million euro. That figure is more in line with total damage estimates of around 400 million euro determined by Klijn et al. (2004) and van der Klis et al. (2005). One explanation for the higher damages in the Damage Scanner and the HIS-SSM could be the difference in cell size. In the Damage Scanner, the grid cell size is 100x100 meter, instead of 25x25 meter in the integrated flood risk model, which can result in higher damages because of aggregation of multiple land uses in one grid cell, resulting from overestimation of residential land in the aggregation process. This can also be seen when one compares the land use map used in this study with the land use map used in the Damage Scanner; the amount of residential and commercial land-use is 7.5 per cent higher in the Damage Scanner than in the land use map used in this study. This overestimation results from the fact that residential land tends to dominate, but not completely fill cells at a coarser resolution, as has also been observed by Bouwer et al. (2009). Another reason could be that in the integrated flood risk model a greater variety of agricultural land uses are used with much lower maximum damages. For these classes, much lower maximum damages are chosen because it is not likely that inundations of more than 0.5 meter will cause any more damage to agricultural crops. Finally, the HIS-SSM model calculates the damages by using objects. In the integrated flood risk model, objects such as tractors and other agricultural machines are not taken into account, which results in lower damages.

Even though high flood risk values are found for extreme rainfall events in Noord-Beveland, this does not mean that this will also be the case for the rest of the Netherlands. Since Noord-Beveland has much agriculture, not many urban areas and many secondary defenses, it is not very representative for the rest of the Netherlands. For instance the 'Randstad' area (middle west of the Netherlands) is much more urban and will therefore probably have a different comparison of the examined types of flood risk.

6.2 Methodological issues

The model used in this study seemed very useful when determining flood risk for both extreme rainfall events and large-scale flooding. But several methodological issues remain that should be taken into account. First, the aggregation of the built up areas in the land use maps could have been better. There are only three different urban land use classes, while more differentiation would be desirable. Second, there could have been more detailed investigation about the maximum damages for a number of land use classes. For a number of classes, determining the maximum damages was sometimes difficult, even though it was calculated with HIS-SSM damage maps and literature studies. Therefore, more research and investigation is required to provide consistent estimates of the maximum damages. Third, it was difficult to develop the model with different inundation maps as inputs. Several adjustments must been made to fit the different inundation maps in the same model.

Also important to take into account is that the model only has been used for determining the flood risk of a small, specific area. It is interesting to see whether results for larger areas or areas with different land uses are similar to those reported in this study.

Finally, it is always hard to validate the model in a consistent way when observation data is lacking. Since there have not been many large-scale floods or extreme rainfall events, it is hard to test if the model calculates realistic absolute damage estimates. As both types of risk are estimated using the same model, the influence of any bias in, for instance, maximum damages will affect both estimates.

7. Conclusion

The main objective of this study was to create a common methodology to assess flood risk of extreme rainfall and large-scale flooding in the Netherlands. Based on the literature we were able to incorporate both types of flood risk within an integrated model that allowed us to compare the different types of flood risk in a plausible and consistent way.

We then applied the model to analyze flood risk in the 'Noord-Beveland' area. Results show that even though the highest total damages are found to result from inundations of large-scale floods, the flood risk of extreme rainfall events are in general much higher when both are expressed in terms of annual expected damage. The reasons are that extreme rainfall events cause larger areas to inundate and occur with a higher probability, which combines to drive up flood risk. Further investigation should be done in other parts of the Netherlands to test if this is the case for more dike rings.

Our model does not quantify some types of indirect damage, such as human casualties and social disturbances. These should be taken into account to provide an even more consistent comparison. We expect that they would have increase the damage associated with large-scale floods. Nonetheless, we question whether the difference large difference (64 times) would be completely bridged by these additional effects. Aside from these unquantified factors, there are a number of data comparability issues, such as the differences in exposure and the distribution of the damage, which should also be kept in mind when comparing different types of flood risk.

Even though the model requires further refinements our initial results suggest it is possible to compare different forms of flood risk within an integrated model. Our finding that higher flood risk can be associated with extreme rainfall events suggests there is a need for the focus of public policy to shift away from large-scale flooding onto extreme rainfall events.

8. Acknowledgements

As a start, Ylva Peddemors of the Province of Zeeland and Govert Verhoeven of Deltares are thanked for providing the inundation maps. Karin de Bruijn and Olivier Hoes are thanked for valuable discussion during the start-up phase of this project. Stuart Donovan is thanked for considerably improving the use of English. The Netherlands Environmental Assessment Agency (PBL) is thanked for providing the Land Use Scanner model and related data sets. This research was carried out as part of the Dutch National Research Programmes 'Climate changes Spatial Planning' and 'Knowledge for Climate' (http://www.climateresearchnetherlands.nl/).

9. References

Aerts, J., Sprong, T. and Bannink, B.A. (Eds.), (2008). *Aandacht voor Veiligheid.* DG Water Report 2009/2008 (Dutch).

Bouwer, L. M., Bubeck, P., and Aerts, J. C. J. H. (2010). Changes in future flood risk due to climate and development in a Dutch polder area. *Global Environmental Change* 20(3): 463-471.

Bouwer, L.M., Bubeck, P., Wagtendonk, A.J., and Aerts, J.C.J.H. (2009). Inundation scenarios for flood damage evaluation in polder area. *Natural Hazards and Earth System Sciences* 9: 1995-2007, doi:10.5194/nhess-9-1995-2009.

Bouwer, L.M., Crompton, R.P., Faust, E., Höppe, P and Pielke Jr., R.A. (2007). Confronting disaster losses. Science 318: 753.

Brienne, M., Koppert, S., Koopman, A. and Verkennis, A. (2002). Financiele onderbouwing kengetallen hoogwaterschade. VROM, Rijkswaterstaat, Dienst Weg -en Waterbouwkunde (Dutch).

CBS (2011). Statline; online statistical database. Centraal Bureau voor de Statistiek, http://cbs.statline.nl/ (last accessed June, 2011).

de Bruijn, K.M. (2006). *Bepalen van schade ten gevolge van overstromingen.* Q4290. WL | Delft Hydraulics, Delft (Dutch).

de Bruijn, K.M. (2007). *Review on existing knowledge and tools for flood risk mapping. Project Q3668.* Delft, the Netherlands: WL | Delft Hydraulics.

de Moel, H., Aerts, J. C. J. H., and Koomen, E. (2011). Development of flood exposure in the Netherlands during the 20th and 21st century. *Global Environmental Change* 21: 620-627.

de Wit, A.J.W. (2003). Land use mapping and monitoring in the Netherlands using remote sensing data. *IEEE international geoscience and remote sensing symposium,* Toulouse.

de Wit, A.J.W. and Clevers, J.G.P.W. (2004). Efficiency and accuracy of per-field classification for operational crop mapping. *International Journal of Remote Sensing,* 25(20): 4091-4112.

Forster, S., Kuhlmann, B., Lindenschmidt, K. E. and Bronstert, A. (2008). *Assessing* flood risk for a rural detention area. *Natural Hazards and Earth System Sciences* 8(2): 311-322.

Grossi, P. and Kunreuther, H. (2005). *An introduction to Catastrophe Models and Insurance. Chapter 2 in: Catastrophe modeling: A new approach to managing risk.* Risk Management and Decision Processes Center, The Warton School, University of Pennsylvania.

Hoes, O.A.C. (2007). *Aanpak wateroverlast in polders op basis van risicobeheer.* Technische Universiteit Delft (Dutch).

IPCC (2007). *Climate change 2007: impacts, adaptation and vulnerability. Contribution of Working Group 2 to the Fourth Assessment Report of the Intergovernmental Panel on Climate Change.* Cambridge University Press, Cambridge.

Jacobs, C., Koomen, E., Bouwman, A. and Van der Burg, A. (2011). Lessons learned from land-use simulation in regional planning applications. Chapter 8 in: Koomen, E. and Borsboom-van Beurden, J. *Land-use modeling in planning practice.* Springer, Heidelberg.

Klijn, F., Baan, P., de Bruijn, K. and Kwadijk, J. (2007). MNP. *Overstromingsrisico's in Nederland in een veranderend klimaat. Verwachtingen, schattingen en berekeningen voor het project Nederland Later.* WL - Delft Hydraulics (Dutch).

Klijn, F., van der Klis, H., Stijnen, J., de Bruijn, K. and Kok, M. (2004). *Overstromingsrisico dijkringen in Nederland. Betooglijn en deskundigenoordeel. Rapport.* HKVLIJN IN WATER & WL nr. Q3503.10, april 2004 (Dutch).

Kok, M. and Klopstra, D. (2009). *Van Neerslag tot Schade.* Eindrapport. HKV Lijn in Water. Universiteit Twente en KNMI, Lelystad (Dutch).

Kok, M., Huizinga, H.J., Vrouwenvelder, A.C.W.M. and Barendregt, A. (2005). *Standaardmethode2004 – Schade en Slachtoffers als gevolg van overstromingen.* DWW-2005-005. 2005. RWS Dienst Weg- en Waterbouwkunde (Dutch).

Koomen, E. and Borsboom-van Beurden, J. (2011). *Land-use modeling in planning practice.* Springer, Heidelberg.

Koomen, E., Loonen, W. and Hilferink, M. (2008). Climate-change adaptations in land-use planning; a scenario-based approach. In: Bernard, L., Friis-Christensen, A. and Pundt, H. (eds.), *The European Information Society; Taking Geoinformation Science One Step Further.* Springer, Berlin: 261-282.

Linde, A.H.T., Aerts, J.C.J.H., Bakker, A.M.R. and Kwadijk, J.C.J. (2010). Simulating low-probability peak discharges for the Rhine basin using resampled climate modeling data. *Water Resources Research* 46.

Merz, B., Elmer, F. and Thieken, A.H. (2009). Significance of "high probability/low damage" versus "low probability/high damage" flood events. *Natural Hazards and Earth System Sciences* 9 (3): 1033-1046.

Merz, B., Thieken, A.H. and Gocht, M. (2007). Flood risk mapping at the local scale: concepts and challanges. In: Begum,S., Stive,M.J.F. & Hall,J.W. (Eds.), *Flood Risk Management in Europe - innovation in policy and practice:* 231-251. Dordrecht, Netherlands: Springer.

Milly, P.C.D., Wetherald, R.T., Dunne, K.A. and Delworth, T.L. (2002). Increasing risk of great floods in a changing climate. *Nature,* 415(6871): 514-517.

NBW (2003). Nationaal Bestuursakkoord Water, (www.nederlandleeftmetwater.nl), Den Haag (Dutch).

Nieuwenhuizen, W., Wolfert, H.P., Higler, L.W.G., Dijkman, M., Huizinga, H.J., Kopinga, J., Makaske, A., Nijhof, B.S.J., Olsthoorn, A.F.M. and Wösten, J.H.M. (2003). *Standaardmethode Schade aan LNC waarden als gevolg van overstromingen; methode voor het bepalen van de gevolgen van overstromingen voor de aspecten opgaande begroeiing, vegetatie, aquatische ecosystemen en historische bouwkunde.* Wageningen, Alterra, Research Instituut voor de Groene Ruimte. Alterra-rapport 709 (Dutch).

Penning-Rowsell, E. C., Fordham, M., Correia, F. N., Gardiner, J., Green, C., Hubert, G., Ketteridge, A.-M., Klaus, J., Parker, D., Peerbolte, B., Pfl¨ugner, W., Reitano, B., Rocha, J., Sanchez-Arcilla, A., Saraiva, M. d. G., Schmidtke, R., Torterotot, J.-P., Van der Veen, A., Wierstra, E., and Wind, H. (1994). *Flood hazard assessment, modeling and management: Results from the EUROflood project, in: Floods across Europe: Flood hazard assessment, modeling and management*, edited by Penning-Rowsell, E. C. and Fordham, M., Middlesex University Press, London.

Riedijk, A., van Wilgenburg, R., Koomen, E. and Borsboom-van Beurden, J.A.M. (2007). *Integrated scenarios of socio-economic and climate change; a framework for the 'Climate changes Spatial Planning' program*, Spinlab Research Memorandum SL-06, Amsterdam.

RWS-DWW (2005). *Flood Risks and Safety in the Netherlands (Floris). Rep. No. DWW-2006-014.*

Sayers, P.B., Gouldby, B.P., Simm, J.D., Meadowcroft, I. and Hall, J. (2002). *Risk, Performance and Uncertainty in Flood and Coastal Defence. A Review.* R&D Technical Report FD2302/TR1 (HR Wallingford Report SR587), Crown copyright, London, U.K. Smits, I., Wijngaard J.B., Versteeg R.P. en Kok M., (2004) *Statistiek van extreme neerslag in Nederland.* Rapport 2004-26, STOWA, Utrecht. (Dutch)

TAW (2000). *Van Overschrijdingskans naar Overstromingskans. Achtergrondrapport.* Technische Adviescommissie voor de Waterkeringen, Delft (Dutch).

Van der Hoeven, E., Aerts, J., Van der Klis, H. and Koomen, E. (2009). An Integrated Discussion Support System for new Dutch flood risk management strategies. Chapter 8. In: Geertman, S. and Stillwell, J.C.H. (eds.), *Planning Support Systems: Best Practices and New Methods.* Springer, Berlin: 159-174.

van der Klis, H., Baan, P. and N. Asselman (2005). *Historische analyse van de gevolgen van overstromingen in Nederland. Een globale schatting van de situatie rond 1950, 1975 en 2005.* WL rapport Q4005.11 (Dutch).

van Oort, P.A.J., Bregt, A.K., de Bruin, S., de Wit, A.J.W. and Stein, A. (2004). Spatial variability in classification accuracy of agricultural crops in the Dutch national land-cover database. *International Journal of Geographical Information Science,* 18(6): 611-626.

van Veen, N. (2005). *Influence of future zoning on flood risks.* ERSA conference papers, European Regional Science Association.

Vanneuville W., Maddens R., Collard Ch., Bogaert P., De Maeyer Ph., Antrop M., (2006) *Impact op mens en economie t.g.v. overstromingen bekeken in het licht van wijzigende hydraulische condities, omgevingsfactoren en klimatologische omstandigheden,* studie uitgevoerd in opdracht van de Vlaamse Milieumaatschappij, MIRA, MIRA/2006/02, Gent. (Dutch)

Vrijling, J. K. (2001). Probabilistic design of water defense systems in The Netherlands. *Reliability Engineering & System Safety,* 74(3): 337-344.

Wind, H.G., Nierop, T.M., de Blois, C.J. and de Kok, J.L. (1999). Analysis of flood damages from the 1993 and 1995 Meuse Floods, *Water Resource. Res.,* 35(11): 3459–3465, doi:10.1029/1999WR900192.

2

Flash Flood Hazards

Dénes Lóczy, Szabolcs Czigány and Ervin Pirkhoffer
Institute of Environmental Sciences, University of Pécs
Hungary

1. Introduction

Climate change research has revealed that the frequency of extreme weather phenomena with increasing damage to human assets has been gradually growing worldwide (Intergovernmental Panel on Climate Change [IPCC], 2007). The likelihood of increasing frequency of heavy precipitation events is assessed as 'likely' for the last four decades of the 20th century and 'very likely' for the 21st century. This also means that over most regions of the Earth's land surface an ever growing proportion of total precipitation will fall in the form of heavy rainfalls (Burroughs, 2003). The intensification trend of tropical cyclone activity, observed in some regions since 1970, will probably also continue in the 21st century. As a consequence, rainfall events concentrated in time and space are expected to lead to serious local flooding in many parts of the world.

Floods are remarkable hydrometeorological phenomena and forceful agents of geomorphic evolution in most physical geographical belts and, from the viewpoint of human society, among the most important environmental hazards. Except for extreme environments, floodplains and the immediate surroundings of streams are usually densely inhabited areas and, therefore, they are of high vulnerability to floods. According to the European Environment Agency (EEA, 2010), floods rank as number one on the list of natural disasters in Europe over the past decade. Authors of the report claim that "the events resulting in the largest overall losses were the floods in Central Europe (2002, over EUR 20 billion), in Italy, France and the Swiss Alps (2000, about EUR 12 billion) and in the United Kingdom (2007, over EUR 4 billion)" (p. 8.). With accumulating knowledge on the water regime of major rivers, the inundation hazard from riverine floods can be defined with some precision. To estimate the magnitude of this hazard in small catchments, however, poses more problems.

2. Flash flood research

2.1 Definitions and approaches

Flash floods (synonym: storm-driven floods) can be defined from various aspects: as hydrometeorological phenomena, natural hazards or geomorphic agents. Inundations can be referred to four basic classes: riverine floods, excess water (from rising groundwater table), coastal floods and flash floods (Lóczy, 2010). Although riverine floods along major rivers remain to be the most severe natural hazard which threaten to inflict serious damage to human life and property, recently the latter classes have also attracted more attention in scientific circles.

From a hydrometeorological aspect, flash floods are best described as events involving "too much water in too little time" (Grundfest & Ripps, 2000). This means that exceptionally high amounts of rainfall, combined with very efficient and rapid runoff on relatively small catchments, are typical of flash floods. A flash flood immediately follows the inducing storm event. The term 'flash' itself indicates a sudden rapid hydrological response of a usually small catchment, where water levels may rise to their maximum within minutes or a few hours after the onset of the rain event. Flash floods are highly localized in space: they are restricted to basins of a few hundred square kilometres or less. They are also restricted in time: response times not exceeding a few hours or are even less. Therefore, extremely short time is left for warning (Georgakakos, 1987, 2006; Collier, 2007; Carpenter et al. 1999).

It is often emphasized that heavy rainfall is a necessary but not sufficient condition for inducing flash floods. Since the entire physical environment influences their origin, flash floods are proper subjects for physical geographical investigations (Czigány et al., 2008). For instance, soil moisture conditions prior to the rainfall events are major hydrological controls of flash flood generation (Norbiato et al., 2008; Czigány et al., 2010). It is only with knowledge on the topography, soils and human impact on the catchment (steep slopes, drainage density, impermeable surfaces, saturated soils and land use) that the flood/no flood threshold can be established with some precision. Anthropogenic influences are important because some basins respond particularly rapidly to intense rainfall in the wake of disturbances in the natural drainage (stream channelization, deforestation, housing development, fire etc.) (Norbiato et al., 2008). As hydrometeorological phenomena, flash floods are best characterized by their magnitude (total amount and intensity of inducing rainfall), return interval, total runoff and similar parameters.

As geomorphological phenomena flash floods are short-duration events caused by an abrupt rise in the discharge of a river or stream, which may have remarkable geomorphic impacts through erosion and sedimentation (Reid, 2004). Previously, some geomorphologists restricted this concept to the ephemeral streams of arid and semiarid areas (Reid et al., 1994), but now the view is more excepted that the 'flashy' flood hydrographs of subtropical seasonal climates and even of humid temperate regions can also be covered in the flash flood category. There may be, however, significant differences in runoff generation and geomorphic consequences (Bull & Kirkby, 2002). The geomorphic consequences of flash floods are usually judged from the stream flood hydrograph, sediment load transported and sediment accumulation.

Flash floods are naturally not novel phenomena, the frequency of their occurrence, however, shows an increasing tendency. Until some recent disasters, flash floods have not been so intensively studied as conventional large riverine floods. In some particularly affected countries (e.g. in the United States and the United Kingdom), however, their research dates back to the 1970s and 80s (e.g. Grundfest, 1977, 1987; Georgakakos, 1987; Schmittner & Giresse, 1996; Carpentier et al., 1999; Pontrelli et al., 1999).

In the case of a sophisticated hydrological approach, in addition to precipitation, several environmental factors are also to be considered in flash flood modelling as boundary conditions. Soil characteristics (actual moisture content, permeability, ground surface alterations and vertical soil profile) influence runoff production and help define flash flood prone areas. Various catchment characteristics (e.g. size, shape, slope, land cover) also affect runoff and the potential occurrence of flash floods. Consequently, the approach towards flash flood hazard assessment should be substantially different from that applied for modelling inundations along large rivers, in coastal areas or in valleys and lowlands due to elevated groundwater levels (Czigány et al. 2011b - see also below).

Flash floods are often associated with other natural hazards. Then their damage is partly due to the fact that they often trigger debris flows, i.e. hyperconcentrated flows, where the proportion of sediment load surpasses that of water discharge (Iverson, 1997). The particle size of the sediment swept at rates up to 20 m s^{-1} - confined within narrow valleys or erosion gullies - may range from clay to large blocks. The rainfall that produces flash floods often saturates entire hillslopes and subsequently may induce extensive slumps. Through blocking valleys and impounding stream flow, the slumps create suitable conditions for the next flood.

2.2 Flash floods in the world

Disastrous flash flood events can be cited from almost all continents. The majority of the 'classic', best documented events have been reported from the United States. Every year, riverine, coastal and flash floods are responsible for more fatalities than any other meteorological phenomenon in the Unites States. On the 30-year average, flood-related death toll totals 120 fatalities annually. From 1996 through 2003, 3000 flash flood events were documented in a year on average (Collier, 2007). Although some authors note that remarkable progress in research and warning have been made in the US and in some other countries (e.g. in the UK), flash floods are still among the most dangerous natural phenomena worldwide (Davis, 2001). Detailed documentation is available since the 1970s. In 1972 alone two disastrous floods were recorded in the US: 125 people were killed in Buffalo Creek, West Virginia, as a consequence of the failure of a coal-waste dam (Davies et al., 1972) and 238 people in Rapid City, South Dakota, where 380 mm rain fell within 6 hours (Davis, 2001). One of the best documented flash floods of all time occurred in the Thompson Canyon, Colorado, a small watershed (181 km^2) drained by one of the tributaries of the Colorado River. In 1976, 350 mm rain fell in less than six hours, flooding the narrow canyon floor (Caracena et al., 1979) and when the water level rose suddenly and unexpectedly by 6.5 m (Davis, 2001). 145 people were killed, 418 houses destroyed and 138 damaged. Total material damage amounted to USD 40 million.

A flash flood (well studied and even documented in videos, now on YouTube) took place in England, on the Cornwall Peninsula, in Boscastle, on August 16, 2004 (Golding, 2005). It provided an opportunity for the application of the land surface model called MOSES-PDM (Met Office Surface Exchange Scheme incorporating the Probability Distributed Model) to portray the evolution of soil moisture conditions from meteorological information (radar rainfall and satellite cloud observations). The entire rainfall event lasted for only about seven hours, but was very localized. The total 24-hour cumulative rainfall reached 200.4 mm at one location (Otterham) (Golding, 2005). In places upstream from Boscastle rainfall intensity reached 24 mm within 15 minutes, while in Boscastle 89 mm rain fell in an hour (Golding, 2005). The probability of such a high-intensity rainfall in Boscastle, at least according to the available statistical data, is 1 to 1,300. Antecedent high precipitation also influenced soil moisture and runoff on the higher portions of the catchment. The intense rainfall was followed by a 2-metre rise in the water level of the local Valency Stream, the discharge of which reached 180 m^3 s^{-1}, a value of an estimated return time of 400 years

(Bettess, 2005). During the Boscastle flash flood event, 100 residential homes were destroyed and 75 cars were swept to the sea. Due to the efficient assistance of the available rescue teams, no fatalities happened. This rare event resulted from a combination of hydrometeorological factors (Golding, 2005): unusually high rainfall efficiency (relative to the moisture content of the inflowing air) and the exceptionally long stay of intense storms over the same catchment.

For similar reasons, mountain environments in the Mediterranean are also seriously threatened by flash flooding (e.g. Borga et al., 2007). In the Aragonian Pyrenees, the catchment of the Barranco de Áras stream (only 19 km^2 in area) was affected by enduring (5-hour) rainfall with 500 mm h^{-1} peak intensity and an estimated 243 mm total amount on 6 August 1996. In the Biescas camp-site 87 people died because flash flooding of 600 m s^{-1} discharge was combined with a debris flow transporting 68,000 m^3 debris (Gutiérrez et al., 1998). The tragic underestimation of the capacity of check dams by engineers had also contributed to the disaster. In the French Côte d'Azur, in Draguignan (Var *département*) the 10 June 2010 flash flood killed 37 people in the town and its neighbourhood, caused blackouts and cut away the village from the world (Telegraph, 2010). It was triggered by a huge cloudburst (350 mm within 20 hours), unobserved in the area since 1827. (For a more complete overview of flash floods in Europe see Gaume et al., 2009.)

As it has been mentioned, however, arid and semiarid regions are the most favoured environments for flash flood generation (Reid et al., 1994). According to research in Israel (Cohen & Laronne, 2005), for instance, flash floods of arid regions involve both bedload and suspended sediment concentrations much higher than in the perennial rivers of humid environments. In arid or desert regions storms cut arroyos (intermittent gullies with flat floors and vertical walls). Flash flooding in an arroyo can occur in less than a minute, with enough power to wash away sections of pavement, large boulders, cars and even houses. Although the sediment yield of individual events is large, fortunately, flood events rarely occur and mean annual sediment yields remain low in arid environments (Graf, 2002).

The prediction of heavy rainfall and ensuing flash floods is particularly challenging in the tropical belt, especially on islands with intense, localized and mostly convective rainfalls (e.g. Kodama & Barnes, 1977). Devastating events have been reported from various Caribbean islands (Laing, 2004). During an El Niño winter, on 5–6 January 1992, heavy rainfall produced flash floods in Puerto Rico and caused 23 deaths and 88 million U.S. dollars in damage (National Oceanic and Atmospheric Administration, National Weather Service [NOAA NWS] 1992). At a few stations the amount of the rainfall, associated with a quasi-stationary front at the surface and an upper-level trough, was up to 500 mm (Laing, 2004).

High relief often generates heavy rainfall through orographic lift or by creating persistent low-level convergence which induces new convection (Weston & Roy, 1994). Such hydrometeorological and other environmental conditions were associated with another deadly flash flood, too, that occurred in Jamaica on 3–4 January 1998. The northeastern region was affected by heavy rainfall, which induced both flash floods and mudslides which caused five deaths and more than nine million U.S. dollars in damage to property, agriculture, and infrastructure (Laing, 2004). The situation was aggravated by antecedent rainfall from a strong cold front, currents of moist air masses at lower and higher topographic levels. Similarly to the Puerto Rico event, orographic lifting contributed to the disaster also here. The steep slopes of Blue Mountain ridge enhanced and localized convection over the region. In addition to antecedent rainfall, land-use practices (e.g. floodplain encroachment) also increase flood hazards.

Some examples of flash floods can also be cited from a continent infamous of rather extreme spatial and temporal variations in weather conditions, Australia. The most frequent cause of flash flooding is slow-moving thunderstorms. These systems, related to the El Niño-Southern Oscillaton (ENSO) circulation pattern, can involve strong updrafts of air which suspend huge amounts of rain before releasing a deluge onto the ground (Allan, 1993). Water in creeks, drains and natural watercourses can rise at dangerous rates. On the evening of 26 January 1971, seven people died in Canberra as flash flood waters from a nearby thunderstorm flooded roadways near a drainage channel. It was estimated that around 95 mm of rain fell in one hour during the event. On another occasion, in Sydney on 7 November 1984, 127 mm fell in one hour leading to damage of around AUD 128 million (in July 1996 terms) (Australian Government, Bureau of Meteorology [AG BoM]). In the drier ('outback') regions of Australia flash floods are more common, but - the vulnerability being lower - their documentation is not so good.

Flash floods are increasingly observed in urban areas, where the surface is unable to absorb large amounts of water in a short period. In urban areas the hazard is exacerbated by various - and not exclusively physical - contributing factors and vulnerability is significantly higher - often because of the sudden rise of water levels. For instance, many cities of Latin America show uncontrolled and disorganized urban growth (Stevaux & Latrubesse, 2010) with infrastructure and production systems (railways, roads, plants) concentrated in densely populated valleys. Disasters are produced by a combination of tropical storms inducing flash floods and landslides and urban occupation of the valleys. It is often emphasized that the increased proportion of impervious urban surfaces and the limited drainage capacity are responsible for flooding. A storm on December 14–16, 1999, caused catastrophic landslides and flooding along a 40-km coastal strip north of Caracas, in the coastal state of Vargas, Venezuela with its extremely steep and rugged topography (mountains 2700 m high within about 6–10 km of the coast) (Brandes, 2000). The rivers and streams of this mountainous region drain to the north and emerge from steep canyons onto alluvial fans before emptying into the Caribbean Sea. Damage to communities and infrastructure was so serious because here little flat area is available for development with the exception of the alluvial fans. In Vargas state probably almost 50,000 people were killed, more than 8000 individual residences, and 700 apartment buildings were destroyed or damaged and total economic losses are estimated at USD 1.79 billion (Wieczorek et al., 2001). On average, at least one or two major flash-flood or landslide events per century have been recorded in this region since Spanish occupation in the 17th century.

2.3 Flash floods in Hungary

Until very recently flood hazard research in Hungary had focused on riverine floods, particularly those along the two major rivers, the Danube and its main tributary, the Tisza (Lóczy & Juhász, 1996; Lóczy, 2010). There are relatively few papers published on flash flood events in Hungary (Gyenizse & Vass, 1998; Fábián et al., 2009). In the mainly lowland and hill environments of Hungary flash floods do not appear a major hazard. Another reason is the lack of appropriate monitoring systems in the flash flood affected catchments (Vass, 1997). In the wake of the events of the first decade in the 21st century, however, flash flood related disasters and their consequences have been appearing more and more frequently in the Hungarian media.

Some recent events that made news took place in the Mátra Mountains (North-Hungary) in 1999 (Koris & Winter, 1999) and again on 18 April 2005 (Horváth, 2005). The rainfall resulted from an atmospheric complex of several convective cells transporting moist air like a conveyor belt against the mountain slopes. Huge boulders of volcanic rock were transported by the local stream (Fig. 1). As an aftermath of the flood, slumps in 500 m length along the undercut bank are regularly generated.

Fig. 1. Deposits of debris flow after the Mátrakeresztes flash flood (by permission of the Nógrád County Disaster Prevention Directorate)

Some of the most disastrous events in Hungary occurred on 15–16 May 2010, when a strong cyclone reached the Carpathian Basin. Hitherto virtually unknown stream names (e.g. Hábi Canal, Bükkösd Stream and Baranya Canal) appeared in the media. The ensuing floods caused significant economic losses in Southern Transdanubia (Southwest-Hungary), a region of mostly dissected hill topography and a dense drainage network (Fig. 2). Daily precipitation amounts, intensities and stream stages broke records and cumulative precipitation locally exceeded 300 mm in the Kapos drainage basin during May and June. In Csikóstőttős village 65 people were evacuated. A one metre high flood swept away a children's camp in Szekszárd, where firemen assisted to evacuate the campers. On 16 June 2010 182 mm of rainfall fell on the village of Iklódbördőce in the Zala Hills (Southwest-Hungary) and caused a mudflow. Estimated by the insurance companies, the May and June events caused ca HUF 100 billion (EUR 360 million) economic losses, at least 3,100 residential homes were damaged and the agricultural damage totals ca HUF 30 billion (EUR 110 million). A summary of water-related damage recorded by insurance companies shows the distribution of insurance events in Southern Transdanubia between 1980 and 2005 (Fig. 3). In the light of the 2010 floods, the number of events presented here seems to be underestimated (property insurance was probably not comprehensive), but the map is informative of the zones of highest flood risk.

3. Flash flood modelling

3.1 Objectives

The above mentioned events directed the attention of water management experts and of the wider public to flash flood hazard. Consequently, flood prevention needs to be also

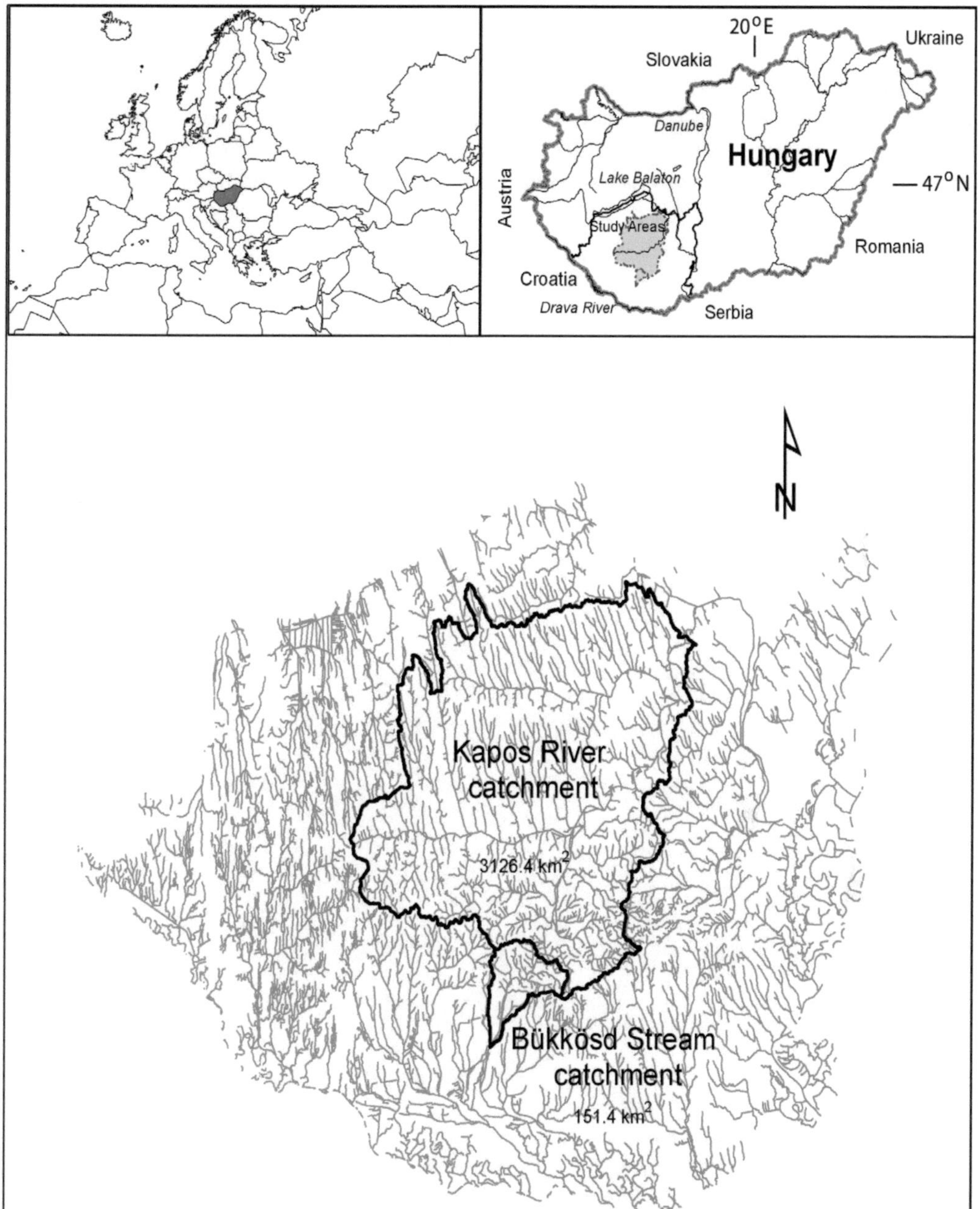

Fig. 2. The drainage network of Southern Transdanubia with the catchments studied and an inserted location map (from the river network database of Hungary)

extended to the previously neglected small mountainous catchments at relatively low elevations, which cover ca 30 per cent of the entire land area of Hungary, while the cumulative length of streams total more than 20,000 km (Kaliczka, 1998). It is recognized that flood assessment and prevention measures, such as the presently planned and constructed nationwide Flood Risk Information System (abbreviated from the Hungarian as ÁKIR), a model and software-based flood prediction system, also have to cover minor catchments potentially affected by flash floods (Pászthory & Szigeti, 2009).

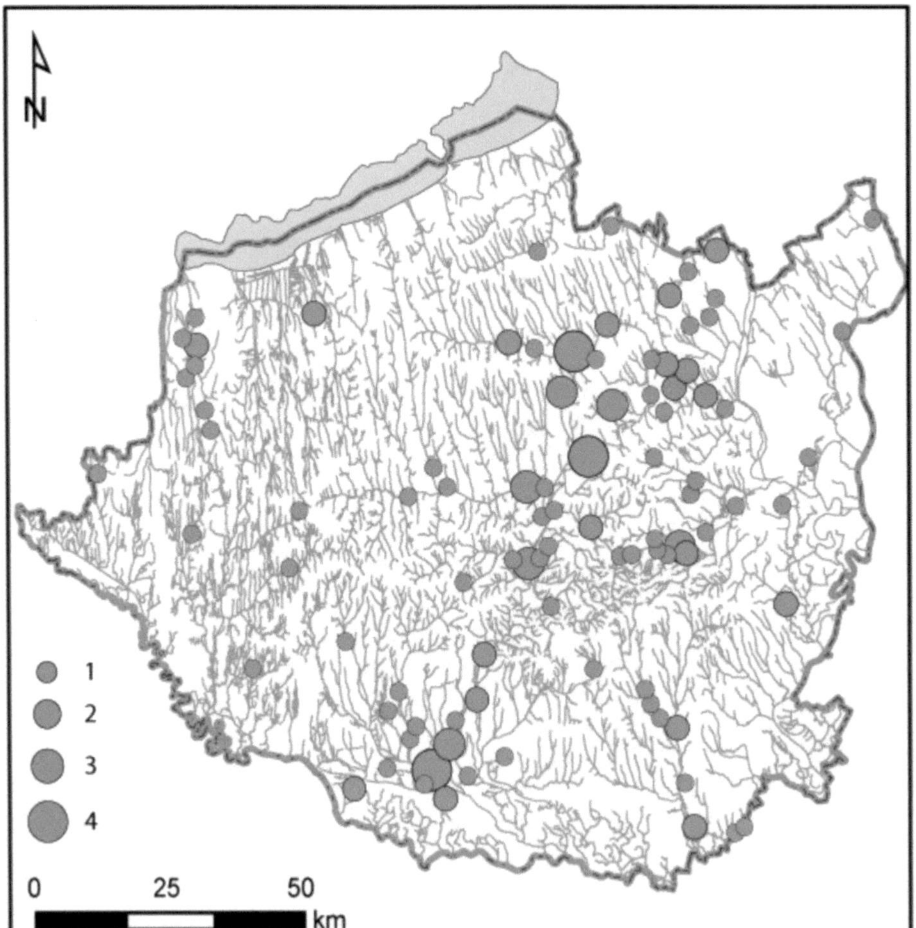

Fig. 3. Water-related events with damage to property in Southern Transdanubia, 1980–2005, based on insurance data (after Varannai, 2005). The map shows the number of occurrences

In the following part of this chapter a proposed flood risk assessment mapping procedure and numerical flood forecasting system are outlined. Our objective is to identify flash flood risk in order to promote the development of a flood warning system and to mitigate flood-related life and property losses. The screening of the country's territory for flash flood hazard and rating the risk in the various regions would also help insurance companies in estimating expectable damage.

3.2 International models

The modelling of flash flood hazards requires a more complex approach than that of large riverine floods as more environmental factors have to be considered and regularly monitored. Flood modelling serves flood forecasting, i.e. the estimation of future flood conditions, while flood warning means the information of the public on the timing and location of a flood event allowing them sufficient time to take preparatory actions. A Decision Support System (DSS) in flood management assists authorities to make decisions based on forecast information (the expected characteristics of the flood, the number of inhabitants threatened and the evacuation infrastructure available) (Maarten et al., 2007).

Given the significance of catchment properties in the generation of floods, distributed hydrological models seem to be best suited for the purpose of flash flood prediction. These models of various levels of complexity are built on a grid-based network, small subbasins or triangulated irregular networks (TINs). Some frequently used examples of physically based, distributed hydrological models (cited by Paudel 2010, see references there) are

- Gridded Surface/Subsurface Hydrologic Analysis (GSSHA), developed by the US Army Corps of Engineers Engineering Research and Development Center (USACE-ERDC);
- Modular Modeling System (MMS) – US Geological Survey;
- Hydrology Lab's Research Modeling System (HLRMS) – National Weather Service Hydrology Laboratory;
- Vflo™ – Institute of Environmental Sciences, University of Oklahoma – Institute of Environmental and Natural Sciences, Lancaster University, UK;
- MIKE-SHE – Danish Hydraulic Institute;
- Hydrologic Research Center Distributed Hydrologic Model (HRCDHM) – Hydrologic Research Center;
- Soil and Water Assessment Tool (SWAT) - US Department of Agriculture, Agriculture Research Service (USDA-ARS)
- TIN-based Real-Time Integrated Basin Simulator (tRIBS) – Massachusetts Institute of Technology;
- Variable Infiltration Capacity (VIC) – University of Washington.

The US Army Corps of Engineers (USACE) has played a vital role in the development and application of hydrological models in the Unites States since the early 1960s. USACE models are extensively used throughout the world. As a well-established standard model, HEC-HMS is widely used in the United States and worldwide for the simulation of surface runoff. For the mapping of inundated areas the models HEC-RAS, HEC-GeoRAS and ArcGIS 9.1 are useful. The HEC-HMS is designed to simulate the precipitation-induced runoff processes in catchments of dendritic drainage pattern. It is applicable in a wide range of geographical areas, equally for large drainage basins and small catchment for solving the widest possible range of problems (water availability, urban drainage, flow forecasting, future urbanization impact, reservoir spillway design, flood damage reduction, floodplain regulation etc.) (US Army Corps of Engineers, 2005).

The default model used in the European Flood Forecasting System (EFFS) is LISFLOOD (De Roo et al. 2000), a physically based catchment model, developed for the European river basins. As a rainfall-runoff model it has inputs of data on topography, precipitation amounts and intensities, antecedent soil moisture, land use type and soil type in the form of maps (topography, drainage network, CORINE land cover, soil depth, soil class). The meteorological variables required are rainfall, potential evaporation (for bare soil, closed canopy and water surfaces) and daily mean air temperature (De Roo et al., 2000).

3.3 Methods

3.3.1 Differences between the modelling of riverine and flash floods

Conventional large riverine floods and flash floods do not only differ in their general characteristics but also in their time of concentration (T_c or lag time) and duration of flood peaks. As it has been mentioned, for flash floods T_c is not more than 6 hours (NOAA definition). This extremely short lead time makes warning and prevention very difficult. In unexplored and ungauged catchments the single source of information is field surveys. Flood reconstruction is often only possible from the falling limb of the hydrograph or by assessing the aftermaths of the event (including the study of deposits - Costa, 1983).

As far as the triggering process is concerned, flash floods are generally associated with intense and convective rainfalls - often further enhanced by orographic effects (Horváth, 2005). Large riverine floods, on the other hand, are often preceded by days of incessant rainfall over hundred or thousand square kilometres affecting several drainage basins. It is to be noted here that in humid continental environments flash floods do not only occur in the summer, but rain-on-snow events may also generate winter flash floods (as shown for Southwest-Hungary by Pirkhoffer et al., 2009a,b; Czigány et al., 2010).

For flash flood modelling and forecasting usually an area of 10 to 200 km^2 is selected, i.e. about one or two orders of magnitude less than for large riverine floods. During flash floods peak flow may exceed baseflow several hundred times - although peak discharge only lasts for a few hours. Moreover, as flash floods are usually triggered on the upper reaches of the stream, where the channel is narrow, stage increase is even more pronounced than flow changes. Figure 4 shows an idealized, rapidly rising and slowly attenuating hydrograph, typical of many hill regions, where T_c is extremely short. The second flow peak, triggered by a moderate rainfall, is due to higher soil moisture. The time of residence in the reservoirs (e.g. canopy or surface storage) of the hydrological cycle is usually much shorter for flash floods. Rainfall intensity largely exceeds infiltration rate, and, thus, excess runoff into intermittent streams reach the beds of permanent rivers, where a rapid rise of water will be observed (Fig. 4.).

Numerical modelling is further complicated by a plethora of environmental factors to be considered during the simulation process. Judging from the available data on documented flash floods, it can be claimed that built-up areas of the highest risk are located along the boundaries of areas of higher elevation and adjacent lowlands as well as at the abrupt narrowing of river valleys (bottlenecks). As periodically water-filled gullies often function as preferential flow paths during convective rainstorms, adequate knowledge on topography is essential for highly accurate flash flood prediction.

3.3.2 Runoff modelling

In case studies and pilot catchments we used 50-m and 10-m Digital Elevation Models (DEMs) based on topographic maps. In the field TOPCON HiPER Pro RTK GNSS high-precision GPS and SOKKIA surveying instruments were employed to improve the spatial resolution of the generated DEM to 1 m. This, however, was achieved only locally - usually in the immediate vicinity of watercourses.

Surface runoff was simulated using HEC-HMS, which has the advantage of working with distributed precipitation data available from weather radar and a continuous soil-moisture-accounting model (Pirkhoffer et al., 2009b). Radar images and various meteorological data were obtained from the Hungarian Meteorological Service (OMSz), while hydrological data (e.g. water stage and discharge) were received from the Research Institute for Environmental Protection and Water Management (VITUKI Rt.), the South-Transdanubian Environmental Protection and Water Management Directorate (DDKÖVIZIG) and the MECSEKÉRC Rt., a successor enterprise to the former uranium mining company. To obtain field data we monitored soil moisture (using Time Domain Reflectrometry technique), canopy cover and precipitation at 14 monitoring stations in a 1.7 km^2 pilot catchment (Pirkhoffer et al., 2009b; Czigány et al., 2010). Runoff output data were then compared with observed flow.

3.3.3 Rapid screening and GIS-based risk assessment

The first comprehensive, but least detailed type of approach to flash flood modelling is rapid screening that usually employs ARC GIS and SGA GIS softwares (Pirkhoffer et al., 2009b; Czigány et al., 2011a). The input data for this analysis comprise various topographical, geological, soil and land use parameters. Rapid screening models serve to delineate the area with a natural hazard or rate vulnerability and risk in that area (Cobby et al., 2009; Czigány et al., 2008) in order provide a general overview of its level for experts, decision-makers and the public.

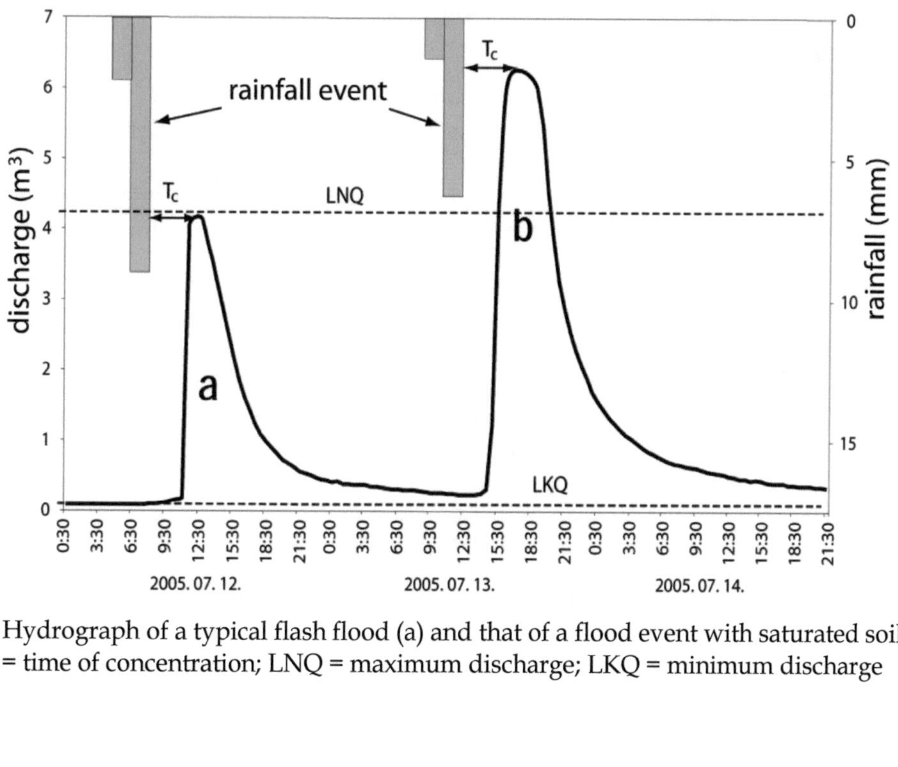

Fig. 4. Hydrograph of a typical flash flood (a) and that of a flood event with saturated soils (b). T_c = time of concentration; LNQ = maximum discharge; LKQ = minimum discharge

First, the catchments potentially affected by flash floods are identified. In the next step, risk assessment has to be carried out individually for each catchment as catchment properties influence flood level and stream behaviour. The impacts of floods are most pronounced along the watercourses and at the outflow point of the catchment. Therefore, all catchments are assigned with a unique ID number and the outflow points (usually in built-up areas) receive the same ID.

There are two approaches available for flood risk assessment: the first is based on passive factors, i.e. those which do not change significantly with time, while the second method focuses on the active factors, i.e. those which show significant variations with time (precipitation, canopy cover and soil moisture content). Passive environmental factors are determined with relatively high accuracy; spatially and temporally correct data on active factors, however, are difficult to obtain (Bálint & Szlávik, 2001).

The environmental factors incorporated in the 1:100,000 risk map are classified into five categories: topographical parameters (derived from the DEM), drainage netwok (from the river network database of Hungary), land cover (from CORINE Land Cover 2000), soils (from the AGROTOPO Hungarian soil database) and hydrological conditions. The three topographical properties were average slope, slope range and valley density for the catchment. Four soil parameters, which influence surface runoff, infiltration and interception were considered: soil depth, physical soil type, the ratio of barren/vegetation-covered surfaces (in limestone areas). Data on the hydrological factors contributing to flash flood generation were borrowed from the river network database of Hungary, created in accordance with the Water Framework Directive of the European Union. As confluences (number of tributary rivers) are prone to enhance the magnitude of flash floods (also proven during the Mátrakeresztes flash flood event - Horváth, 2005), first the number of stream confluences per unit area (1 km^2) were determined. Then drainage density was incorporated in the model.

GIS-based risk mapping represents a transitional type of models. It is closely related to rapid screening, but already points towards numerical analyses. The basic difference from rapid screening is that flooding in this case is not directly associated with a given rainfall event. Rainfalls are incorporated in a rather hypothetical manner: the extent of flooding is determined from a threshold height above the valley floor or the mean stream stage.

GIS models are primarily based on topography: all parameters, including runoff, T_c and drainage network, are derived from a topographic map or a DEM (Digital Elevation Model). The spatial resolution is at least 5 or 10 m. (Errors tend to be significant: between the calculated and the actual watercourses may reach 100 m.) The models which ignore infiltration and the canopy effect and define runoff direction and volume exclusively from topographic models are called impervious surface (IS) models. However, to obtain a true picture of runoff behaviour, the impact of soils and land use (Fig. 5a) cannot be neglected. Figure 5b clearly show the differences in runoff according to IS models with light colours, while the black zones indicate runoff also influenced by soils and land use.

Channel widths vary greatly in areas of high relief, ranging from 0.5 m to dozens of metres. When the spatial resolution of the topographic map exceeds channel width, a channel as a physical entity will not be shown on the final output map but a theoretical centerline represents the stream (valley inundation model). In this case, however, we have to define the valley floor through visual interpretation, wherever possible, including bottlenecks and broader floodplains (Fig. 5c). As it has been mentioned flood levels are approximated by height above the channel (Fig. 5d).

Flood risk was determined by the complex, superimposed impact of the 50-m resolution input grid databases of passive factors through appropriate weighting. Figure 5 summarizes the major elements of a GIS based runoff model and its mapping possibilities. Obviously, the number of included input parameters will determine the accuracy of the output vulnerability map.

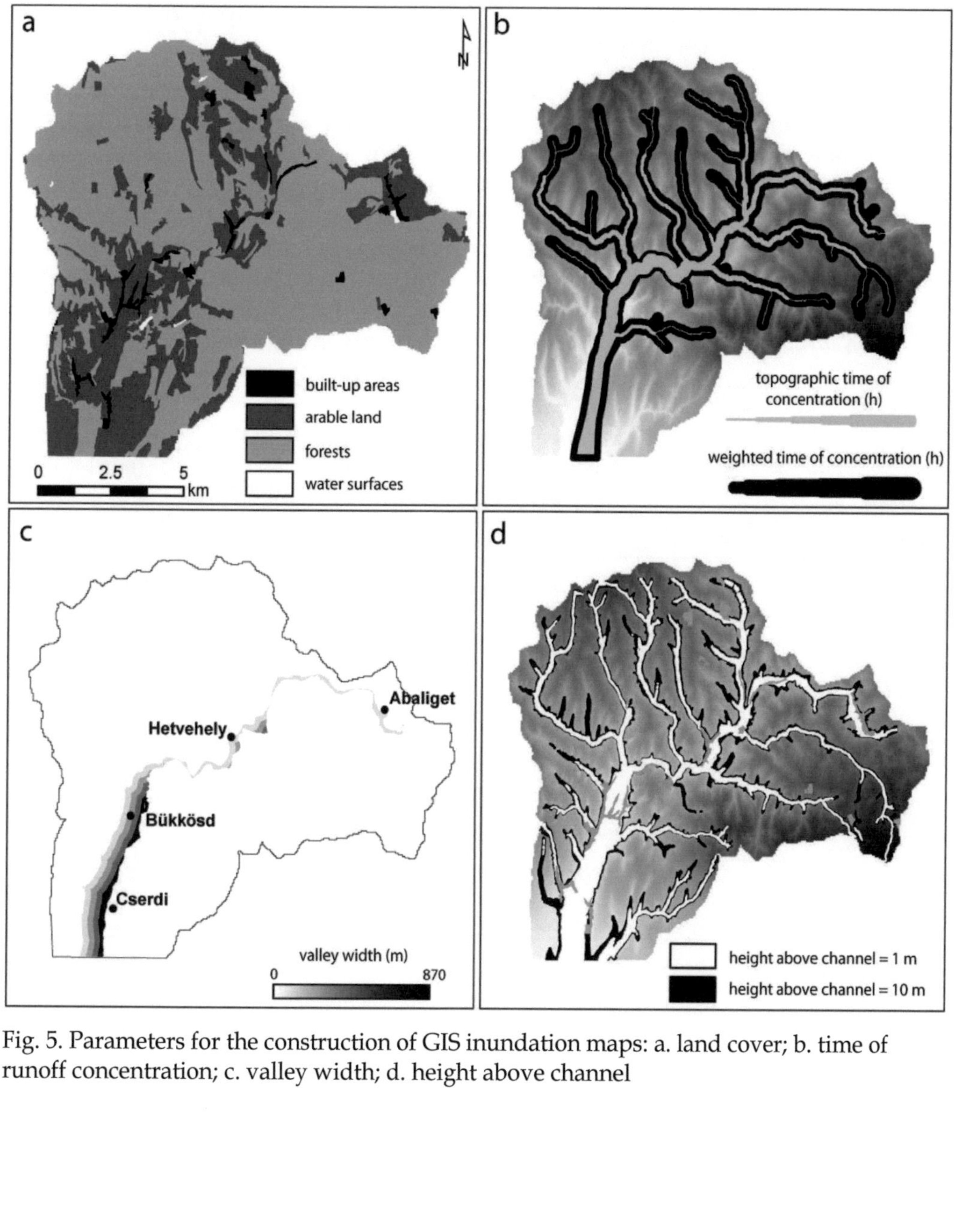

Fig. 5. Parameters for the construction of GIS inundation maps: a. land cover; b. time of runoff concentration; c. valley width; d. height above channel

3.4 Discussion

3.4.1 Correlation between rainfall and flood levels

As mentioned above, the primary triggering factors of flash floods are high-intensity convective rainfalls that are often associated with supercells. Below we discuss the spatial and temporal features of the heaviest rainfall event (in the second half of May and early June 2010) in Southern Transdanubia. The attention this event attracted helped us collect the necessary input data for the analyses.

Insurance only covers property damage in Hungary if rainfall events exceed 30 mm daily precipitation, officially confirmed by the Hungarian Meteorological Service (Varannai, 2005). The average of at least one event exceeding 30 mm occurs each year in the study area. The actual number of events in this category is shown in Table 1.

A persistent waving low-pressure system dominated in the central and western part of the Mediterranean and Central and Eastern Europe in mid-May and stayed in this region for three to four days. Similarly, the Carpathian basin was affected by moist air masses generating extensive, prolonged and relatively high-intensity precipitation on 14 to 17 May. The second half of May was characterized by local but intense showers and downpours. The 15 and 16 May flash floods were typical from a hydrological viewpoint, but unusual from a meteorological aspect as typical convective cells were not observed in this period. However, the soils were saturated in the upper and steep portions of the catchments of the Baranya and Hábi Canals and the Bükkösd Stream prior to the event, in early May.

Meteorological station	Number of rainy days above 30 mm precipitation	Number of rainy days	Cumulative precipitation (mm)
Siófok	3	21	257.9
Sellye	4	23	274.5
Sátorhely	2	22	177.3
Sármellék	2	23	204.2
Pécs	3	25	253.6
Árpádtető	5	24	385.0
Nemeskisfalud	3	24	273.2
Nagykanizsa	2	22	251.0
Kisbárapáti	2	25	185.1
Keszthely	3	24	385.5
Kaposvár	3	22	226.7
Iregszemcse	2	26	226.7
Iklódbördőce	4	22	285.6
Homokszentgyörgy	1	22	175.1
Fonyód	3	27	278.3
Bátaapáti	3	25	308.0

Table 1. Selected rainfall properties of the studied area between May 1 and June 16, 2010

Therefore, the soils acted as an impervious surface triggering extreme surface runoff. Soil moisture content only slightly decreased in the following two-week period, thus the second storm with less cumulative rainfall induced flash floods again on 31 May and 1 June. Over the period of 1 May to 16 June the cumulative number of rainy days reached at least 21 at all rain gauges operated by the Hungarian Meteorological Service in Southwest-Hungary (Table 1 and Fig. 6). Groundwater tables in the observation wells of the area indicated a mean rise of 1 to 1.2 m over the entire region (DDKÖVIZIG, 2010).

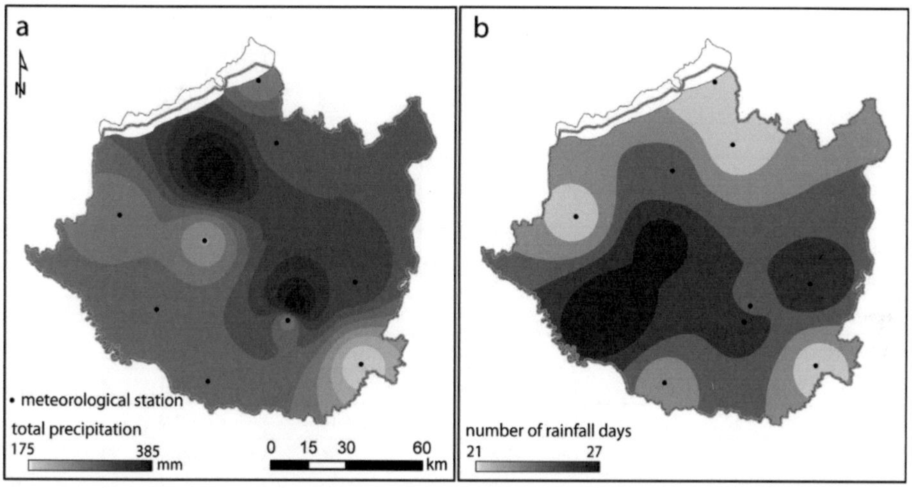

Fig. 6. Total cumulative rainfall (a) and number of rainy days (b) in Southern Transdanubia between 1 May and 16 June 2010 (data provided by the Hungarian Meteorological Service)

Table 2 clearly illustrates the extreme precipitation characteristics of the mentioned 47-day period. At many rain gauges in the study area precipitation reached or even exceeded 50% of the mean annual rainfall. The long-term average May precipitation in Pécs is 84 mm, while the cumulative precipitation in May 2010 was nearly threefold higher. The return time of such precipitation is estimated at 400 years.

The extremity of rainfall is also clearly reflected in the actual intensity values. For short periods, intensity values reached 30 mm h^{-1}, while 10-minute intensity was 51.6 mm h^{-1} at the Keszthely main meteorological station. For small mountainous catchments it is essential to know the areal extent of the rainfall zone. Due to the scarcity of rain gauges, we have to rely on radar images. Convective cells are around 5 to 10 km across, thus radar images of adequate (at present 2 by 2 km) resolution are extremely helpful in the estimation of the areal extent of precipitation for modelling purposes. Heavy rainfall characterized the settlements of Sásd and Csikóstőttős on 15 May 2010 (Fig. 7) and maximum rainfall and intensity were observed basically in the same area on the following day (16 May 2010).

On 15 May 86 mm of rain fell on the upper catchments of the Baranya Canal, where T_c is shortest within the catchment, with similar flood stages. As a consequence, rapidly rising flood stages were just slightly off from the previous records (Fig. 8). South of the divide, in the mountainous Bükkösd Stream catchment, the rainfall was much more prolonged and high water stages persisted longer at the Szentlőrinc stream gauge than at the gauges upstream (Fig. 9).

Meteorological station	Total precipitation in the study period (mm)	Annual mean precipitation, 1941-1970 (mm)	Total precipitation in % of the mean of 1941-1970	Annual mean cumulative precipitation, 1961-1990 (mm)	Total precipitation in % of the mean of 1961-1990
Bátaapáti	308.0	741	41.57	593.0	51.94
Fonyód	278.3	730	38.12	561.2	49.59
Homokszentgyörgy	175.1	773	22.65	648.2	27.01
Iklódbördőce	285.6	...	...	688.0	41.51
Iregszemcse	226.7	640	35.42	617.0	36.74
Kaposvár	225.8	746	30.27	578.6	39.02
Keszthely	385.5	664	58.06	526.9	73.16
Kisbárapáti	185.1	688	26.9	559.3	33.09
Nagykanizsa	251.0	743	33.78	726.0	34.57
Nemeskisfalud	273.2	...	...	648.8	42.11
Pécs. Ifjúság u. 6.	331.6	741	44.75	...	...
Pécs Pogány	253.6	666	38.08	620.0	40.90
Pécs, Árpádtető	385.0	839	45.9	729.6	52.77
Sármellék	204.2	...	...	585.3	34.89
Mohács, Sátorhely	177.3	631	28.10	588.0	30.15
Sellye	274.5	725	37.86	695.6	39.46
Siófok	257.9	615	41.93	577.0	44.70

Table 2. Cumulative rainfall amounts of selected settlements in Southwest-Hungary between May 1 and June 16, 2010, compared to the long-term annual average

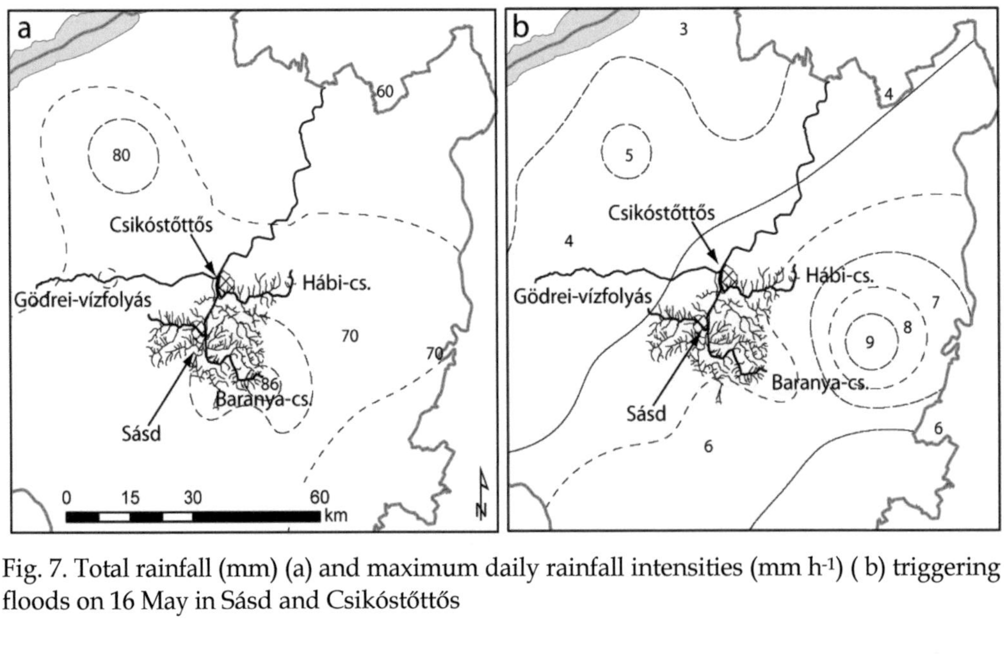

Fig. 7. Total rainfall (mm) (a) and maximum daily rainfall intensities (mm h^{-1}) (b) triggering floods on 16 May in Sásd and Csikóstőttős

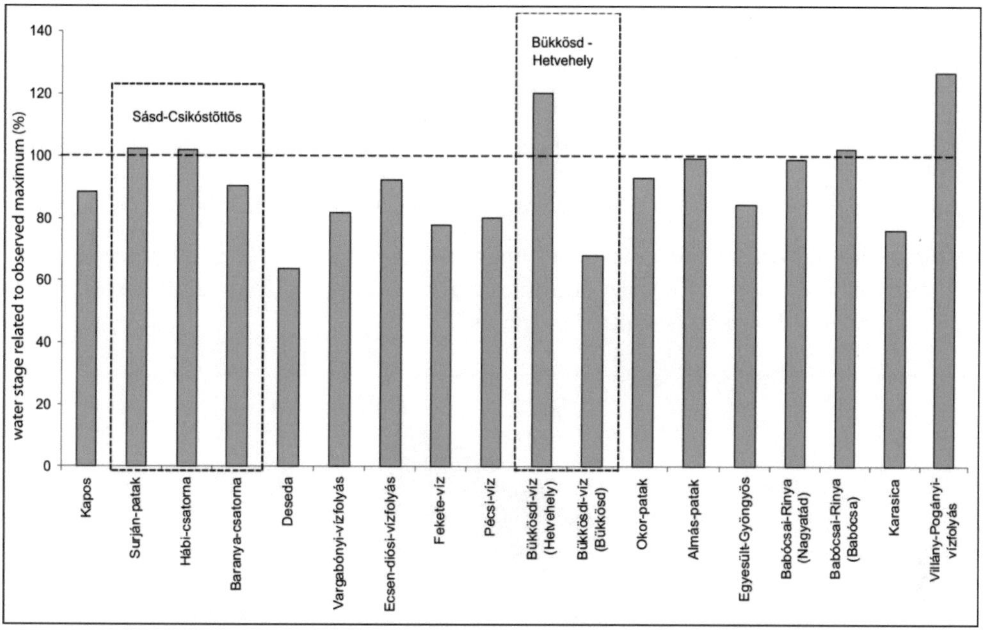

Fig. 8. Stages of selected Southern Transdanubian watercourses between 15 and 18 May 2010, in percentage of the highest stage observed to date. Crucial settlements are marked

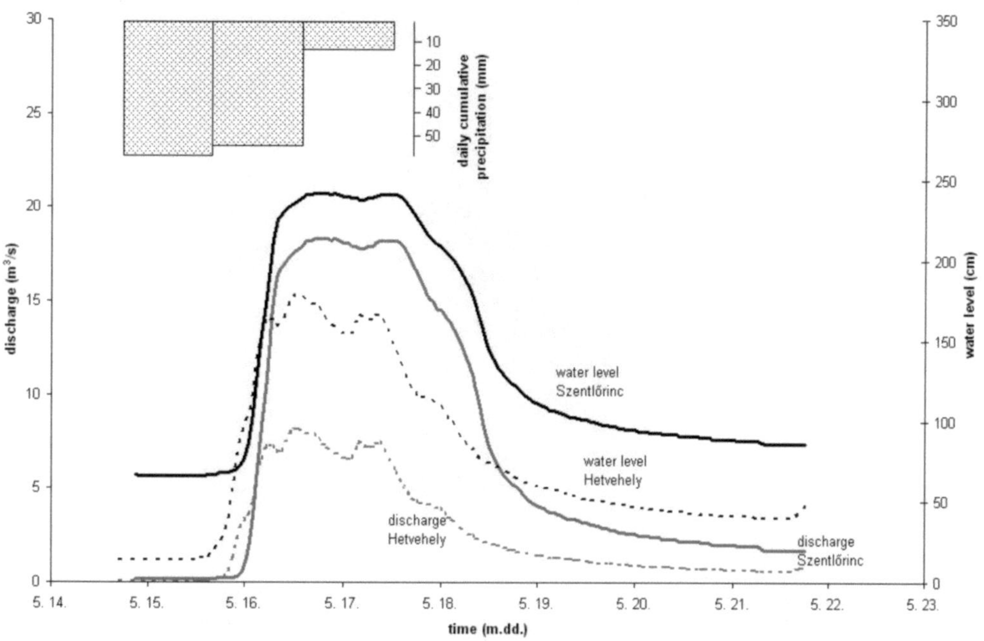

Fig. 9. Discharge and water level curves of the Bükkösd Stream at the Szentlőrinc and Hetvehely stream gauges, showing cumulative rainfall amounts from 15 to 22 May 2010 (data from Institute of Hydrology, Research Institute for Environmental Protection and Water Management [VITUKI])

3.4.2 Results of risk mapping and hydrological modelling

Altogether 210 catchments were delineated in the study area. Their average size is 42 km^2, the smallest is 2 km^2, while the largest is 300 km^2 in area. Figure 10 shows the catchments delineated using the above described method. Considering the combined effect of hydrology and precipitation pattern in Southern Transdanubia, there is a risk of flash flood with a return period of maximum 10 years in almost all mountains and hills of the region.

The categories shown on the output risk map indicate a relatively good correspondence with observed locations of flooding and inundations during the May, 2010 events and the map seems reliable for risk assessment purposes (Fig. 10).

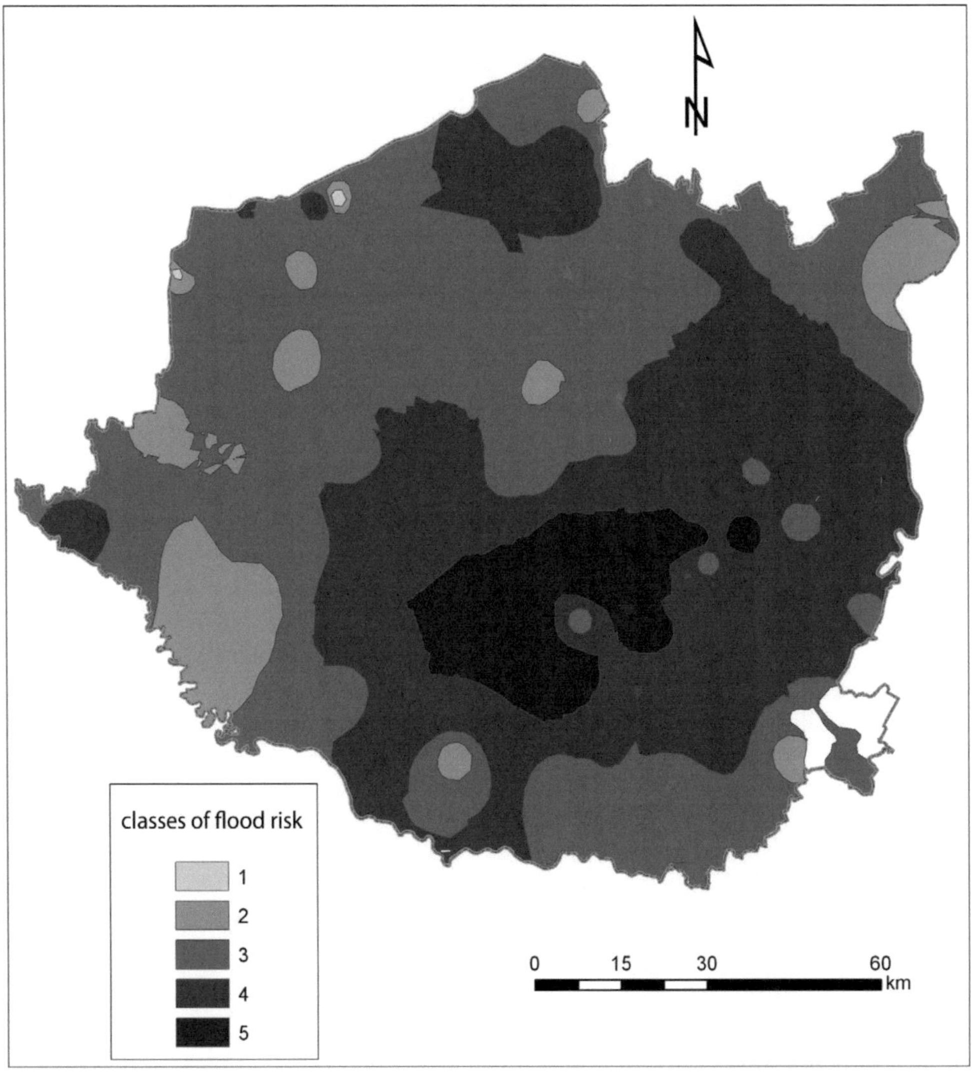

Fig. 10. A flood risk map of Southern Transdanubia prepared using the rapid screening technique. 1 = lowest risk; 2 = highest risk

The basically static approach of GIS-based modelling (focusing on passive factors of inundation risk) is supplemented by hydrodynamic modelling, which expresses basic physical and hydrological relationships with mathematical equations (Maddox et al. 1979). Runoff is represented in critical flow or stage value, which is further analyzed with a flood transformation model. If appropriate data of sufficient spatial resolution are available, the HEC software environment is also suitable for the estimation of the extention of potentially inundated areas. Thus, it can also fulfil a verification function.

Firstly, the HEC-HMS model determines the actual discharge responding to critical rainfall for the catchment under study. However, the output data verification will only be feasible if stream gauge data are available for the catchment. If the simulation is carried out on an unexplored catchment, total runoff (flow) has to be estimated by empirically based equations (Koris, 2002).

Threshold precipitation values, i.e. those that trigger floods with a given return period are determined for various flood levels. In our investigations, based on observed rainfall, a 400-year return period (during which probably a series of undocumented flash flood events occurred) had to be taken into consideration. In this case, in addition to the actual rainfall values, we have to acquire comprehensive knowledge on the entire hydrological cycle, including information on elements like the hydraulic conductivity and infiltration rate of soils, canopy and surface storage. The numerical models also involve topographical analyses, but they are focused on the study of cross-sections. Valley cross-sections are established at predetermined spacing and analyzed along the whole length of the watercourse (Fig. 11). The actual width of the cross-section is designed with regard to the critical flood level above the valley floor or the mean long-term water stage. River flow or stage values are then determined for each cross-section (Fig. 11).

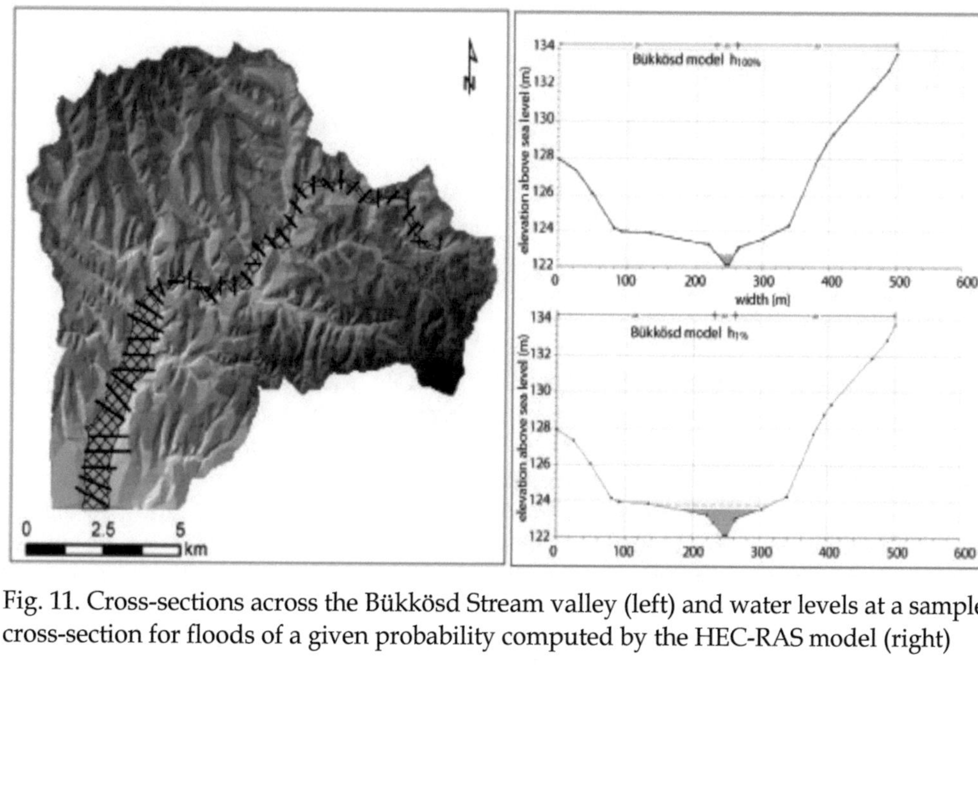

Fig. 11. Cross-sections across the Bükkösd Stream valley (left) and water levels at a sample cross-section for floods of a given probability computed by the HEC-RAS model (right)

Today numerical models are widely applied tools to simulate the areal extent of inundations and flooding along a watercourse (i.e during riverine floods) (Gaume et al., 2004). They are also suitable for flood simulation in urban environments, where the proportion of permeable surfaces are limited and impervious paved surfaces are widespread (e.g. Xia et al., 2011). Numerical modelling is particularly suitable for the analyses of risk scenarios, such as dam breaching, and also capable of the exact localization and parameterization of the elements of the channel and drainage systems (e.g. bridges, levees and culverts) and even appropriate for the 3D representation of these structures.

4. Potential prediction models

The prediction of flood occurrence is the ultimate goal of modelling. Prediction models are classified into two categories: real-time direct forecasting (similar to weather forecasting – e.g. Doswell III, 1996, 1998) and flowchart-type modelling or scenario building (similar to climate change prediction). The models of the first type of forecasting do not seem useful in flash flood prediction in our case since no appropriate monitoring system exists in the study area. Applying the second approach scenarios or flowcharts can be designed for the most endangered catchments in the region (e.g. Alkema, 2003). Flowchart-type modelling takes advantage of the results of the aforementioned rapid screening analysis and risk assessment. The most probable environmental change scenarios are generated with preselected boundary conditions. The boundary conditions incorporated in the model include soil moisture content, relief, surface storage, canopy cover, cumulative rainfall and rainfall intensity. To validate the suitability of this model type we need to verify it with hindcast modelling, i.e. we perform simulation backward in time to see whether it reconstructs the observed event.

For a flowchart analysis data on preceding rainfall events have to be collected to see whether the previous rainfall event was followed by flood warning. Precipitation data originates from meteorological data usually with a 3-hour lead time. These rainfall prediction schemes determine whether a heavy convective, or a prolonged and relatively low-intensity rainfall is expected. All the scenarios in a flow chart model contain a unique code. An analytical software investigates the resemblance of the present scenario to all the predetermined scenarios and finally selects the most adequate output scenario. Finally, it supports the decision of authorities on issuing a flood warning or not.

5. Conclusions

Being an increasingly dangerous environmental hazard, flash floods are intensively investigated worldwide. Information on floods in remote areas is accumulating. Probably related to climate change, flash floods are also becoming a common phenomenon in the mountainous and hilly region of Southern Transdanubia. At least 433 settlements are located and 700,000 inhabitants live in areas potentially affected by flash floods – although the return period of severe inundations is likely to exceed 100 years for most localities. Unfortunately, new developments often ignore this potential hazards and during prolonged dry periods agricultural cultivation frequently extends over floodplains. However, a flood of long return time may cause economic losses of millions of Euros. The investigations in Southwest-Hungary helped to identify areas where, despite the available long-term statistics, settlements, residential areas and farmlands are potentially affected by flooding and serious damage from floods can be predicted. The choice of methodological approaches to the topic is rapidly broadening. The combination of rapid screening methods, GIS-based risk assessment with numerical hydrological modelling and the flowchart analysis of probable scenarios opens new vistas in this research field.

6. Acknowledgements

This research was supported by the Baross Gábor Program (Grant No. REG_DD_KFI_09/PTE_TM09), the Hungarian Science Fundation (OTKA, Grant No. T 68903) and the Bolyai János Scholarship. The authors are grateful for the data and support provided by the Hungarian Meteorological Service, the South-Transdanubian Water Management Directorate (DDKÖVIZIG), the VITUKI Rt. and the Mecsekérc Zrt. The authors are especially indebted to Ákos Horváth, Gábor Horváth, András Varannai, Gergely Resitcky and Roland Vendégh for their contribution to the present work.

7. References

AG BoM (2010). *Facts on Flash Floods in NSW.* Australian Government, Bureau of Meteorology, Melbourne, Victoria, Australia Available from http://www.bom.gov.au/nsw/sevwx/flashfact.shtml Accessed on 10.02.2011

Alkema, D. (2003). Flood risk assessment for EIA (environmental impact assessment): an example of a motorway near Trento, Italy. *Studi Trentini di Scienze Naturali*, Vol.78, pp. 147-153, ISSN 2035-7699

Allan, R.J. (1993). Historical fluctuations in ENSO and teleconnection structure since 1879: Near-global patterns. *Quaternary Australasia,* Vol.11, No.1, 17–27, ISSN 0811-0433

Bálint, G. & Szlávik, L. (2001). Hegy- és dombvidéki kisvízfolyások szélsőséges árvizeinek vizsgálata (Investigation of extreme floods on streams in mountains and hills), *International Conference "Water and Nater Conservation in the Danube–Tisza River Basin"*, pp. 341–353, Debrecen, Hungary, 19–21 September 2001, Hungarian Hydrological Society, Budapest (in Hungarian)

Bettess, R. (2005). *Flooding in Boscastle and North Cornwall, August 2004: Phase 2 Studies Report,* HR Wallingford Report EX5160 Available from http://www.hrwallingford.co.uk/downloads/publications/EX5160_Boscastle_floodings.pdf Accessed on 06.06.2011

Borga, M.; Boscolo, P.; Zanon, F. & Sangati, M. (2007). Hydrometeorological analysis of the August 29, 2003 flash flood in the eastern Italian Alps. *Journal of Hydrometeorology,* Vol.8, No.5, pp. 1049–1067, ISSN 1525-755X

Brandes, H. (2000). The Venezuela flash floods and debris flows of 15-16 December 1999. *Landslide News,* Japan Landslide Society, Vol.13, No.1, pp. 5-7, ISSN 0919-5629

Bull, L.J. & Kirkby, M.J. (Eds.). (2002). *Dryland Rivers: Processes and Management in Mediterranean Climates,* John Wiley and Sons, ISBN 0-471-49123-3, Chichester, UK

Burroughs, W.J. (Ed.). 2003. *Climate into the 21st century,* World Meteorological Organization, Cambridge University Press, ISBN 9780521792028, Geneva, Switzerland – Cambridge, UK

Caracena, F.; Maddox, R. A.; Hoxit, L. R. & Chappell, C. F. (1979). Mesoanalysis of the Big Thompson storm. *Monthly Weather Review,* Vol.107, No.1, pp. 1–17, ISSN 0027-0644

Carpenter, T.M.; Sperfslage, J.A.; Georgakakos; K.P., Sweeney, T. & Fread, D.L. (1999). National threshold runoff estimation utilizing GIS is support of operational flash flood warning systems. *Journal of Hydrology,* Vol.224, No.1, pp. 21–44, ISSN 0022-1694

Cobby, D.; Falconer, R.; Forbes, P.; Smyth, P. & Widgery, N. (2009). Potential warning services for groundwater and pluvial flooding, In: *Flood Risk Management: Research and Practice,* Samuels, P., Huntington, S., Allsop, W. & Harrop, J., (Eds.), pp. 1273–1280, Taylor & Francis Group, ISBN 978-0-415-48507-4, London, UK

Cohen, H. & Laronne, J.B. (2005). High rates of sediment transport by flashfloods in the Southern Judean Desert, Israel. *Hydrological Processes,* Vol.19, No.8, pp. 1687–1702. doi: 10. 1002/hyp.5630, online ISSN 1099-1085

Collier, C. (2007). Flash flood forecasting: what are the limits of predictability? *Quarterly Journal of the Royal Meteorological Society,* Vol.133, No.622A, pp. 3–23, doi: 10.1002/qj.29, online ISSN: 1477-870X

Costa, J.E. (1983). Paleohydraulic reconstruction of flash-flood peaks from boulder deposits in the Colorado Front Range. *Geological Society of America Bulletin,* Vol.94, No.8, pp. 986–1004, doi: 10.1130/0016-7606(1983)94<986:PROFPF> 2.0.CO;2, ISSN 0016-7606

Czigány, S.; Pirkhoffer, E. & Geresdi, I. (2008). Environmental impacts of flash floods in Hungary, In: *Flood Risk Management: Research and Practice,* Samuels, P., Huntington, S., Allsop, W. & Harrop, J., (Eds.), pp. 1439–1447, Taylor & Francis Group, ISBN 978-0-415-48507-4, London, UK

Czigány, Sz.; Pirkhoffer, E. & Geresdi, I. (2010). Impact of extreme rainfall and soil moisture on flash flood generation. *Időjárás,* Vol.114, No.1, pp. 79–100, ISSN 0324-6329

Czigány, Sz.; Pirkhoffer, E.; Nagyváradi, L.; Hegedűs, P. & Geresdi, I. (2011a). Rapid screening of flash flood-affected watersheds in Hungary. *Zeistschrift für Geomorphololgie,* Vol.55, Supplementary Issue 1, pp. 1-13, ISSN 0372-8854

Czigány, Sz.; Pirkhoffer, E.; Fábián, Sz. Á. & Ilisics, N. (2011b). Flash floods as natural hazards in Hungary, with special focus on SW Hungary. *Riscuri şi catastrofe, Cluj-Napoca, Romania,* Casa cartii de stiinta. Vol.8, No.1, pp. 117–130, ISSN 1584-5273

Davis, R.S. (2001). Flash flood forecast and detection methods. *Meteorological Monographs,* American Meteorological Society, Vol.28, pp. 481–526, doi: 10.1175/0065-9401-28.50.481, ISSN 0065-9401

Davies, W.; Bailey, J. & Kelly, D. (1972). West Virginia's Buffalo Creek Flood: A Study of the Hydrology and Engineering Geology. *US Department of the Interior, Geological Survey Circular* No.667, pp. 1-32, ISSN 0364-6017

DDKÖVIZIG (2010). *Az esőzések miatt kialakult helyzet Dél-Dunántúlon (The situation caused by heavy rainfall in Southern Transdanubia).* Available from http://www.ddkovizig.hu/magyar/hirek/az_esozesek_miatt_kialakult_helyzet_del-dunantulon (in Hungarian) Accessed on 01.06.2010

De Roo, A.P.J.; Wesseling, C.G. & Van Deursen, W.P.A. (2000). Physically based river basin modelling within a GIS: The LISFLOOD model. *Hydrological Processes,* Vol.14, No. 11-12, pp. 1981–1992, online ISSN 1099-1085

Doswell III, C. A.; Brooks, H. E. & Maddox R. A. (1996). Flash flood forecasting: An ingredients-based methodology. *Weather Forecasting,* Vol.11, No.4, pp. 560–581, ISSN 0882-8156

Doswell III, C. A.; Brooks, H. E.; Romero, R. & Alonso, S. (1998). A diagnostic study of three heavy precipitation episodes in the western Mediterranean region. *Weather Forecasting,* Vol.13, No.1., 102–124, ISSN 0882-8156

EEA (2005). *Climate Change and Floods in Europe.* Report of the European Environmental Agency. Copenhagen, Denmark Available from http://www.eea.europa.eu/hu/publications/briefing _2005_1 Accessed on 01.06.2010

Fábián, Sz. Á.; Görcs, N. L.; Kovács, I. P.; Radvánszky, B. & Varga, G. (2009). Reconstruction of a flash flood event in a small catchment: Nagykónyi, Hungary. *Zeitschrift für Geomorphologie,* Vol.53, No.2, pp. 215–222, ISSN 0372-8854

Gaume, E.; Livet, M.; Desbordes, M. & Villeneuve, J.P. (2004) Hydrological analysis of the river Aude, France, flash flood on 12 and 13 November 1999. *Journal of Hydrology,* Vol. 286, No.1, pp. 135–154, ISSN 0022-1694

Gaume, E.; Bain, V.; Bernardara, P.; Newinger, O.; Barbuc, M.; Bateman, A.; Blaškovičova, L.; Bloschl, G.; Borga, M.; Dumitrescu, A.; Daliakopoulos, I.; Garcia, J.; Irimescu, A.; Kohnova, S.; Koutroulis, A.; Marchi, L.; Matreata, S.; Medina, V.; Preciso, E.; Sempere-Torres, D.; Stancalie, G.; Szolgay, J.; Tsanis, I.; Velascom, D. & Viglione, A. (2009). A compilation of data on European flash floods. *Journal of Hydrology.* Vol. 367, No.1, pp. 70–78, ISSN 0022-1694

Georgakakos, K. P. (1987). Real-time flash flood prediction. *Journal of Geophysical Research,* Vol.92, No.D8, pp. 9615–9629, ISSN 0148-0227

Georgakakos, K. P. (2006). Analytical results for operational flash flood guidance. *Journal of Hydrology,* Vol.317, No.1-2, pp. 81–103, doi: 10.1016/j.jhydrol.2005.05.009, ISSN 0022-1694

Golding, B. (2005). Meteorology of the Boscastle flood. *National Hydrology Seminar 2005,* Galway, 15 November 2005, Irish National Committee, International Hydrological Programme Available from http://www.opw.ie/hydrology/data/speeches/c_gold~1.pdf

Graf, W. L. (2002). *Fluvial Processes in Dryland Rivers,* Reprint of 1st edition (1988), Blackburn Press, ISBN 978-1930665514, Caldwell, New Jersey, USA

Grundfest, E. (1977). What people did during the Big Thompson Flood. *Working Paper No. 32,* Natural Hazards Research and Applications Information Center, Boulder, Colorado, USA

Grundfest, E. (1987). What we have learned since the Big Thompson Flood. *Proceedings of the Tenth Anniversary Conference,* 17–19 July 1986, Special Publication No. 16. Natural Hazards Research and Applications Information Center, Boulder, Colorado, USA Available from http://www.uccs.edu/~geogenvs/flood/ Accessed on 10.05.2011

Grundfest, E. & Ripps, A. (2000). Flash floods, In: *Floods. Vol. 1,* Parker, D. J. (Ed.), pp. 377–390, Routledge, ISBN 978-0415172387, London and New York,

Gutiérrez, F.; Gutiérrez, M. & Sancho, C. (1998). Geomorphological and sedimentological analysis of a catastrophic flash flood in the Arás drainage basin (Central Pyrenees, Spain). *Geomorphology,* Vol.22, No. 3-4, pp. 265-283, ISSN 0169-555X

Gyenizse, P. & Vass, P. (1998). A természeti környezet szerepe a Nyugat-Mecsek településeinek kialakulásában és fejlődésében (The role of the physical environment in the development of settlement in the Western Mecsek Mountains). *Földrajzi Értesítő*, Vol.47, No.1–2, pp. 131-148, ISSN 0054-1503

Horváth, Á. (2005). A 2005. április 18-i mátrakeresztesi árvíz meteorológiai háttere (Background to the 18 April 2005 flood at Mátrakeresztes, North-Hungary). *Légkör*, Vol.50, No.1, pp. 6–10, ISSN 0133-3666 (in Hungarian with English summary)

Iverson, R. M. (1997). The physics of debris flows. *Reviews of Geophysics*, Vol.35. No.3, pp. 245–296, ISSN 8755-1209

Kaliczka, L. (1998). *Hegy és dombvidéki vízrendezés (Water management in mountains and hills)*, Manuscript lecture notes. Eötvös József Technical College, Baja (in Hungarian) Available from:
http://levelezo.atw.hu/Jegyzet/hegydombvizrend.pdf Accessed on 03.05.2011

Kodama, K. & Barnes, G. M. (1997). Heavy rain events over the south-facing slopes of Hawaii: Attendant conditions. *Weather Forecasting*, Vol.12, pp. 347–367, ISSN 0882-8156

Koris, K (2002). A hazai hegy- és dombvidéki kisvízgyűjtők árvízhozamainak meghatározása (Determining flood discharges on small cathcment in the mountains and hills of Hungary). *Vízügyi Közlemények*, Vol.84, No.1, pp. 64–77, ISSN 0042-7616 (in Hungarian)

Koris, K. & Winter, J. (2000). Az 1999. évi nyári rendkívüli árvizek a Mátra és a Bükk déli vízgyűjtőjén (Extreme summer floods in the southern catchments of the Mátra and Bükk Mountains, North-Hungary, in 1999). *Vízügyi Közlemények*, Vol.82, N.2, pp. 199–219, ISSN 0042-7616 (in Hungarian)

Laing, A. G. (2004). Cases of heavy precipitation and flash floods in the Caribbean during El Niño winters. *Journal of Hydrometeorology*, Vol.5, (August 2004), pp. 577–594, doi: 10.1175/1525-7541(2004)005<0577:COHPAF>2.0.CO;2, ISSN 1525-7541

Lóczy, D. & Juhász, Á. (1996). Hungary, In: *Geomorphological hazards of Europe*, Embleton, C. and Embleton, Ch., (Ed.), 243-262, Elsevier, ISBN 0-444-88824-1, Amsterdam, The Netherlands

Lóczy, D. (2010). Flood hazard in Hungary: a re-assessment. *Central European Journal of Geosciences*, Vol.2, No.4, pp. 537–547, doi: 10.2478/v10085-010-0029-0, ISSN 18961517

Maarten, R.; Erlich, M.; Versini, P.-A.; Gaume, E.; Lumbroso, D.; Asselman, N.; Hooijer, A. & de Bruijn, K. (2007). *Review of flood event management Decision Support Systems*. FLOODsite Project Report T19-07-01 Available from:
http://floodsite.net/html/cd_task17-19/docs/reports/T19/Task19_report_M19_1 review_v1_4.pdf Accessed on 03.06.2011

Maddox, R. A.; Chappell, C. F. & Hoxit, L. R. (1979). Synoptic and mesoalpha-scale aspects of flash flood events. *Bulletin of American Meteorological Society*, Vol.60, pp. 115–123, ISSN 1520-0477

NOAA, National Weather Service (1992). *Puerto Rico flash floods, January 5–6, 1992*. Natural Disaster Survey Report, Silver Spring, Maryland, USA, 92 pp.

Norbiato, D.; Borga, M.; Degli Esposti, S.; Gaume, E. & Anquetin, S. (2008). Flash flood warning based on rainfall thresholds and soil moisture conditions: An assessment for gauged and ungauged basins. *Journal of Hydrology* Vol.362, Nos. 3–4, pp. 274–290, doi:10.1016/j.jhydrol.2008.08.023, ISSN 0022-1694

Pászthory, R. & Szigeti, F. (2009). Árvízi Kockázati Információs Rendszer (Flood Risk Information System), *Conference paper at the Conference of the Hungarian Hydrological Society*, Baja (in Hungarian) Available from: http://www.hidrologia.hu/mht/index.php Accessed on 01.06.2010

Paudel, M. (2010). *An examination of distributed hydrologic modeling methods as compared with traditional lumped parameter approaches*. PhD Dissertation, Brigham University, Provo, Utah, USA Available from: http://contentdm.lib.byu.edu/ETD/image/etd3708.pdf Accessed on 06.07.2011

Pirkhoffer, E.; Czigány, S. & Geresdi, I. (2009a). Impact of rainfall pattern on the occurrence of flash floods in Hungary. *Zeitschrift für Geomorphologie*, Vol.53, No.2, pp. 139–157, ISSN 0372-8854

Pirkhoffer, E.; Czigány, Sz.; Geresdi, I. & Lovász, Gy. (2009b). Environmental hazards in small watersheds: flash floods - impact of soil moisture and canopy cover on flash flood generation. *Riscuri şi catastrofe, Cluj-Napoca*, Casa cartii de stiinta, pp. 117–130, ISSN 1584-5273

Pontrelli, M. D.; Bryan, G. & Fritsch, J. M. (1999). The Madison County, Virginia, flash flood of 27 June 1995. *Weather Forecasting*, Vol.14, No.4, pp. 384–404, ISSN 0882-8156

Reid, I. (2004). Flash flood, In: *Encyclopedia of Geomorphology Vol.1.*, Goudie, A.S., (editor-in-chief), Routledge, pp. 376–378, ISBN 0-415-32737-7, London, UK

Reid, I.; Laronne, J.B.; Powell, D.M. & Garcia, C. (1994). Flash floods in desert rivers: studying the unexpected. *EOS, Transactions, American Geophysical Union*, Vol.75, No.39, p. 452, doi: 10.1029/94EO01076 ISSN 0096-3941

Schmittner, K.E. & Giresse, P. (1996). Modelling and application of the geomorphic and environmental controls on flash flood flow. *Geomorphology*, Vol.16, No.4, pp. 337–347, doi:10.1016/0169-555X(96)00002-5, ISSN 0169-555X

Stevaux, J.C. & Latrubesse, E. (2010). Urban Floods in Brazil, In: *Geomorphology of Natural Hazards and Human Exacerbated Disasters in Latin America*, E. Latrubesse, (Ed.), pp. 245-266, Elsevier, ISBN 9780444531179, Amsterdam, The Netherlands

US Army Corps of Engineers (2005). *Hydrologic Modeling System HEC-HMS*. User's Manual, Version 3.0.0. USACE Hydrologic Engineering Center, Davis, California, USA

Vass, P. (1997). Árvizek a Bükkösdi-patak felső szakaszán (Floods in the headwaters of the Bükkösd Stream), In: *Földrajzi tanulmányok a pécsi doktoriskolából I.*, Tésits, R. & Tóth, J. (Eds.), pp. 261–285, Bornus Nyomda, Pécs, Hungary (in Hungarian, English summary)

Weston, K. J. & Roy, M. G. (1994). The directional-dependence of the enhancement of rainfall over complex orography. *Meteorological Applications*, Vol.1, No.3, pp. 267–275, doi: 10.1002/met.5060010308, ISSN 1469-8080

Wieczorek, G. F.; Larsen, M. C.; Eaton, L. S.; Morgan, B. A. & Blair, J. L. (2001). Debris-flow and flooding hazards associated with the December 1999 storm in coastal

Venezuela and strategies for mitigation. *U. S. Geological Survey Open File Report 01-0144*, Available from:
http://pubs.usgs.gov/of/2001/ofr-01-0144 Accessed on 02.04.2011
Xia, J.Q.; Falconer, R.A.; Lin, B.J. & Tan, G.M. (2011). Modelling flash flood risk in urban areas. Water Management, Vol.164, No.6, pp. 267–282, doi: 10.1680/wama.2011.164.6.267, ISSN 1741-7589

3

Change of Groundwater Flow Characteristics After Construction of the Waterworks System Protective Measures on the Danube River – A Case Study in Slovakia

František Burger
Slovak Academy of Sciences/Institute of Hydrology
Slovak Republic

1. Introduction

The waterworks construction affects the hydrological regime of the flow of groundwater in the river alluvia that is usually in the hydrodynamic relation to the regime of the fluctuation of the surface watercourse level. To elaborate the prognosis of the changes in the regime of groundwater means to determine their sequence in time for the entire period of their creation until the final stable status is reached. It has to be made on the basis of the knowledge of hydrogeological situation within the territory and the contemporary regime of groundwater. The creation of such changes may be invoked by natural changes or also anthropogenic interventions into the water situation within the territory. The above implies this is the unsteady flow task from hydrodynamic point of view. It is natural since groundwater flow has always somehow the character of unsteady flow. It is implied by natural conditions of their supply and drainage. The regime of supply and drainage of groundwater in water-bearing collectors depends upon the factors not changed in time, so in general, the mode of fluctuation of groundwater level is affected by the changes in time as well and therefore it is unsteady. However, if the conditions of supply and drainage of groundwater are changed in time negligibly, or if the area of interest of the water-bearing collector is located in a certain sufficient distance from the source of supply and drainage point, the flow of groundwater may be practically considered to be stable. The time slope of the forecast changes within the determined area of flow then shall, in addition to other conditions, depend upon time changes in surface and groundwater at its edges, i.e. in the areas of supply or drainage of groundwater. The forecast time changes shall then depend upon the character of the peripheral impacts, thus they shall be different in the case of the natural changes in hydrological conditions and different in the case of artificial structural interventions into contemporary hydrological conditions. In particular, the morphological changes in the river bed, contents of suspended sediments in the watercourse and natural colmatage of the watercourse belong amongst the natural changes in the hydrological conditions of the territories affecting the groundwater regime (Velísková, 2010; Komboš, 2008). All the technical structural measures amending the conditions of supply and drainage of groundwater belong amongst the interventions into the waterworks conditions of the territories affecting the groundwater regime (Šoltész & Baroková, 2004). In many cases, it is necessary to know not only the final condition achieved by the groundwater level after the implementation of any technical measure, but also the time after what the final condition is reached, or the time procedure of settlement of the new groundwater level status. That means, in general, the task regarding the long-term prognosis in the changes in the groundwater regime must be compiled as the task of unsteady flow of groundwater, where the reached final steady condition is the extreme case (Duba, 1964).

The protective measures on the Dunube River

The designed waterworks complex consists of Gabčíkovo waterworks and Nagymaros waterworks which are, in terms of hydraulic, navigation, and energy distribution, a single operating system. The multipurpose hydroelectric project was built together with Hungary, according to an interstate Treaty signed in 1977. The waterworks complex on the Danube was designed to have an additional level at Nagymaros, consisting of a reservoir 95 km long and the Nagymaros power plant. This level was to be located between the Hungarian towns of Nagymaros and Visegrad and its purpose was to use the gradient of the reservoir for production of electricity and to allow ships to pass. When the Gabčíkovo Project was 90% completed, Hungary stopped fulfilling its treaty obligations in 1989 and tried to end the Treaty in 1992 (www.gabcikovo.gov.sk).

In 1992, the Slovak party put into operation the Gabčíkovo waterworks using an alternative solution on the territory of the Slovak Republic (so called "C" variant) and it wholly completed the works on the object "Protective measures of the Nagymaros waterworks storage reservoir". The necessity of the construction of the object was implied by the reason of the prevention of an unfavourable impact of the dammed level of the Danube River by the Nagymaros step on the territory of the Slovak Republic. This was the reinforcement of the Danube River dam on the territory of the Slovak Republic, the Váh river dam, the Hron river dam and the Ipeľ river dam. The backwater of the Danube River would prevent the gravitation outflow if the internal waters into the Danube River. The erected underground walls in the protective dams prevent the gravitation outflow of internal waters from the territory of the Slovak Republic in the Komárno - the Ipeľ river estuary section, even when there is no backwater in the Danube River. For that reason the internal waters of the territory must be pumped into the Danube River through the erected pumping stations. The administrator of the river basis incurs increased costs related to the activity without their compensation. At the time of the decision of the Hungarian party on the termination of the works on the Nagymaros waterworks, the majority of the protective measures had already been implemented or in the uppermost stage of progress. Subsequently, their scope was minimised and they completed the objects related to

- the flood protection of the territory and
- the diversion of internal waters.

Protective measures against the level impoundment in the reservoir Nagymaros were built on the Slovak territory during the construction of Gabčíkovo waterworks. These consist of renovation of existing dams with newly built underground sealing walls, reinforcement of banks and building seepage canals. Protective measures include the construction of drainage channels and pumping stations and channels. Since the Hungarian side does not build up the lower reservoir, the operation will only be the Gabčíkovo waterworks and the protective measures established in the Slovak Republic that have been running for the maintenance of surface and ground water management at each water stage in the Danube River.

The Patince - Štúrovo section, RK 1751.8 to 1716.0

The construction of the underground walls in the Kravany nad Dunajom section RK 1746.6 to 1722.5 and in the Štúrovo section RK 1722.5 - 1716.0 took place in 03/1985 - 06/1996. There are two so called "windows" omitted in the non-permeable underground wall. The entire construction was carried out before 11/2002 (the data provided by Vodohospodárska výstavba š.p. Bratislava).

2. Modelling and numerical simulation of groundwater flow in the Čenkov reparian alluvial aquifer

2.1 The long-term minimal anthropogenic disruption in natural conditions in of the study area

The solution of groundwater flow features assessments, which are due to later anthropogenic investigations into the area are ranked as almost natural, are going out from the evaluation of former groundwater regimes based on observations in the nature, knowledge of geological structure of the area and hydrogeologic conditions, which is serving as a base to the water-level regime assessment and the subsequent assessment of the main groundwater flow directions.

The aim of the task to be solved is to create a numerical model for a steady groundwater flow in the reparian alluvial aquifer of the Čenkov plain, and its calibration, verification and obtaining of results by a simulation that is at groundwater level, using filtration velocity vectors, groundwater paths by particle tracking and the water budget. One assumption is that long-term minimal anthropogenics disrupted the natural conditions of the study area. As the date of the simulation was chosen on the day of 29 September 1954, because of the steady state of water flow through the study area, and also of the Danube low stage and because of existence of solving similar task by other methods in the past (Duba, 1964) and thereby available data needed for modelling and simulation.

2.1.1 Description of study area

The Čenkov plain is situated in the eastern part of the Danubian lowland, west apart from Štúrovo town. It is the fluvial plain of the Danube, which borders in the south on a 23 kilometre long river section between RK 1722 and 1745 and in the north in an arc stretching across the terrace platform, where on its boundary lies the Moča village, the Búč village, the Júrsky Chlm village, the Mužla village and the Obid village. The fluvial plain is from 2.5 km up to 6 km wide and has an overall area of 66 km^2. Its surface is flat. The heights of the terrain vary from 106 up to 108 m a.s.l. The lowest-situated section under the terrace is on height level 105m a.s.l. and the highest situated section in the Čenkov wood is on the middle of the area 108 - 110m a.s.l. (Fig. 1 and 2).

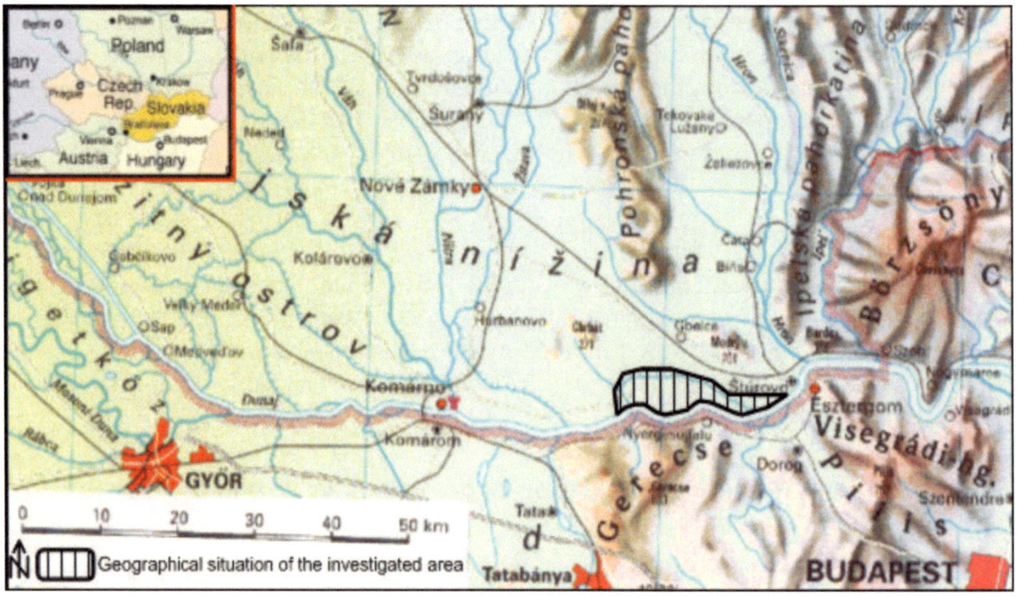

Fig. 1. Geographical situation of the Čenkov plain

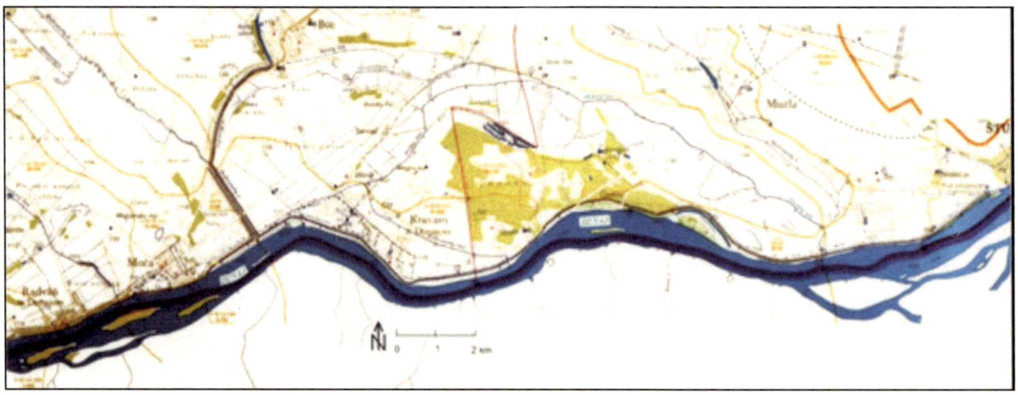

Fig. 2. Water management map of the Čenkov plain

2.1.2 Evaluation of natural conditions

2.1.2.1 Climate

From a climatic point of view the study area belongs to a warm locality in the scope of the south-eastern part of the Danubian lowland, where it has a warm and dry climatic zone with a mild winter. First, the temperature characteristics and yearly air temperature average shows that the south of Slovakia is the warmest locality of the republic. The average 10.4 °C at Štúrovo is convincing. Uniformity of moisture conditions is clear already from the yearly relative air moisture average, which varies from 74 - 81 % and is the lowest in the bottom most parts of the Danubian lowland (Štúrovo 74 %, Komárno 75 %). In the territory of the West-Slovakia district, which is the most productive agricultural locality, the precipitation has significant importance. The centre of this locality is the Danubian lowland, which is indeed the warmest but also has the driest locality. In a series of long-term observations, the lowest annual precipitation totals vary in terms of 300 - 400 mm and minimum monthly precipitation totals in particular months do not even reach (except for July) 5 mm precipitation, whereupon significant dry periods are more often in summer half-year than in winter half-year. The lowest July precipitation totals do not drop under 10 mm. On the other hand wet (precipitation) periods are lasting here mostly from 18 to 20 days, and their appearance is relatively more rare than the appearance of dry periods and it occurs mostly in spring and autumn periods. The highest annual precipitation totals could reach 900 mm, even in singular cases up to 1000 mm of precipitation.

2.1.2.2 Hydrogeology and geology

The Danube fluvial plain at observed river sections is built by sediment deposits of the Danube River, where their thickness varies irregularly between 5 - 12 m and the most frequent thicknesses are between 6 - 9 m. Gravel and sand dominate soil layers, which are in the highest part covered by alluvial loams. Gravel–sand fillings of the Danube fluvial plain's bed in this section belong to Würm, and the cover of sandy loam is Holocene.

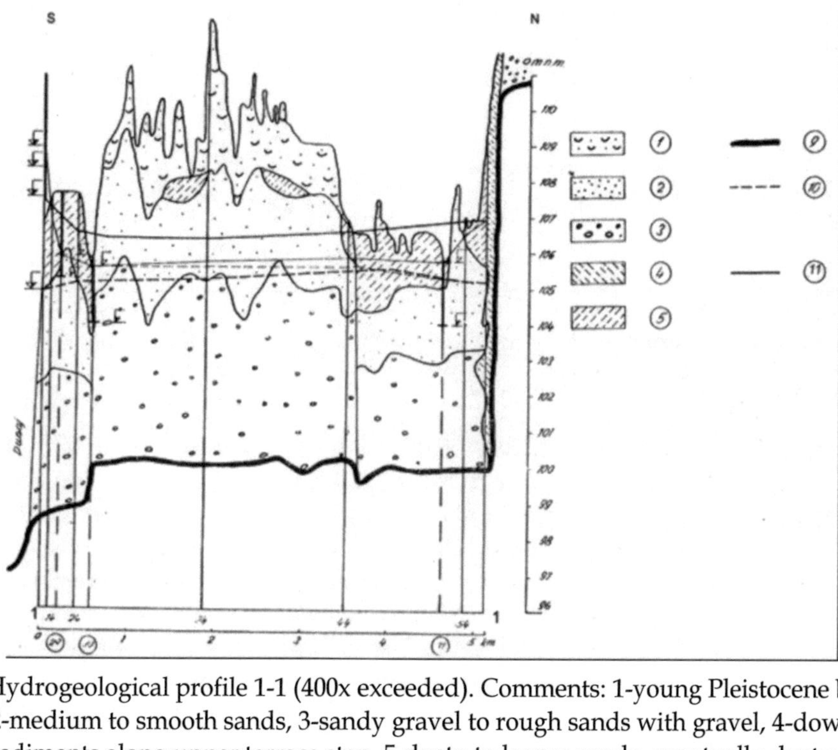

Fig. 3. Hydrogeological profile 1-1 (400x exceeded). Comments: 1-young Pleistocene blown sands, 2-medium to smooth sands, 3-sandy gravel to rough sands with gravel, 4-downhill loamy sediments along upper terrace step, 5-dusty to loamy sands, eventually dusty - sandy loams, 9- marking of the tertiary base surface, 10-groundwater level on 29 Sept.1954, 11-the highest groundwater level in years 1954 - 1956 (Duba, 1964)

It is possible to observe their partial subtilization in longitudinal profiles of gravel sand alluvia (Fig. 3 and 4) in the Danube direction, although the appearance of heavier gravel layers is possible in the whole profile. The left edge of the Danube's fluvial plain is lined by a markedly terraced step with relative height approximately 15 m and base 3 m above the Danube water level. Absolute height of the base is around 110 m a.s.l. it is slowly descending from the Chotin village to the Štúrovo town. Hydrogeological conditions of the terrace were proofed only by a few boreholes, after which hydraulic conductivity of gravel varies from 6.6E-05 $m.s^{-1}$ (the Chotín village) up to 2.0E-03 $m.s^{-1}$ (the Štúrovo town - the Nana village). Groundwater recharge happens entirely from precipitation in locations where permeable blown sands or loamy sands and sandy loams are located in hanger. Ground-water from the terrace is drained on its edge to the lower step, partly on contact as it comes up to the surface and it is taken away by the drainage channels. The ground-water level in the alluvia is mainly influenced by the surface stream of the Danube River and then on other side by water seeping down from an adjacent terrace and through precipitation. The ground-water flow direction according to bilateral relation of the Danube water level and ground-water level was either to the aquifer or to the Danube.

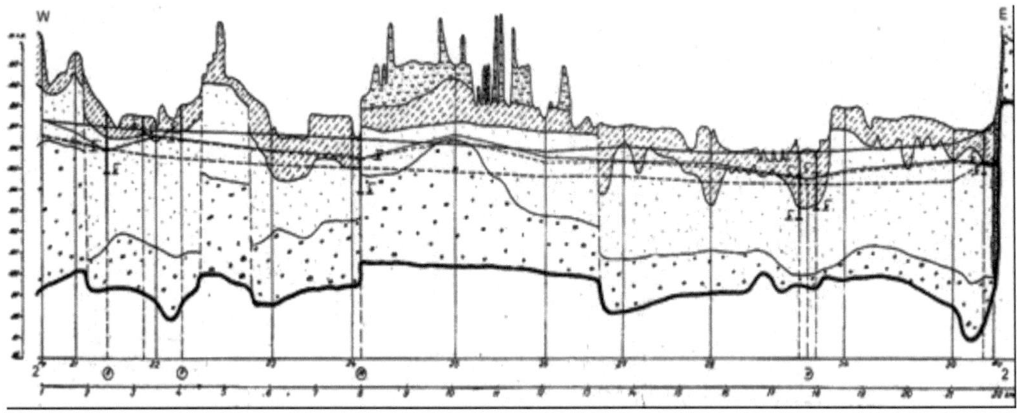

Fig. 4. Hydrogeological profile along the Danube bank (400x exceeded), (Duba, 1964)

2.1.2.3 Hydrology

Through the hydrological characteristics of the study area and the description of surface flows in an objective time it is necessary to concentrate on the Danube River, which has here first- rated importance. Slovak Danube river reach belongs to the upper part of middle part of the river. Danube is keeping its alpine character in Slovak reach, in its upper part it has considerable slope around 0.4‰, it is flowing in its own alluvia and it is creating multiple systems of river arms. Water stages are first of all dependent on the water supply from the Alps. Maximum water stage reaches the Danube in June at the time of alpine snow and glacier melting. From June it comes to permanent decrease and minimum water stages are reached in December and January. The Danube water stage on 29 September 1954 in RK 1742.9 (Radvaň nad Dunajom) was 105.20 m a.s.l. Other surface flows in study area are rather small and short and their discharges are low. The maximum occurs in spring months, and in summer their discharge is considerably decreased. Such streams are Modrianský potok (creek) (from Veľká Dolina), its left-hand side tributary Vojnický potok (creek), and Mužliansky potok (creek). Main channels: the Obidský, the Búčský, the Kraviansky and the Krížny channel belong to the system, as well as the large amount of side drainage channels without any name.

3. Methods and material

3.1 Modelling

3.1.1 Mathematical model of groundwater flow

Three-dimensional groundwater flow of constant density through porous earth material may be described by a partial differential equation (McDonald, M.G. & Harbaugh A.W., 1988):

$$\frac{\partial}{\partial x}\left(K_{xx}\frac{\partial h}{\partial x}\right)+\frac{\partial}{\partial y}\left(K_{yy}\frac{\partial h}{\partial y}\right)+\frac{\partial}{\partial z}\left(K_{zz}\frac{\partial h}{\partial z}\right)-W=S_s\frac{\partial h}{\partial t} \tag{1}$$

Where

x, y, z are Cartesian coordinates in the direction of main axis of hydraulic conductivity K_{xx}, K_{yy}, K_{zz}, K_{xx}, K_{yy}, K_{zz} are the values of hydraulic conductivity in the direction of the axis of Cartesian coordinates x, y, z, which are assumed that they are parallel with major axis of hydraulic conductivity [L T^{-1}], h piezometric pressure head [L], W volumetric flux per unit volume, which represents sources and (or) sinks of water [T^{-1}], S_s specific storage of the porous material [L^{-1}], and t time [T]. In general, S_s, K_{xx}, K_{yy}, and K_{zz} may be the functions of space ($S_s = S_s\,(x, y, z)$ a $K_{xx} = K_{xx}\,(x, y, z)$, etc. and h and W could be the functions of space and time ($h = h\,(x, y, z, t)$, $W = W(x, y, z, t)$) which means that equation (1) describes groundwater flow for unsteady conditions in a heterogeneous and anisotropic medium, provided that the principal axes of hydraulic conductivity are aligned with the coordinate directions. Equation (1), together with the specification of flow and (or) head conditions on aquifer boundaries and specification of initial head conditions, creates a mathematical model of groundwater flow. A solution of equation (1), in an analytical sense, is an algebraic formula which indicates $h\,(x, y, z, t)$, so that when the derivatives of h, with respect to space and time are substituted into equation (1), the equation and its initial and boundary conditions are satisfied. Besides these very simple systems, it is possible to reach an analytical solution of equation (1) only rarely, so therefore it is necessary to use numerical methods for solution. One of the methods is the finite difference method, where the continuous system of equations (1) is substituted by the finite set of discrete points in the space and time, and the partial derivatives are substituted by terms calculated from the differences in head values at these points. Such an approach leads to the system of linear algebraic differential equations. Values of head in specific points in time are obtained by their solutions. These values represent approximation of the time-variable distribution of piezometric head, which could have been obtained by analytical solution of equation (1).

3.1.2 Three-dimensional modular model of groundwater flow" MODFLOW"

The finite difference model originally published by McDonald & Harbaugh (1988), in the form of later modifications and addendums, and its modular computer program was utilized by the solution of the mentioned task. The modular structure consists of the "main program" and a series of independent subroutines called "modules". The explanation of physical and mathematical concepts, on which the model is based, and an explanation on how the modules are implemented into the structure of computer program, is listed in detail in the mentioned work. Ground-water flow in hydrogeological ground-water body is simulated by the use of a finite difference block-central method. The solution of systems of simultaneous linear equations is possible to obtain by various methods.

3.1.3 Conceptual groundwater modelling

3.1.3.1 Definition of model's boundaries

Northern to western boundaries of the modelled area are chosen regarding to the demarcation of hydrogeological groundwater body of the Čenkov plain from the northern side by the higher old-Würm terrace step. The Danube River creates the southern to eastern border (Fig. 2).

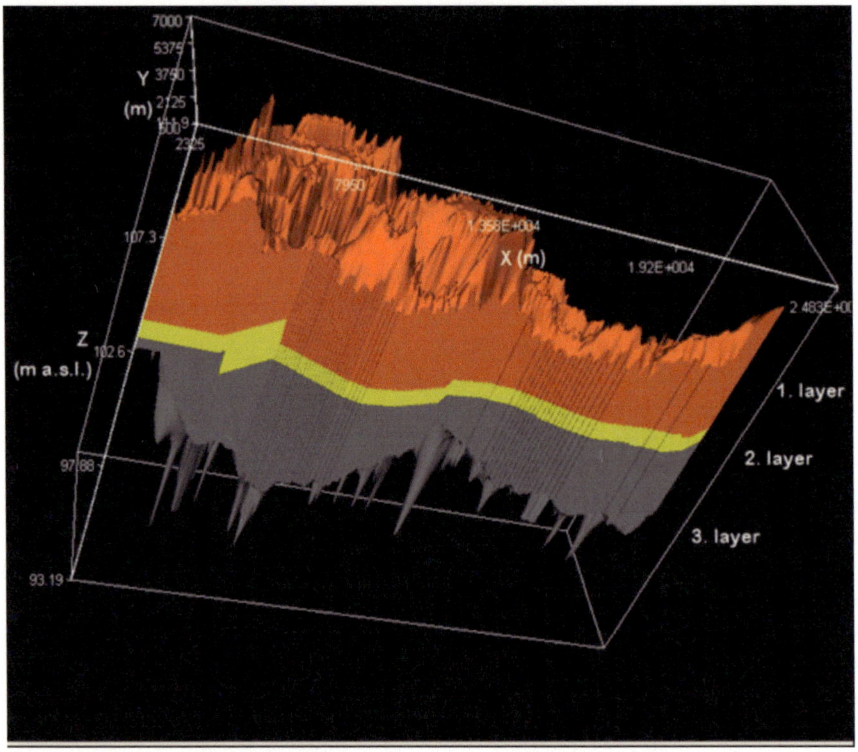

Fig. 5. The aquifer of the Čenkov plain

3.1.3.2 Transformation of the hydrogeological system - vertical schematization

The topography of the surface terrain was processed into digital format from maps with the scale of 1: 10 000. For purposes of modelling, the above described hydrogeological system was transformed into a three-layers system, which consists of an upper covering soil layer (1st layer), of a medium to soft sand layer (2nd layer) and of a sandy gravel to rough sands with a gravel layer (3rd layer) (Fig. 5). These type of layers were selected for modelling as follows: 1st layer: unconfined, 2nd layer: confined/unconfined (transmissivity = const.), 3rd layer: confined/unconfined (transmissivity = const.). Thicknesses of particular layers were defined from geological and hydrogeological data from the survey. The surface of the tertiary base is considered impermeable.

3.1.3.3 Discretization in the space and time

Hydrogeological systems are divided into a mesh of blocks called cells, the locations of which are described in terms of rows, columns, and layers. Footprint dimensions are picked so that the whole area of the Čenkov plain is covered with a smooth overlay. Dimensions of cells are: $\Delta x = \Delta y = 50$ m. Geometry of the model is: 22.5 km x 6.5 km, (130 rows and 450 columns); 3 layers. Grid orientation was picked in the direction of the general groundwater flow and coordinate axis x, y, z are approximately parallel to the main hydraulic conductivity axis. Groundwater flow has always had a certain measure of unsteady flow. This results from natural conditions of recharge and drainage of groundwater. However, if the recharge and drainage groundwater conditions are changing in the time slightly, the flow is quasi-steady and practically represents certain boundary status. From the modelling target point of view a steady status of groundwater flow was considered to the date 29 September 1954.

3.1.3.4 Filtration parameters of the aquifer

Following filtration parameters were necessary for the modelling of this case: horizontal hydraulic conductivity, transmissivity, vertical hydraulic conductivity, effective porosity and coefficient of vertical leakage. *Horizontal hydraulic conductivity* of the groundwater body was obtained from the results of the hydropedological and hydrogeological survey in the study area. Data from pumping tests in probes and boreholes were globally processed by the means of interpolation method of kriging. Values vary from 7.48E-07 m s^{-1} to 3.99E-03 m s^{-1}. *Transmissivity* of the layers was calculated as a multiple of horizontal hydraulic conductivity and thickness of the layer. *Vertical hydraulic conductivity*: by the modelling applications the usual ratio of the horizontal to vertical hydraulic conductivity is from 1 to 10 (Anderson & Woessner, 1992). For the first and second layer ratio 1.0 was selected in compliance with results of the field research and for the third layer the ratio 2.0. *Effective porosity* is the feature of an aquifer to receive and to send out fluid in order to build hydrostatic pressure in the layer and through to the groundwater level. Quantitatively it is expressed by the coefficient of flexible storage and coefficient of free water level storage. The value of the coefficient of the free water level storage depends on hydraulic conductivity and also on grain size distribution of sediments and varies around 0.05 up to 0.15 for loamy sands, 0.15 for soft granulated to dusty sands, 0.19 for soft granulated sands, 0.22 for medium granulated sands and 0.24 for rough granulated sands, gravels etc. Estimated values of flexible storage for unit volume of the groundwater body are stated in the work of Mucha & Šestakov (1987). *Vertical leakage* is required in the case of multiple layers groundwater body and represents the resistance to the water leakage at adjacent layers.

3.1.4 Calibration and verification of the model

Calibration is a process, when the initial input model parameters are adjusted until output (dependent) model parameters at most approach the values measured in the terrain. Calibration of the model is an inverse-model process, i.e. the problem of parameter estimation is an inverse problem. Calibration of the model or the inverse model process could be performed either repetitively, either on a manual basis by way of trial and error, or by using a special computer program. The calibration was executed by the means of the special computer program PEST with manual tuning of some zones. Calibration results for the status up to date 29 September 1954 are shown in Fig. 6. The difference between measured groundwater stages in probes from the Hydrometeorological Institute's observing network have calculated groundwater stages have a maximum value of 0.17 m and the regression coefficient has a value near to one, which refers to high correspondence of calculated results with measured results. The calibrated model of steady groundwater flow was verified at the low water stage in the Danube up to date 7 Aug. 2002 and at high water stage in the Danube up to date 7 May 2000. For both cases a very good accordance of measured and calculated groundwater levels was reached.

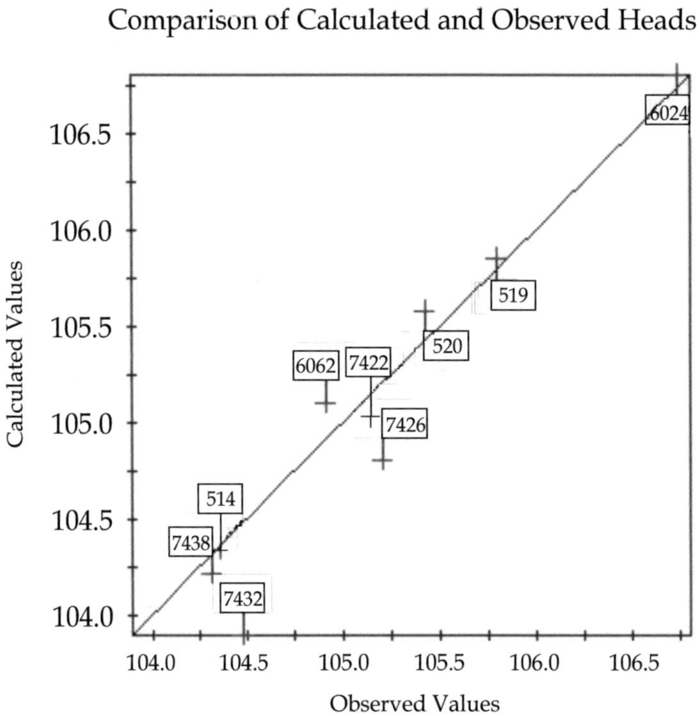

Fig. 6. Plot of calculated versus observed heads

3.1.5 Simulated steady head distribution and flow lines

Data about groundwater fluctuation has shown that the basic factors which are influencing groundwater level changes are atmospheric precipitation, bank filtration from the Danube river and underground inflow from the upper terrace direction north-west and on the northern edge of the area, then evapotranspiration, and underground outflow to the Danube (draining effect of the Danube) and outflow through the drainage canals.

On Fig. 7 we can see that the drainage effect of the Danube is from RK 1745 to RK 1733 and from RK 1727 to RK 1722, and this presents altogether 17 km of bank filtration drainage systems in the river. Bank filtration recharge is from RK 1733 to RK 1727 and that is 6 km bank length of the Danube. The line of direct drainage influence of the Danube goes from the Moča village along the edge of the terrace step to the Búč village, where it turns to the south approximately to the Mária farmstead whence it continues in a southern-easterly

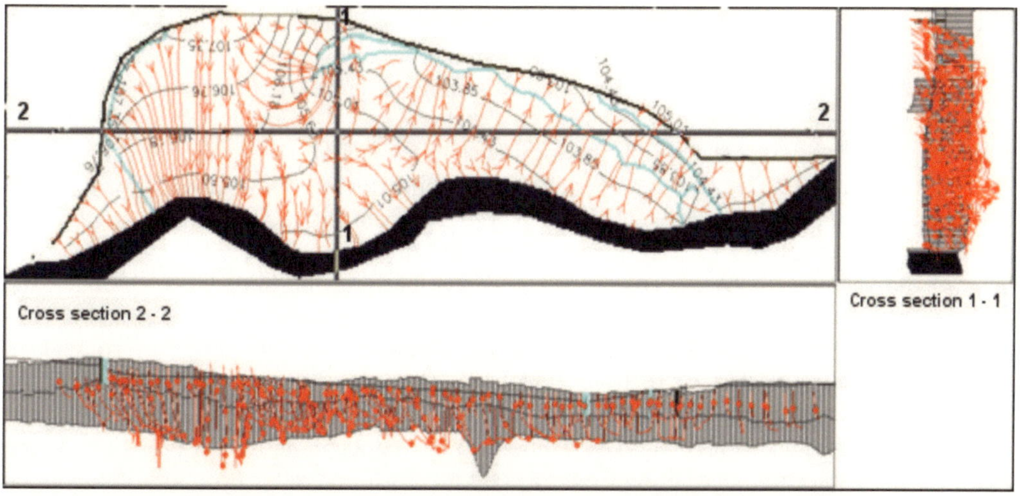

Fig. 7. Simulated steady head distribution and flowlines in the second model layer

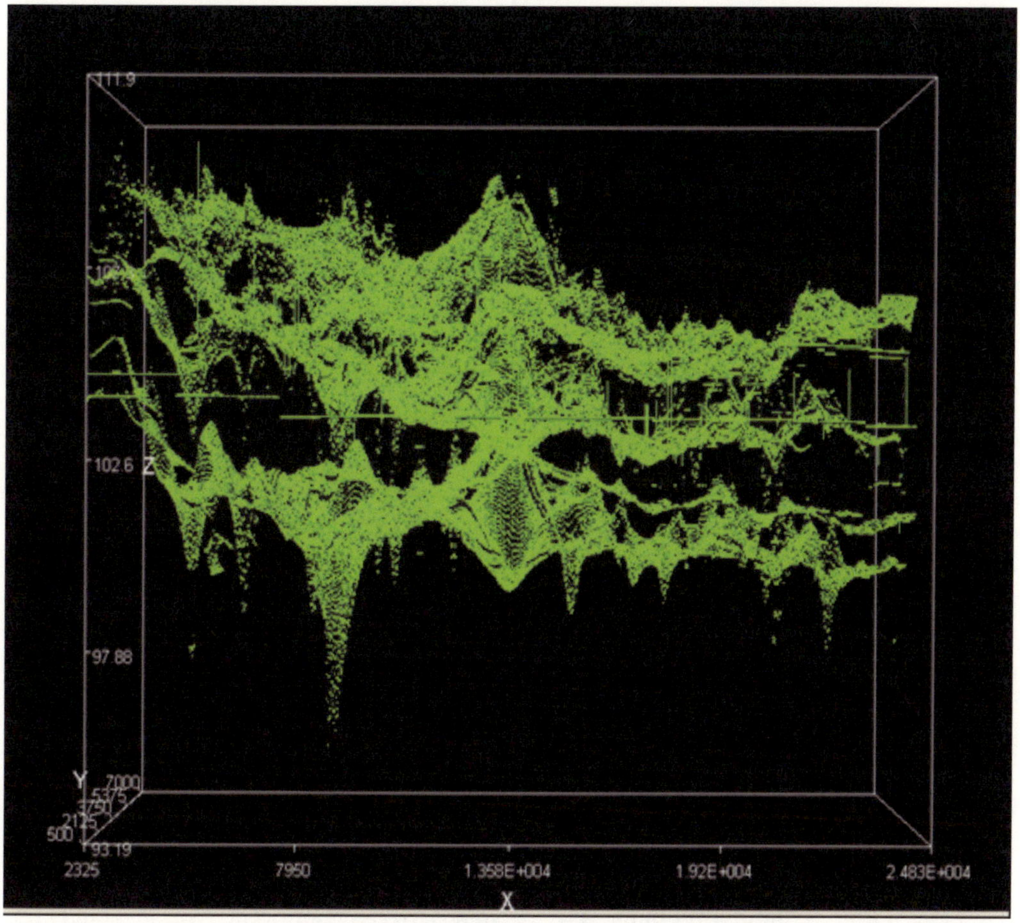

Fig. 8. 3-D visualization of filtration velocity vectors - view from south

direction to the river in the Čenkov locality. The second drainage area is bounded from the south from the Obid canal river mouth to the Danube up to the Štúrovo town and from the north by the "Pod kopanicami" drainage canal. The feeding effect of the river is decreased by the drainage of lower parts of the Mužliansky creek and the Obid canal. The remaining part of the study area is drained by the local system (the Obid, Krížny and Búčsky canal). On Fig. 8 there is a 3D visualization of the whole groundwater body, with velocity vectors in the second layer of the groundwater body where there are the highest filtration velocities along the Obid canal and Mužliansky creek and at their river mouths, and also along the bank of the Danube (drainage) around RK 1742. Similarly in the third layer, there are the highest filtration velocities along open streams and along the Danube bank from RK 1744 to RK 1742 and around RK 1729. The range of filtration velocity values in the whole groundwater body varies from 1E-10 $m.s^{-1}$ to 1.19E-03 $m.s^{-1}$. Particle tracking is used for the tracing of the groundwater flow directions, which means creating flow lines by carrying out the tracing of infinitely small imaginary elements movement situated in the flow field. In the reach from the Moča village to the Čenkov settlement groundwater flow into the Danube except the northern part in which groundwater flows to the drainage canals. From the Obid canal mouth to the Danube River (RK 1727) up to the Štúrovo town groundwater flow direction is again into the Danube. The volume water budget for the whole model at the end of the simulation is calculated in order to control the results. It indicates acceptation of the numerical solution. Continuity has to be preserved also for the total model inflow and outflow or sub region of the layer. The difference in the water budget of the study area is 0.01 % the difference should be in ideal case smaller than 0.1%, what is fulfilled. In general an error up to 1% is accepted (Konikow & Bredehoeft, 1978).

4. The hydrological classification of groundwater regimes

The basic regime specificities of the fluctuation of groundwater level in the alluvial plains and terraces of rivers in dependence upon the fluctuation of the level in a watercourse may be observed only if the natural conditions create the possibility of the hydrodynamic continuity between the level in the surface watercourse and groundwater. If such a relation is possible, then three zones may be earmarked according to the regime symptoms of the groundwater level fluctuation on the riverine territory:

- In the first zone, that is called the narrower riverine zone, every fluctuation of the level of the surface watercourse corresponds with the fluctuation of the groundwater level, of course with the higher distance from the watercourse characterized by the increase in the phase shift and reduction of amplitude. In a certain distance from the watercourse, due to the increased resistance of the porous environment (it is proportional to the length) lower fluctuations in the watercourse level stop influencing the fluctuation of the groundwater level and the next zone is entered.
- In second zone, called wider riverine zone, only greater fluctuations of the watercourse level or the water conditions with longer duration affect the fluctuation of the groundwater level. Its external demarcation (in the direction from the river to the outskirts of the territory) is considered to be the join of the points in which the horizontal straight line plotted at the height of the maximal river level crosses the highest underground level in the riverine valley cross-sections. The river level may not directly influence the groundwater level behind this line via bank filtration.

- In third zone, called the external riverine zone, the groundwater level is always higher than the watercourse level. Despite that it is considered to be the riverine zone, since the gradual direct increase in the groundwater level in narrower and wider riverine zone increases the base for the groundwater drainage of the external riverine zone, in which their level may raise as a consequence of the fact the underground inflow has more difficult outflow conditions and vice versa. Then also the territory is indirectly affected by the outflow river regime - its level fluctuation.

It is clear the boundary between narrower and wider riverine zone, similarly as between wider and external riverine zone is conventional to the certain degree. It must be understood to the intent that its determination pursuant to the stated principles is based upon a certain length of observation time, during which all somehow extreme situations need not to occur. The most distant demarcation of the external riverine zone shall be the edge of the terraced step of the bottom land or other its demarcation at the contact with other hydrogeological units.

If the natural conditions are anthropogenically influenced and create a limited possibility of a hydrodynamic link between the level of the surface stream and groundwater, as it is in the case of the constructed underground non-permeable seating wall (the "NSW") between the Danube River and hydrogeological collector with omitted sections in the NSW, so called "windows", the demarcation of the boundaries between the zones is more difficult. The coefficients of determination for the individual boreholes are considered, while they are very important for the assessment of the degree of dependence between the level of the Danube River and the groundwater level. The boundary between narrower and wider riverine zone shall be determined using two-dimensional models of groundwater flow, displaying the isolines of piezometric groundwater heads and the vectors of filtration speed. The boundary is changed in dependence upon the level condition in the watercourse. The average width of narrower riverine zone of the Danube River at the left side of the lower Váh River in the proximity of an "window" is approximately 2500 m and it is approximately 2300 m on the Čenkov plain. Wider riverine zone is earmarked by the boundary of hydrogeological region Q 057 in both cases.

5. Analyses and comparison of representative groundwater regime of the territory before, and after construction of protective measures

The processing of the observed data and creation of numerical models enables the clarification of the laws of the groundwater regime, in particular to determine its fundamental characteristics, which are: the level heights and main directions of the groundwater flow, depth of the groundwater level under the terrain, the fluctuation of the underground level, the lines of development of changes of the groundwater level in time, volumetric budget and hydrogeological profiles.

The height of levels and main directions of groundwater flow shall be determined using the isolines of the piezometric heights of groundwater level (piezometric contours; for a free level ground water table contours) as the basic document. Firstly they are constructed for the characteristic conditions of the factors that may affect the groundwater regime, according to the knowledge of the hydrogeological and geomorphological conditions of the territory and the preliminary assessment of the observed data. They are the extreme cases of the occurrence of the meteorological factors, such as the periods after extraordinary heavy rainfall or after prolonged rain-free periods. In addition, they are construed also for the periods of the occurrence of the extreme and average conditions of the groundwater for the observation period. Such processing produces the basic data on the conditions of supply and drainage of the groundwater within the territory.

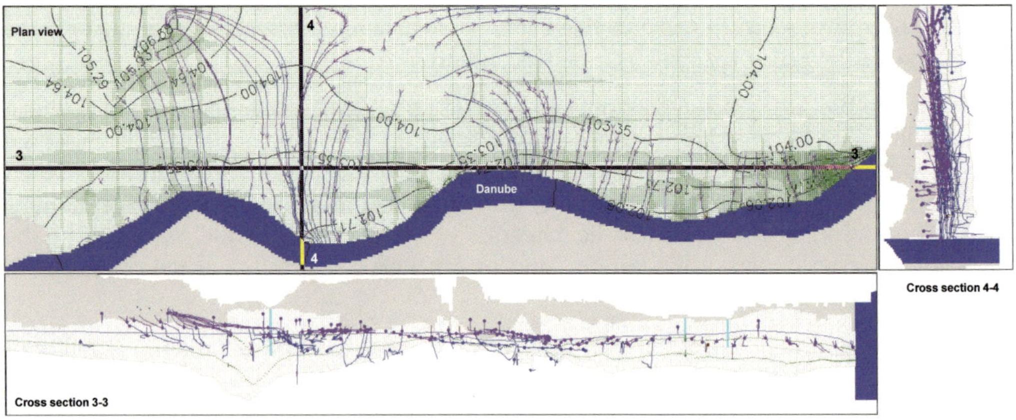

Fig. 9. Steady state head distribution (black lines, m a.s.l.), flowlines (blue lines in the third model layer, purple lines in the second model layer) and velocity vectors(green, m.s^{-1}) in the 3rd model layer before the construction of protective measures - minimum water stage of the Danube

Contour levels: min. 101.42 m a.s.l., max. 104.43 m a.s.l. Flowlines: the direction of the flow of groundwater is from the aquifer to the Danube River lengthwise. Velocity vectors: maximal speeds of the groundwater flow are at the Vojnice brook beyond the effluent from the Búčsky pond in second model layer and somehow smaller speeds are in the Eastern part of the territory of interest at Štúrovo, at Čenkov and at the proximity of the Modranský brook in the third model layer. Maximum value of horizontal pore velocity is 5.11E-04 m.s^{-1} and maximum vertical pore velocity is 3.89E-07 m.s^{-1} (Fig. 9).

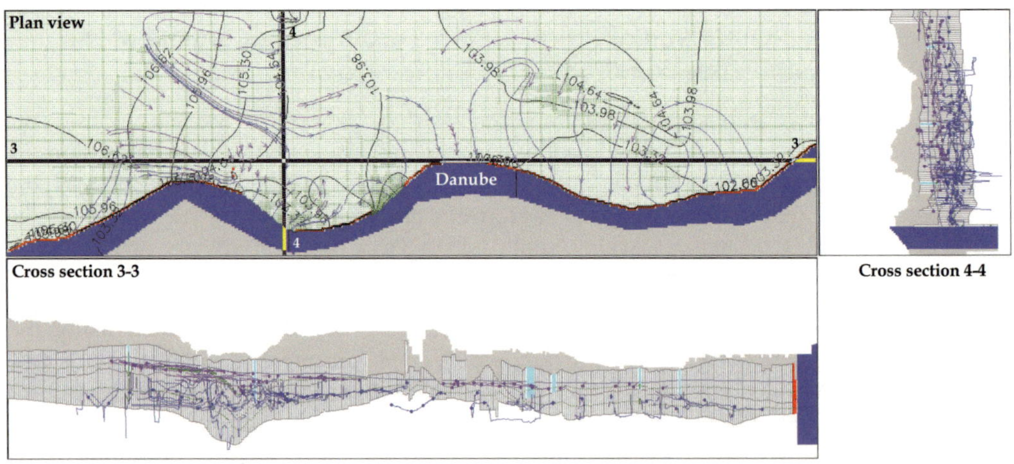

Fig. 10. Steady state head distribution, flowlines and velocity vectors in the 3rd model layer after the construction of protective measures - minimum water stage of the Danube

Contour levels: min. 101.34 m a.s.l., max. 107.94 m a.s.l. Flowlines: the groundwater flows from the aquifer from the Búčšsky pond to the Danube River via the "window" next to Kravany. Smaller amount of groundwater flows to the Danube River via the "Window" at Čenkov. Velocity vectors: the maximal speeds of the groundwater flow are in the Eastern zone of the water source of Kravany towards the wells. Maximum value of horizontal pore velocity is 1.23E-03 m.s-1 and maximum vertical pore velocity is 8E-08 m.s-1 (Fig. 10).

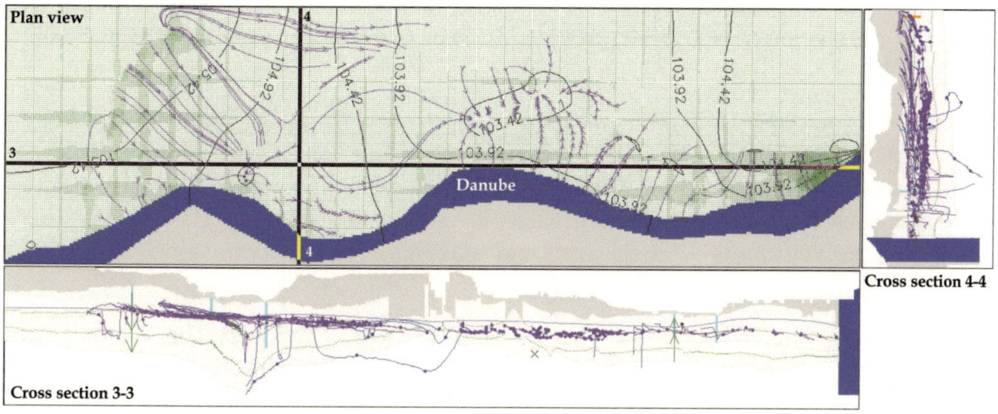

Fig. 11. Steady state head distribution, flowlines and velocity vectors in the 3rd model layer before the construction of protective measures - average water stage of the Danube

Contour levels: min. 103.32 m a.s.l., max. 107.91 m a.s.l. Flowlines: the Western half of the territory of interest is drained by the Danube River. The water source of Kravany drains the circular zone up to the Kravany channel. The aquifer is supplied from the settlement of Čenkov up to the pumping station Obid by the Danube River. The interior is drained by and water is conducted away by the drainage channels, in particular the Obid and Mužľa ones. Velocity vectors: the maximal speeds of flow are around the water source of Kravany in second model layer and somehow lower ones are in the Eastern part of the territory of interest in the section of Obid - Štúrovo in third model layer. Maximum value of horizontal pore velocity is 4.32E-04 m.s-1 and maximum vertical pore velocity is 4.21E-09 m.s-1 (Fig. 11).

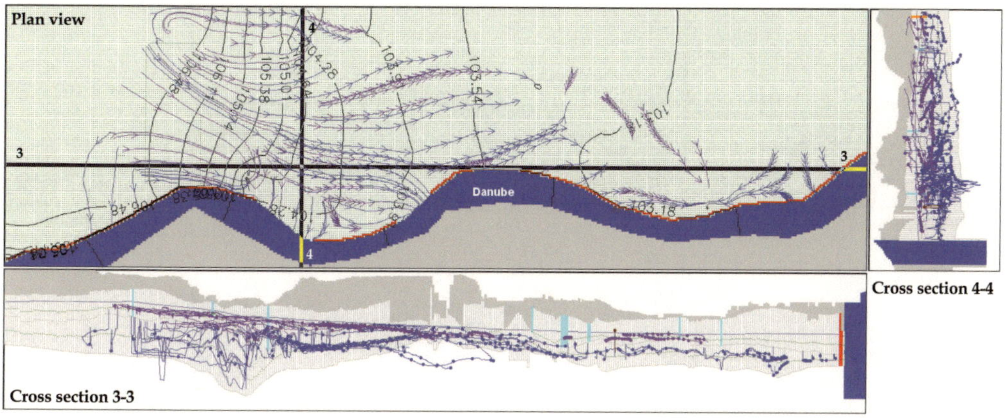

Fig. 12. Steady state head distribution, flowlines and velocity vectors in the 3rd model layer after the construction of protective measures - average water stage of the Danube

Contour levels: min. 102.79 m a.s.l., max. 107.95 m a.s.l. Flowlines: the aquifer drains the Danube River via the window at Kravany approximately from the territory of the intravillain of the village of Kravany. The aquifer drains significantly more via second window at Čenkov, where it takes out the groundwater from the water reservoir at the village Búč. The interior is drained by and water is conducted away by the system of drainage channels, in particular the Obid channel. Velocity vectors: the highest speeds of flow re in second model layer around the water source of Kravany. Maximum value of horizontal pore velocity is 1.13E-03 $m.s^{-1}$ and maximum vertical pore velocity is 1.06E-07 $m.s^{-1}$ (Fig. 12).

Contour levels: min. 104.00 m a.s.l., max. 109.53 m a.s.l. Flowlines: the Danube River fills the aquifer lengthwise. The groundwater is drained in the interior by and conducted away by the system of drainage channels. Velocity vectors: the maximal speeds of flow are between the Kravany channel and the Danube River in second model layer. Maximum value of horizontal pore velocity is 9.42E-03 $m.s^{-1}$ and maximum vertical pore velocity is 5.82E-07 $m.s^{-1}$ (Fig. 13).

Contour levels: min. 102.50 m a.s.l., max. 109.53 m a.s.l. Flowlines: The Danube River supplies the entire aquifer via both windows. The groundwater is drained in the interior by and conducted towards the pumping stations by all drainage channels. Velocity vectors: the maximal speeds of flow are in the area between the Kravany channel and the Danube River in second model layer. The highest speeds of flow of the groundwater in third model layer are in the intravillain of the village of Kravany and at Čenkovo. Maximum value of horizontal pore velocity is 7.67E-04 $m.s^{-1}$ and maximum vertical pore velocity is 6.70E-08 $m.s^{-1}$ (Fig. 14).

Depth of the groundwater level under the terrain is conditioned by its height and morphology of the area. The significance of the processing of depth at analogous water stages as the isolines of piezometric heights lies in the fact they enable to assess the possibility of supply of groundwater from rainfall or their drainage by evapotranspiration in dependence upon the

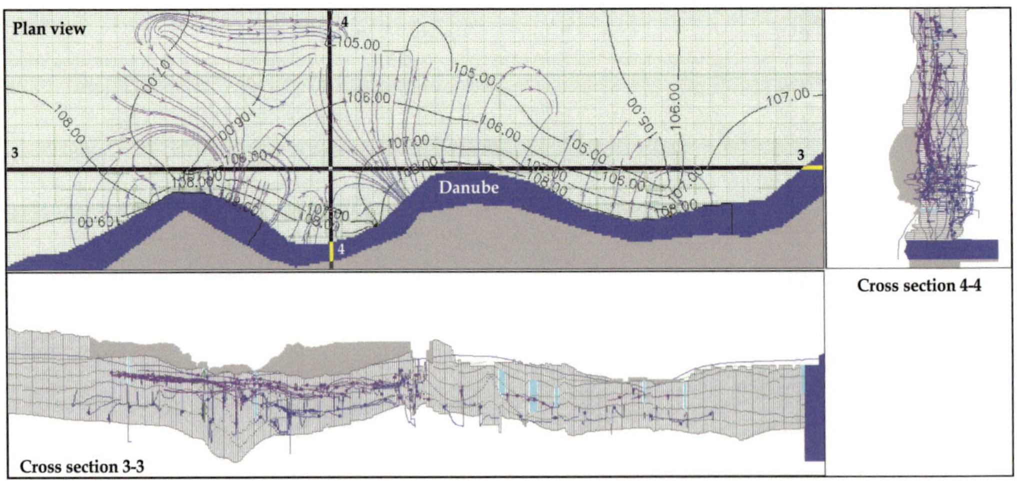

Fig. 13. Steady state head distribution, flowlines and velocity vectors in the 3rd model layer before the construction of protective measures - maximum water stage of the Danube

composition of the surface deposits. When comparing the depths with the thickness of layer of the surface deposits, in the case of their little permeability they enable to assess the groundwater flow regime, thus to earmark the areas or periods with the occurrence of tense or free level of groundwater.

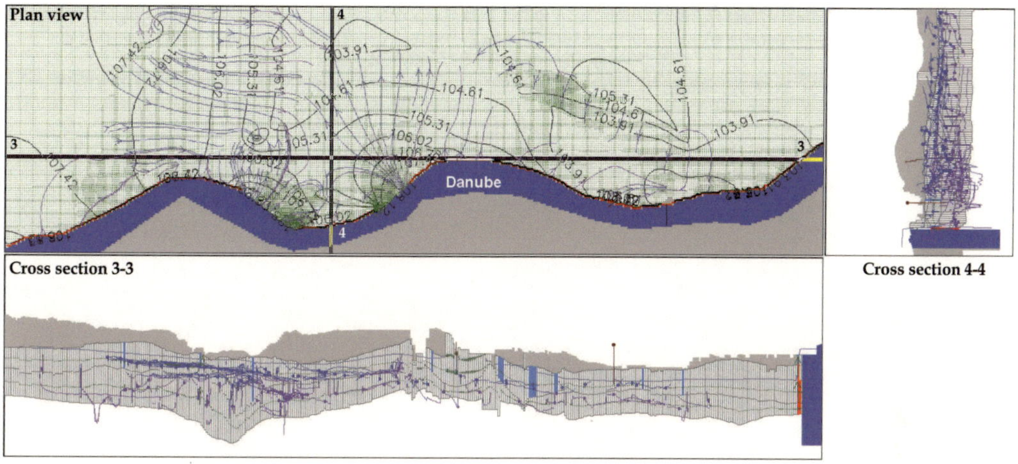

Fig. 14. Steady state head distribution, flowlines and velocity vectors in the 3rd model layer after the construction of protective measures - maximum water stage of the Danube

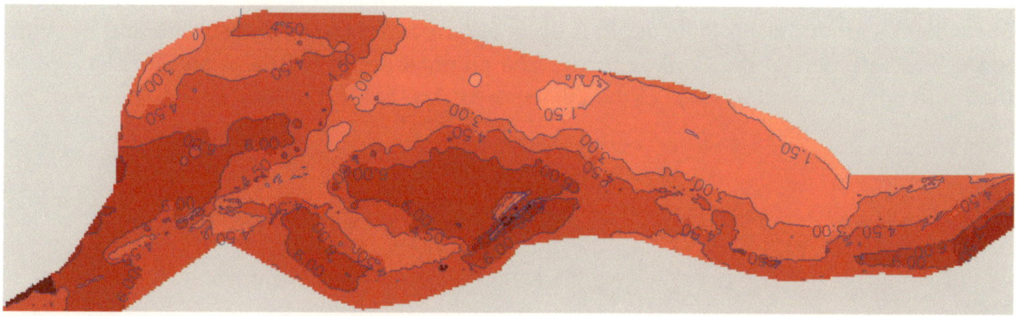

Fig. 15. Depth of groundwater level below the ground surface at minimum water stage of the Danube before the construction of protective measures (m)

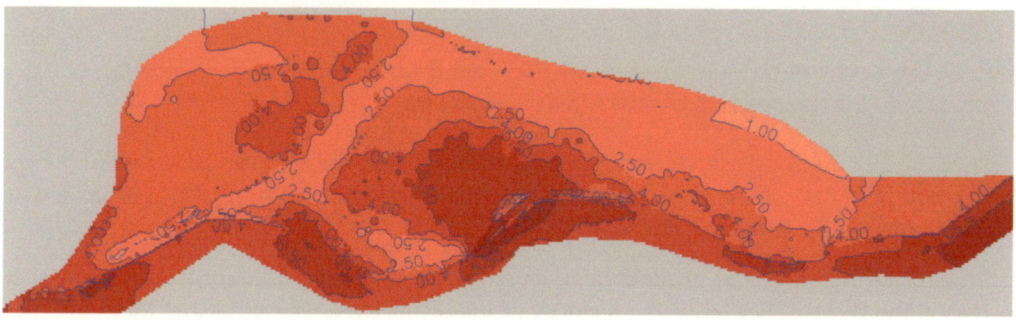

Fig. 16. Depth of groundwater level below the ground surface at minimum water stage of the Danube after the construction of protective measures (m)

The maximal groundwater level depth is up to 8.5 m under the terrain surface and it occurs at the Western border of the territory of interest between the village of Moča and the Modriansky brook. The minimal depth is 2.00 m and it is located at the drainage channels, in particular the Obid, the Krížny and Mužliansky brooks (Fig. 15).

The maximal groundwater level depth is up to 8.0 m under the terrain surface and it occurs in the route of the window in the underground non-permeable wall at Čenkov. The minimal depth is 0.00 m and it is located at the boundary of the territory to the East of the village of Mužla (Fig. 16).

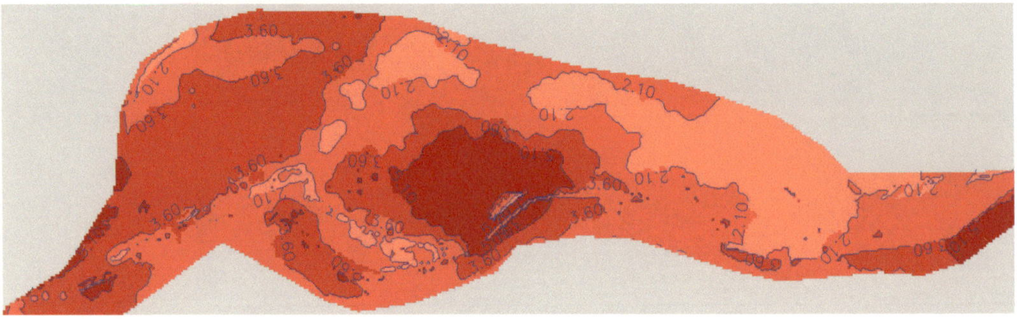

Fig. 17. Depth of groundwater level below the ground surface at average water stage of the Danube before the construction of protective measures (m)

The maximal groundwater level depth is up to 6.30 m under the terrain surface and it occurs in the Southern part of the Čenkov forest. The minimal depth is 0.60 m and it is located to the South next to the intravillain of the village of Mužla (Fig. 17).

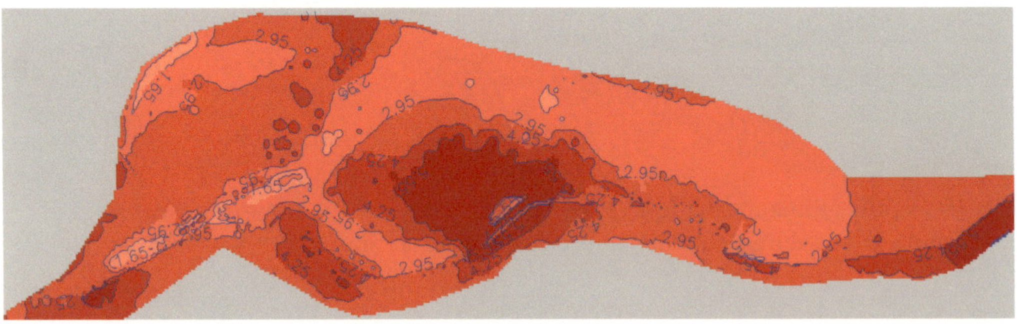

Fig. 18. Depth of groundwater level below the ground surface at average water stage of the Danube after the construction of protective measures (m)

The maximal groundwater level depth is up to 6.86 m under the terrain surface and it occurs in the Southern part of the Čenkov forest. The minimal depth is 0.34 m and it is located at the beginning of the Kravany channel (Fig. 18).

The maximal depth of the groundwater level is 4.37 m under the terrain surface and it occurs at the Western border of the territory of interest between the village of Moča and the Modriansky brook. The piezometric pressure head reaches the value of 3.29 m above the terrain surface in the proximity of the pumping station Obid (Fig. 19).

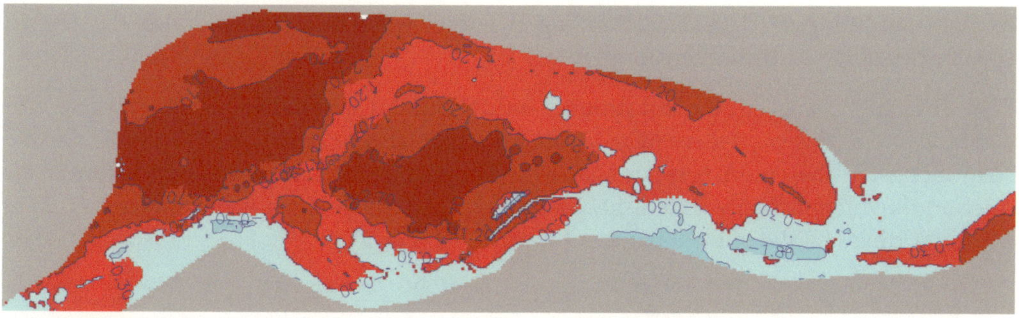

Fig. 19. Depth of groundwater level below the ground surface at maximum water stage of the Danube before the construction of protective measures (m)

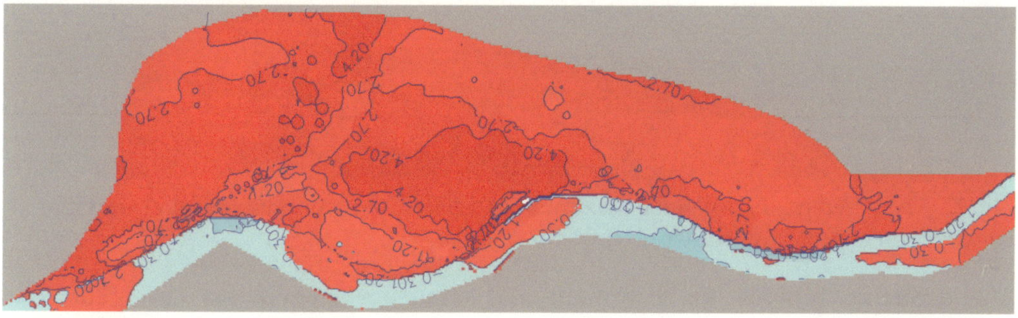

Fig. 20. Depth of groundwater level below the ground surface at maximum water stage of the Danube after the construction of protective measures (m)

The maximal groundwater level depth is up to 5.76 m under the terrain surface and it occurs in the Southern part of the Čenkov forest. The piezometric pressure head reaches the value up to 0.5 m above the terrain surface in the intravillain of the village of Kravany (Fig. 20).

Fluctuation of groundwater levels, plotted as the difference between the extreme heights or depths under the terrain for the observation period, determines the maximal amplitude of the fluctuation on the particular observation spot that may be depicted using the lines with the same fluctuation. From the isolines, it is then possible to determine, when comparing with the values of the fluctuation in the surface recipients and the values of rainfall, as well

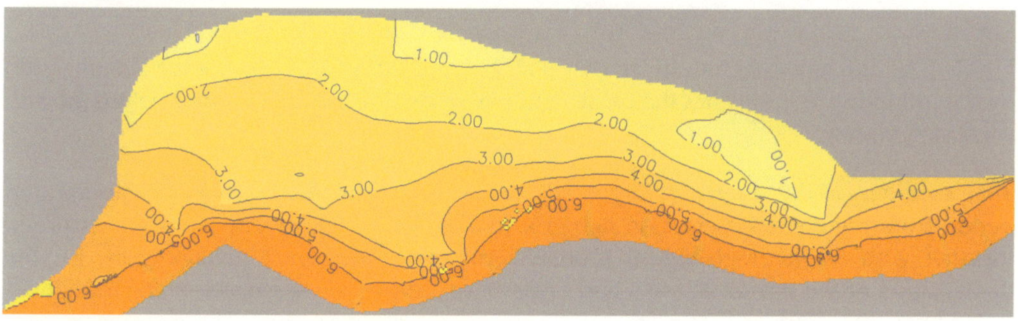

Fig. 21. The difference between the maximum and minimum groundwater level before the construction of protective measures (m)

as the meteorological data characterizing the total evapotranspiration, also the regime specificities of the particular territorial units.

The difference between the maximum and minimum groundwater level before the construction of protective measures reaches the maximal values up to 6.48 m on the narrow strip alongside the entire non-permeable underground wall. To the North towards the interior, the differences are diminished and they reach the minimal values down to 0.00 m on the Northern border of the territory of the interest (Fig. 21).

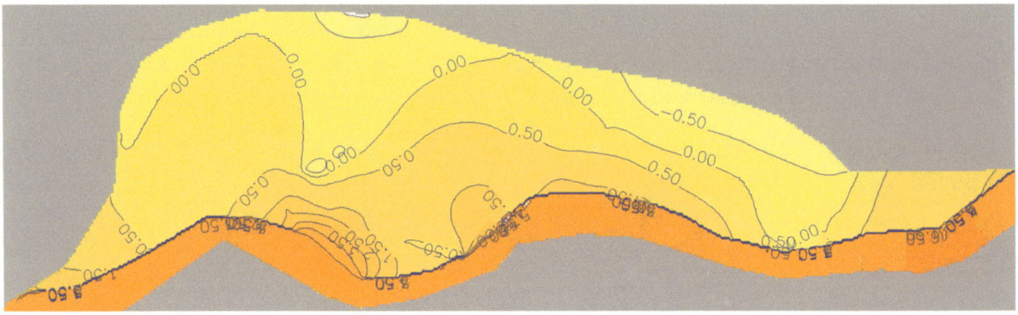

Fig. 22. The difference between the maximum and minimum groundwater level after the construction of protective measures (m)

The difference between the maximum and minimum groundwater level after the construction of protective measures reaches the maximal values up to 6.6 m in the window at Kravany and negative values down to 2.38 m on the Northern border of the territory (Fig. 22).

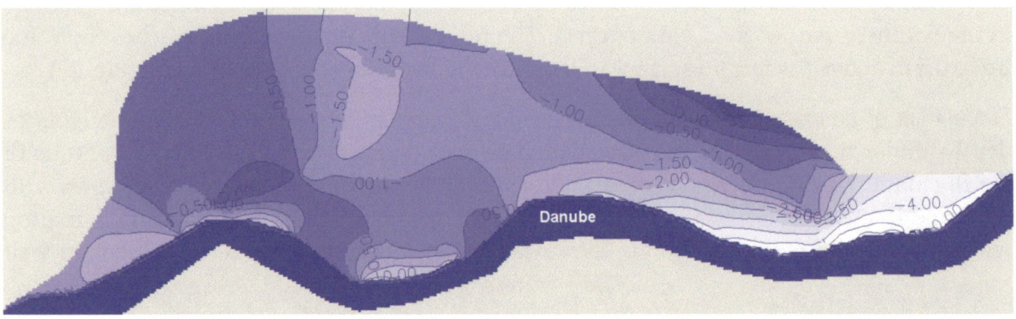

Fig. 23. The difference of groundwater level at the maximum water stage of the Danube (this means groundwater level after the construction of protective measures minus before the construction) (m)

Time slope lines of the changes of the groundwater levels, created for a prolonged observation period from the entire observation network (the SHMÚ, or other boreholes on purpose), together with the hydrogeological profiles form the fundamental preconditions for the demarcation of the territorial units and areas with the prevailing impact of the individual influences, inducing the supply or drainage of groundwater. These impacts are: underground inflow from the rivers or any other surface water recipients, underground inflow from the neighbouring hydrological or hydrogeological units, surface inflow and outflow, surface outflow to the rivers or reservoirs and other surface recipients, underground outflow into other catchment areas or hydrogeological units, rainfall infiltrating on the particular territory, the overall evapotranspiration reducing the stock of groundwater in the case of shallow saturated collectors.

Volumetric budget. A summary of all inflows and outflows to a region is generally called a water budget. In this case, the water budget is termed a volumetric budget because it deals with volumes of water and volumetric flow rates; thus strictly speaking it is not a mass balance. A water budget provides an indication of overall acceptability of the solution. The system of equations solved by the model actually consists of a flow continuity statement for each model cell. The water budget is calculated independently of the equation solution process, and in this sense may provide independent evidence of a valid solution (McDonald, M.G. & Harbaugh A.W.,1988).

The displayed exponential dependence on Fig. 24 from the results of calculations of volume budget show that the approximate limit of the change in the groundwater outflow from the aquifers to the Danube River is an average water stage of the Danube of 104.5 m a.s.l. Water stages of the Danube River exceeding the limit cause low, approximately the same outflow of groundwater from the aquifers to the Danube River both before and after the construction of the protective measures. At water stages of the Danube River below the specified limit the differences in the outflow are increased and at the minimal water stage of the Danube River of 102.5 m a.s.l. the groundwater outflow ($m^3.s^{-1}$) after the completion of the protective measures is approximately five times lower than it was before the completion of the protective measures.

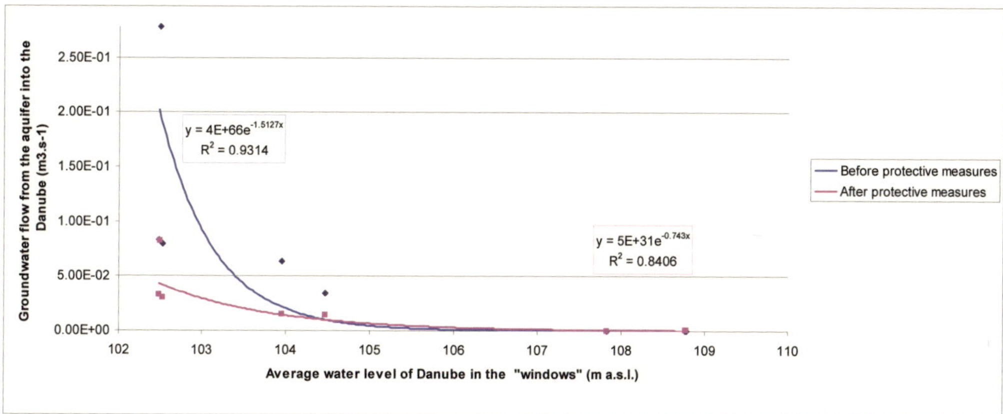

Fig. 24. Groundwater flow from the aquifer into the Danube River ($m^3.s^{-1}$)

Similarly, for the flow of water from the Danube River to the aquifer the relation may be expressed using the exponential dependence (Fig. 25). At the minimal water stage of the Danube River the dependence is almost the same before and after the completion of the protective measures. The differences in the flows are exponentially increased between the average and maximal water stage of the Danube River and at the high water stage of the Danube River after the completion of the protective measures the flow of water from the Danube River to the aquifer is more than fivefold lower than it was before the completion of the protective measures.

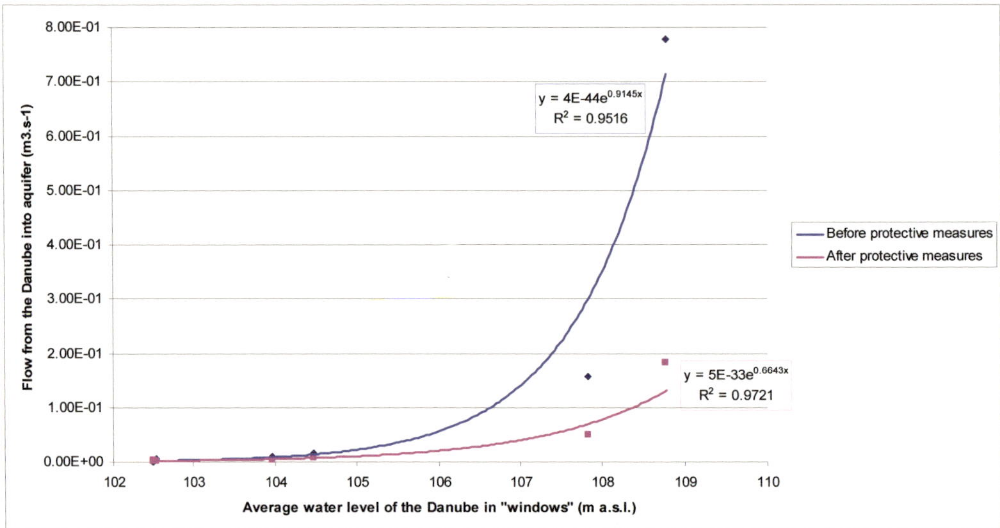

Fig. 25. Flow from the Danube River into the aquifer ($m^3.s^{-1}$)

Hydrogeological profiles (Fig. 4 and 5) with displayed characteristic levels of groundwater, when they are plotted perpendicularly to the surface water recipients, allow to specify the assessment of the impact of the immediate influence of the fluctuation of their level onto the fluctuation of the groundwater level, they allow to determine the distance of the drainage effect of rivers, reservoirs, the inclination of the groundwater level and underground inflow to the observed territory or the underground outflow from it. The hydrogeological profiles alongside the rivers are important for the calculations of the overall bank filtration inflow and outflow. They allow to determine the flow regime, whether it is done with a free level or tense level. Finally, they are very graphic prove for the demarcation of the areas with the intensive inflow and outflow of groundwater, i.e. the areas of their accumulation, or drainage.

6. Conclusions

The processing of the observed data and creation of numerical models enables the clarification of the laws of the groundwater regime, in particular to determine its fundamental characteristics, which are: the level heights and main directions of the groundwater flow, depth of the groundwater level under the terrain surface, the amplitude of the underground level, the lines of development of changes of the groundwater level in time, volumetric budget and hydrogeological profiles. For the purpose of the assessment of the change of the characteristics of the groundwater flow after the construction of the protective measures of the Nagymaros waterworks the condition before the construction of the protective measures was analysed and compared with the condition after the construction of the protective measures (the PMs.) The results imply that:

1. The groundwater level is after the construction of the protective measures:
   - at the minimal water stage of the Danube River higher than before the erection of the PMs on the prevailing portion of the territory of interest (max. by 3.45 m), It is lower on the location of Búčšska lúka and Pod kopanicami (max. by 1.45 m),

- at the average water stage f the Danube River in the Western third of the territory (max. by 1.6 m) and also on the location of Kendeleš (max. by 0.17 m) higher than before the completion of the PMs. Lower (max. by 2.25 m) on the remaining territory,
- at the maximal water stage of the Danube River on the Northern border at the village of Mužla (max. by 1.15 m) and in the proximity of the Kravany channel (max. by 0.35 m) higher than before the completion of the PMs.

2. The main directions of the flow after the completion of the protective measures:
   - at the minimal water stage of the Danube, the change of the direction of the groundwater flow is significant in the Western half of the territory, from the Northern border of the territory to both "windows",
   - at the average water stage of the Danube River, the groundwater from the area of Kravany flows to the Danube River via both "windows" and not to the location of Kendeleš,
   - at the maximal water stage of the Danube, the aquifer is supplied from the Danube not alongside its bank length, but only via the "windows" in the underground wall.
3. The groundwater level depth under the terrain surface after the construction of the protective measures:
   - at the minimal water stage of the Danube the maximal depth of groundwater level was reduced by 0.5 m and the minimal depth reached the level of the terrain surface,
   - at the average water stage of the Danube the maximal depth of groundwater level was increased by 0.56 m and the minimal depth was reduced by 0.36 m,
   - at the maximal water stage of the Danube River the maximal depth of the groundwater level was increased by 1.39 m. The piezometric pressure head above the terrain surface was reduced by 2.79 m.
4. The fluctuation of the groundwater level after the construction of the protective measures:
   - maximal value of the fluctuation was increased by 0.12 m. The minimal value was increased by 2.38 m.
5. Volumetric budget:
   - At water stages of the Danube River below 104.5 m a.s.l. the differences in the outflow of groundwater from the aquifer to the Danube River are increased and at the minimal water stage of the Danube River of 102.5 m a.s.l. the groundwater outflow ($m^3\ s^{-1}$) after the completion of the protective measures is approximately five time lower than it was before the completion of the protective measures. Roughly exponential relation applies here.
   - The differences in the flows are exponentially increased between the average and maximal water stage of the Danube River and at the high water stage of the Danube River after the completion of the protective measures the flow of water from the Danube River to the aquifer is more than fivefold lower than it was before the completion of the protective measures.

Future research should focus on numerical simulations of the underground dam function in the riparian alluvial aquifer. Underground dam belongs to the management types of artificial hydrogeological groundwater body feeding. It is built in shallow alluvial sediments in order to restrain the immediate underground outflow from the groundwater body. It consists of impermeable wall situated along surface flow, which is dropped to the neogene. Artificial groundwater body feeding, which results from integrated surface and groundwater utilization and long lasting sub-surface accumulation, is preferred where it is possible. Artificial feeding has important role by repeated water utilization, because it gives also quality advantages (water clarifying in soil and in groundwater bodies). In order to utilize the underground reservoir for the storage of significant water amount with the intention to utilize it in later period, it is necessary to discover potential accumulation capacity of the groundwater reservoir as well as its convenience for feeding from surface water and easy pumping in the case of necessity. Groundwater reservoir should show sufficient free space between surface terrain and groundwater level for the water storage and water reservation from feeding during the period when the water is not necessary.

7. Acknowledgment

Author would like to express thanks to the Grant Agency of Slovak Academy of Sciences VEGA for the financial support from projects No 2/0123/11 and No 2/0130/09.

8. References

Anderson, M.P. & Woessner, W.W. (1992) *Applied groundwater modelling*. Academic press, Inc., California

Duba, D. (1964) Solution of changes in groundwater level caused by Nagymaros dam construction. *Geologické práce*, Zprávy 32, Bratislava, pp. 91-104 (In Slovak)

Gomboš, M. (2008). Water storage dependability in root zone of soil. *Cereal Research Communications*, Vol.36, No.1, pp. 1194-1194, ISSN 0133-3720

Chiang, W.H. & Kinzelbach W. (2001) *3D-Groundwater Modelling with PMWIN. A Simulation System for Modelling Groundwater Flow and Pollution*, Springer-Verlag Berlin Heidelberg, ISBN 3-540-67744-5

Konikow, L.F., & Bredehoeft, J.D. (1978) Computer model of two-dimensional solute transport and dispersion in groundwater.U.S. Geological Survey Techniques of Water-Resources Investigations, Book 7, chap. C2, 90 p.

McDonald, M.G. & Harbaugh A.W. (1988) *A modular 3-D finite difference groundwater flow model*. USGS, U.S. Geological Survey Open-File Report 83-875, Book 6

Mucha, I. & Šestakov, V.M. (1987) *Groundwater Hydraulics*. ALFA-SNTL, Bratislava-Praha (In Slovak.)

Silva, W.P. & Silva, C.M.D.P.S. (1999-2010) LAB Fit Curve Fitting Software (Nonlinear Regression and Treatment of Data Program) V 7.2.47 online, available from http:/www.labfit.net

Šoltész, A. & Baroková, D. (2004) Analysis, prognosis and control of groundwater level regime based on means of numerical modelling. In: *Global Warming and other Central European Issues in Environmental Protection: Pollution and Water Resources*, Columbia University Press, Vol.XXXV, Columbia, pp.334-347, ISBN 80-89139-06-X

Velísková, Y. (2010) Changes of water resources and soils as components of agro-ecosystem in Slovakia. *Növénytermelés*, Vol. 59, suppl., pp. 203-206, ISSN 0546-8191

http://www.gabcikovo.gov.sk

4

Changes in Groundwater Level Dynamics in Aquifer Systems – Implications for Resource Management in a Semi-Arid Climate

Adelana Michael
Department of Primary Industries/Future Farming Systems Research
Australia

1. Introduction

Groundwater has long been and continues to serve as a reliable source of water for a variety of purposes, including industrial and domestic uses and irrigation. The use of generally high-quality groundwater for irrigation dwarfs all other uses (Burke, 2002); and there are a number of aspects of water quality that have to be managed in such circumstance (e.g salinity, Sodium Absorption Ratio, nutrients, depending on the circumstances of the irrigation). As such there is the need to understand the various implications for use in the management of groundwater resources.

Effective management of groundwater is highly dependent on appropriate reliable and up-to-date information (Adelana, 2009) as may be contained in a groundwater database (GDB). According to FAO (2003a), there are currently thousands of local and personal databases storing key technical and licensing data in a very unsatisfactory manner (mostly in terms of usable formats). Hence, the hard evidence required for the assessment of global trends in groundwater depletion and aquifer degradation is still lacking. It is therefore difficult to assess the extent to which global food production could be at risk from either over-abstraction or from groundwater quality deterioration.

A study on groundwater and food security conducted by FAO (2003a) revealed that compiling reliable groundwater-level and abstraction data (to determine depletion rates) was fraught with problems of coverage, consistency and reliability. Therefore obtaining reliable time-series data on groundwater levels in specific aquifers in many countries may be key to assessing global trend and invariably future impact on food security. The complete lack of a GDB is seriously constraining the formulation and implementation of effective groundwater management policies in many countries. This reinstates the importance of consistency and reliability of groundwater level monitoring for effective groundwater management. In order to ensure sustainable management groundwater level responses must be considered in relation to climate changes and in response to increased agricultural food production.

In the context of varying climatic conditions and frequent lower than average annual rainfall, observed groundwater responses vary and subsequently reduce recharge, stream flow, and the water balance. For example, over the last ten years, decrease in rainfall amount and rain intensity has been the major factor responsible for the declining groundwater levels across northern Victoria in SE Australia (Reid, 2010; Reid et al., 2007). The prolonged effects are expected to contribute a negative impact on water security, agricultural production and the ecosystem. However, under conditions of reduced groundwater use (with recycled water or inter-catchment water transfer), the impacts of irrigated agriculture on the hydrodynamics of shallow aquifer systems and the quality of the groundwater will also need to be fully quantified. Such impacts have been witnessed in other groundwater systems across Australia (Giambastiani et al., 2009; Kelly et al., 2009; McLean & Jankoski, 2002; McLean et al., 2000; Schaffer & Pigois, 2009) and elsewhere in the world (Abidin et al., 2001; Adelana et al., 2006a, 2006b; Chai et al., 2004; Hotta et al., 2010; Lopez-Quiroz et al. 2009).

This study demonstrates the importance of consistent groundwater level monitoring in relation to (and its implications on) effective and sustainable resource management as well as improved the understanding of climate impacts on groundwater levels. Two case examples are selected from areas at different level of groundwater monitoring, used to illustrate impact of climate variability as well as the importance of reliable and consistent groundwater monitoring database.

2. Background

Water use in both study areas (the Werribee Plains, Western Melbourne metropolitan, South-east Australia (Figure 1) and the Cape Flats, Cape Town metropolitan area, South Africa (Figure 2)) supports year-round irrigation, and is one of conjunctive use, including a channel network fed by releases from reservoirs and recycled water, respectively, and supplementary groundwater extractions. This represents two long established irrigation districts: the Werribee Irrigation District (WID) and the Cape Flats farming areas, both known for their market gardens. At a national scale, the WID is major suppliers of lettuce, cabbage, broccoli and cauliflower (SRWA, 2009), while the Cape Flats, especially the Greater Philippi horticultural area, is an important source of Cape Town's fresh produce (such as lettuce, onions, fresh fruit, bananas, potatoes) and which, at the regional scale produces 70-80% of vegetable sold in the Greater City of Cape Town (Rabe 1992, CCT 2010). For the two areas, the location, the highly productive soils and intensive cropping capability allow for diverse production and all-year-round supply. Moreover the close proximity of the two farming areas to fast growing commercial centres (Melbourne and Cape Town, respectively) provide market advantages and increases the value of the land for urban development.

Active groundwater management of the system in the Werribee Plains was initiated in 1998, at which time a safe yield of 2,400 ML/yr was estimated, compared to the sum of licensed groundwater extraction about 6,000 ML/yr. The installation of meters on all licensed bores occurred in 2004. A 25% restriction in licensed volume was in place (SKM, 2004) and this has since been regularly reviewed. Southern Rural Water Authority (SRWA) is the responsible agency for the management of groundwater resources in this district. Until recently, irrigators have been able to consistently rely on approximately 10,000 ML of water rights from SRWA's water distribution system (predominantly concrete-lined channels) and 5,000 ML of groundwater licences in the underlying shallow Groundwater Management Area (Rodda & Kent, 2004). In the Cape Flats, the Department of Water Affairs (DWA) is responsible for permits, licensing and metering. All information regarding registered groundwater users and licensed volumes are encoded onto WARMS (Water use And Registration Management System section of DWA) database. In practise, the farmers in the Cape Town area irrigate their crops, particularly during the dry summer months and intensely in drier years. As at December 2006, the highest single registered volume was 699.15 ML/yr (Adelana, 2011). From WARMS record in 2006, there were 211 bores used for agriculture, 25 for industry and two for water supply within the City of Cape Town municipality (although a number of unregistered household bores may exist). The City of Cape Town has water restriction and management plan in place since 2002.

In the WID, expected threats to the aquifer include seawater intrusion from the coastline and estuarine portion of the Werribee River, inter-aquifer transfer of saline groundwater, and water level-induced bore failure. Reduced rainfall conditions exacerbate these threats by reduced recharge from both rainfall and channel leakage, increased estuarine length of the Werribee River, and an increased dependency on groundwater (SRWA, 2009). In the Cape Flats aquifer the maximum extent of seawater intrusion into the Cape Flats aquifer has been estimated to be approximately 1,000 m from the coastline (Gerber 1981), although recent studies (Adelana, 2011; Adelana & Xu, 2006) did not confirm inland saltwater movement. Nevertheless, surface water in the Cape Flats is known to be contaminated from various sources (Usher et al., 2004; Adelana & Xu, 2006) and the potential treat to groundwater identified (Adelana & Xu, 2006).

Within the WID, the highest percentage of groundwater extraction is from the Werribee deltaic sediments. Regions of the deltaic aquifer adjacent to the coastline and estuary have exhibited depressed watertable conditions, with hydraulic heads falling below mean sea level and/or at lowest recorded levels. These regions are also exhibiting rising groundwater salinity, particularly in deeper piezometers (SRWA, 2009). In the Western Cape, agricultural sector is one of the largest users of water resources; but rapid economic development and population growth is also generating increased pressure on water supplies. For example, the growth in urban water demand in the Greater Cape Town Metropolitan Area was projected to increase from 243 million m^3 in 1990 to 456 million m^3 in 2010; whereas for irrigation water demand the increase is from 56 million m^3 in 1991 to 193 million m^3 in 2010 (Ninham Shand, 1994). Over 60 % of the annual urban demand and 90 % of the irrigation demand occurs in summer (Adelana, 2011).

3. Study approach

In order to investigate varying climatic conditions and the impact of frequent lower than average annual rainfall on observed groundwater levels the long-term climate data are analysed and compared for both study areas. In the long-term, rainfall, minimum and maximum temperatures are related to climate variability. The climate data obtained were analysed and statistically interpreted. Long-term data are from the South African Weather Service (Station: Cape Town Observatory/Airport) and Bureau of Meteorology (BOM with station in Laverton near Werribee).

The groundwater databases of the Department of Primary Industries (DPI) and Department of Sustainability and Environment (DSE) Groundwater Management System (GMS) were examined to select representative bores tapping the Werribee Delta aquifer. Also, from the National Groundwater Database (NGDB) managed by DWA, a few bores screened in the

Cape Flats aquifer were selected. These bores were investigated by evaluating the groundwater levels and salinity (specifically the electrical conductivity) within shallow aquifers. The criteria for selection were continuous groundwater level record (minimum of 10 years record, with minimal interruptions or errors) and screened within the respective aquifers under this study.

The time-series groundwater data at selected locations within Cape Town area were compared with those of bore network data in the Werribee Plain. The analysis of this data was undertaken using Hydrograph Analysis: Rainfall and Time Trends (HARTT), a statistical tool that analyses groundwater data using the effect of long-term rainfall patterns, determined by accumulative residual techniques (Ferdowsian et al., 2001). This method can differentiate between the effect of rainfall fluctuations and the underlying trend of groundwater level over time. Rainfall is represented as an accumulation of deviations from average rainfall, and the lag between rainfall and its impact on groundwater is explicitly represented. HARTT produces a fitted curve through the groundwater level readings.

According to Ferdowsian et al. (2001), two variables are used to produce this line:

Rainfall variable (X_1); accumulative monthly residual rainfall (AMRR, mm), or accumulative annual residual rainfall (AARR, mm).
Time trend (X_2) (1,2,3 days…from first reading)

At any point along the fitted curve, the following equation holds:

$$Y = c + aX_1(\text{rainfall}) + bX_2\,(\text{time trend}) \tag{1}$$

Where:
c is the intercept.
a and b are coefficients calculated in the multiple regression analysis.
Y is the water level depth at a point along the fitted curve.

So, to calculate the effect of rainfall, the following equation is used:

$$Y' = aX_1(\text{rainfall}) \tag{2}$$

And to calculate the underlying trend, the following equation is used:

$$Y'' = c + bX_2\,(\text{time trend}) \tag{3}$$

The R^2 value (the coefficient of determination) is the degree of fit of the calculated curve compared to the recorded water levels (a value of 1 is a perfect fit; the degree of fit becomes less with decreasing values below 1). The p-value indicates the level of significance of each variable. If the p-value is less than 0.05, then the variable is significant. If it is less than 0.01, then it is highly significant. If the trend is not significant (as determined by R^2) then the rate of rise or fall is not reliable. And if the rainfall variable is not significant then the reliability of the effect and the delay period (in the hydrograph response to effective rainfall) is low (Ferdowsian et al., 2001).

The method improves the estimation of time trends and allows for better interpretation of treatment effects on groundwater levels. The advantage and limitation of this method over other techniques of hydrograph analyses have been highlighted in Cheng et al. (2011).

Access to several unpublished reports has also yielded valuable information. A general overview of the study area is presented with the description of geology and hydrogeological settings in order to first understand the groundwater system in both areas.

4. Description of the study area

The vegetable growing Werribee Irrigation District (WID) lies on Melbourne's rapidly-developing western urban fringe underlain by shallow Delta aquifer. The name of the management area for the Werribee Delta aquifer is the Deutgam Water Supply Protection Area (WSPA). The aquifer is linked to both Port Phillip Bay and the tidal extent of the Werribee River (SRWA, 2009). Deutgam WSPA is located around the Werribee South irrigation area (Figure 1). On the other hand, the fresh fruits and vegetable farm area in Cape Town is located on the Cape Flats, especially the Phillipi-Mitchells Plain Irrigation area. A large portion of the area around Cape Town is the sand-covered coastal plain (Cape Flats) shown in figure 2. The City of Cape Town Management Area (CMA) is largely surrounded by the Atlantic Ocean to the west and south with the most prominent landmass being the Cape Peninsula, attached to the mainland by the sandy plain of the Cape Flats (Schalke, 1973; Theron et al., 1992). The greater portion of the entire sand cover of the Western Cape are been considered in this study, particularly the south-western part of the City of Cape Town and the north-western end (Atlantis), where basic data and bore information are available.

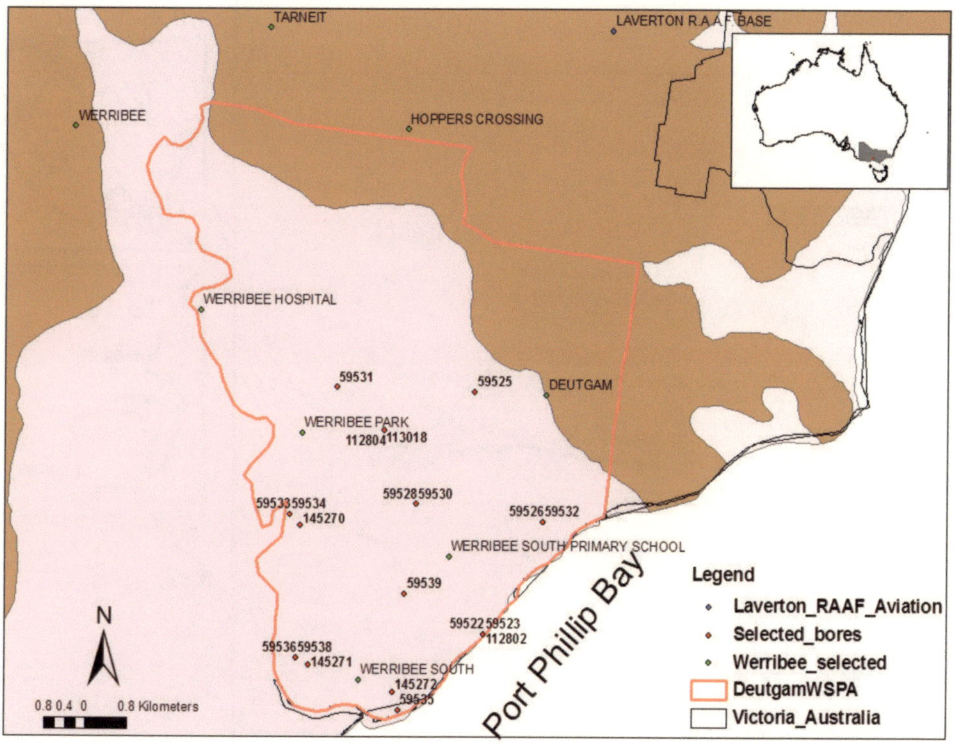

Fig. 1. Location of the Werribee Plains and Deutgam WSPA, western fringe of Melbourne. Inset: Deutgam WSPA (red spot) in Victoria (grey shade) within map of Australia

4.1 Geology and hydrogeology

4.1.1 The Cape Flats

A study of the geological units show the oldest rock in Cape Town and suburbs are the meta-sediments of the pre-Cambrian Malmesbury Group, which occupy the coastal plain between Saldanha and False Bay in the west, to the first mountain ranges in the east (Meyer, 2001). Several erosional windows to this Group are exposed in mainly fault-controlled valleys further to the east and south. Natural features are varied and include narrow flats, kloofs and gorges, cliffs, rocky shores, wave-cut platforms, small bays and sandy and gravel beaches. On the Cape Flats sand dunes are frequent with a prevalent southeasterly orientation; and the highest dunes are only 65 m above sea level (Schalke, 1973; Theron et al., 1992). The sand is derived from two main sources: (i) weathering followed by deposition, under marine conditions, of the quartzite and sandstone of the Malmesbury Formation and the Table Mountain Series; (ii) the beaches in the area, from where Aeolian sand was deposited as dunes on top of the marine sands.

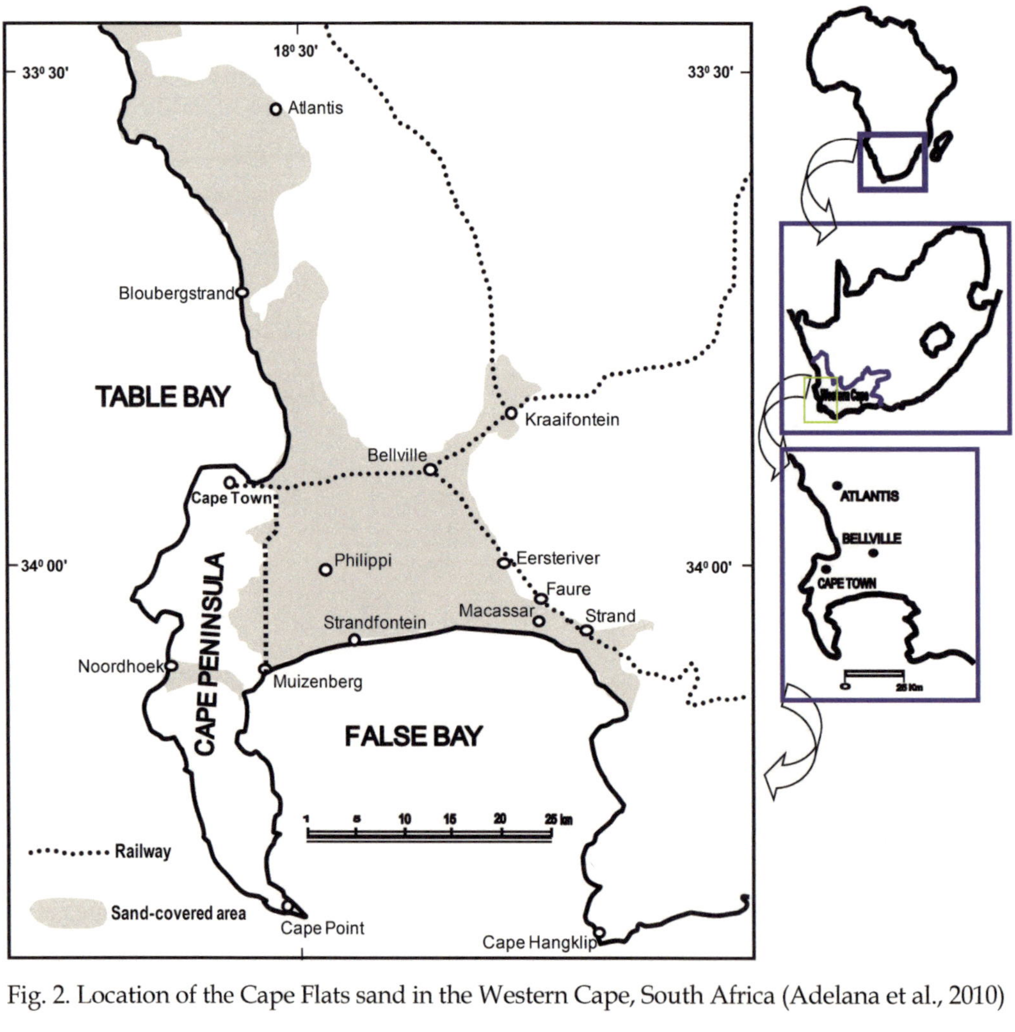

Fig. 2. Location of the Cape Flats sand in the Western Cape, South Africa (Adelana et al., 2010)

According to Meyer (2001) bore yield (from about 497 boreholes in the Sandveld Group) indicates that 41% of boreholes yield 0.5L/s and less while 30% yields 2L/s and more. Transmissivity values range from 32.5-619m^2/d (from recent pumping test data in Adelana, 2011), but typical values between 200 and 350 m^2/d were recorded in Gerber (1981). A detailed description of the hydrogeology of the different geological units is documented in Meyer (2001). The net groundwater recharge to the Cape Flats aquifer in the south-western Cape varies between 15% and 47% of mean annual precipitation (Adelana, 2011). The general aquifer configuration and flow direction in the Cape Flats has been presented as indicating flow from western and south-eastern to the coast. A conceptual model of the aquifer has been developed to indicate all flow is regionally unconfined and two-dimensional with negligible vertical components, although inter-bedded clay and peat layers produce semi-confined conditions in places (Adelana et al., 2010).

4.1.2 The Werribee Plains

The Deutgam WSPA includes all geological units to 40m below the natural surface, encompassing the shallow Werribee Delta sediments (DWSPACC, 2002). An alluvial deposit up to 20 m thick has accumulated in the gorge of the Werribee River. This gorge is the major terrain feature of the Werribee Plains with its alluvial deposit known to be gravely at the base and fumes upwards to become clayey at the surface (Condon, 1951). According to this work and more recent studies (Holdgate et al., 2001, 2002; Holdgate & Gallagher, 2003), the alluvial terraces on the valley floor provide evidence of Pleistocene and Holocene sea level changes. This alluvium, eroded by rejuvenated streams, was deposited (in Late Quaternary times) along the base of the Werribee River. There are prominent intra-volcanic sands within the Newer Volcanics (along the Werribee Plains) while the Older Volcanics were picked in few bores between coal-bearing sediments of the Werribee Formation (Holdgate et al., 2001). In general, the Werribee Formation is disconformably overlain by marine sandstone and mudstone/marlstone (Taylor, 1963 as cited in Holdgate et al., 2002; Holdgate & Gallagher, 2003). Across the Werribee Plains these exceed 120 m in thickness (Holdgate et al., 2001).

The groundwater system used in the Werribee South is called the Werribee Delta aquifer. The Werribee Delta sediments consist of sand and gravel lenses situated within clays and silts. The variable nature of the deltaic sediments resulted in a wide variation in aquifer parameters (SKM, 2002). According to SKM (1998), within the coarser sand horizons the hydraulic conductivity ranges from 10 to 15m/day, with a specific yield of 0.01 to 0.2 but the overall hydraulic conductivity of the aquifer is less than 5m/day with representative specific yield in order of 0.04. Typical bore yields for the Werribee Delta aquifer system are generally less than 5L/s, however yields up to 15L/s have been recorded (SKM, 2002). The selected bores for this study were screened in the Werribee Delta aquifer system, which are mostly sandy or silty clay material at shallow depths but with significant sand and gravel seams at a relatively deeper depth. The Werribee Delta aquifer system is unconfined to semi-confined and groundwater depth varied between 4-7m below ground surface. Recharge to the aquifer system is predominantly from direct rainfall infiltration and surplus irrigation water (SKM, 2002) as well as leakage from the ageing concrete-lined channels (Rodda & Kent, 2004).

4.2 Climate

The study areas (Cape Flats and Werribee Plains) are both under Mediterranean climate. Climate is temperate with warm dry summers and maximum rainfall occurring during winter/spring respectively. Historical average annual rainfall (1913-2009) varies from 1100 mm/yr in the upper north-west of the Werribee catchment to 540mm/yr near Werribee (SRWA, 2009). Historical data (1841-2009) showed there is a variable rainfall gradient in the Greater Cape Town area; rainfall is largely controlled by topography - between 500 mm and 1700 mm on the Cape Peninsula, to 500 mm and 800 mm on the Cape Flats, and ranging from 800 to over 2600 mm in the mountains to the east of the Western Cape region (Adelana, 2011). To the north of the Western Cape, this climate regime grades into semi-desert whereas to the south-east coast the climate becomes less seasonal and tends towards sub-tropical. Drier summer conditions and lower winter temperatures tend to inhibit some plants' growth.

Therefore, rainfall, minimum and maximum temperatures were analysed and compared to show climate variability over the years, and to identify/assess its impact on groundwater levels. Figure 3 show the annual/seasonal rainfall variation in the study areas. There is a similar pattern in the fluctuation of observed annual rainfall being less than the long-term average in many years. Long-term or historical climatic conditions indicate that on average, annual rainfall in the Werribee for the period 1950-1979 exceeded that for the period 1980-2009, with the period 1997-2009 being one of considerably lower than average annual rainfall. For example, during 2004/05, rainfall in the Werribee River catchment was approximately equal to the long-term average; whereas rainfall was about 60% of long term average for 2005/06, although inflows were only 21% of the long term average (SRWA, 2006). Consequently, storage levels fell from an average 34% at the start of the year to 16% at the end of the year and irrigators and diverters in the Werribee system were allocated 80% of their water entitlement (SRWA, 2006).

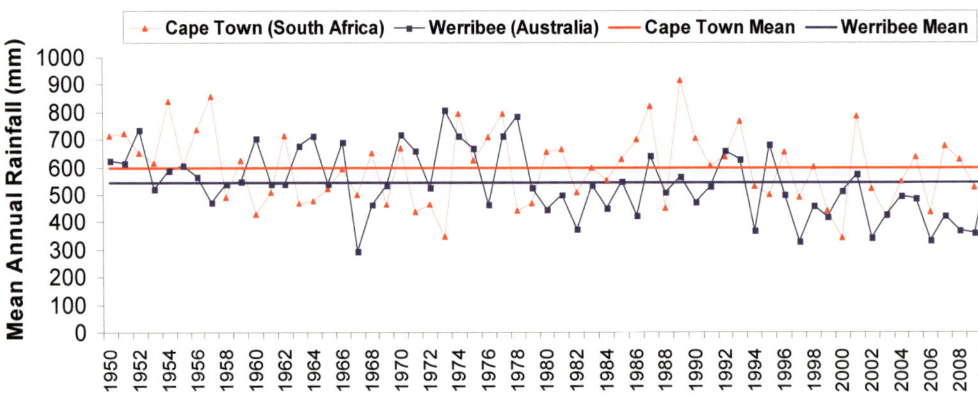

Fig. 3. Annual mean of rainfall in the study areas 1950-2010 (Station: Cape Town Airport and Laverton RAAF Base)

In the Cape Flats from 1958 the trend in rainfall showed continuous decrease up till 1974; 1982-1985 was also a dry period with total average rainfall below annual mean. Since then there has been much fluctuation in the pattern of rainfall in the Cape Town area. This was shown to be comparable to older records (1921-1941) of relatively dry periods; for example

1935 recorded the least annual rainfall (229.4 mm/yr) (Adelana, 2011). A similar situation is observed from 1999-2003, with the exception of year 2001 that showed a relatively wetter record (i.e.784 mm/yr). Based on available information going as far back as the 1960s, Cape Town enters into a drought cycle (i.e. a lower than average rainfall pattern) on average every 6 years (Cape Water Solutions, 2010). The last of such a cycle was in 2003 and 2004 with dry winter and nearly 200mm less than long-term mean of annual rainfall. The consequences include lower dam levels and the imposition of water restrictions.

Seasonal patterns in the Cape Town area show a marked winter rainfall incidence, with June/July typically the wettest month. The general climatic trend throughout the study area

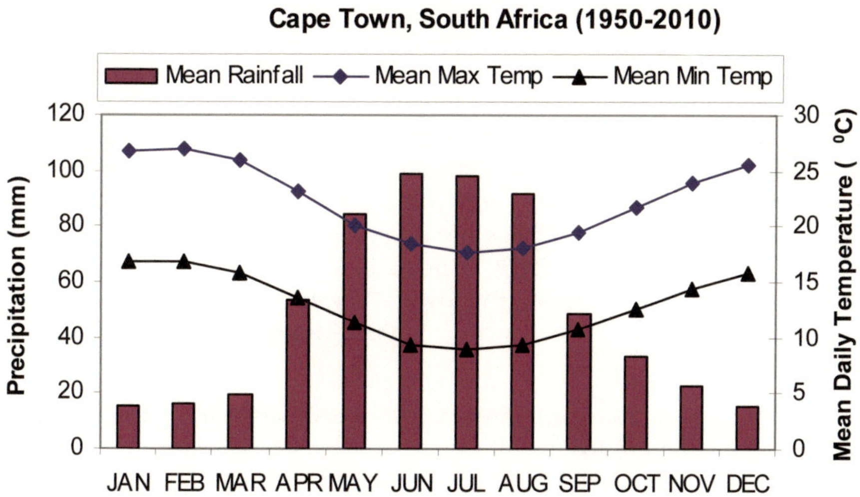

(a)

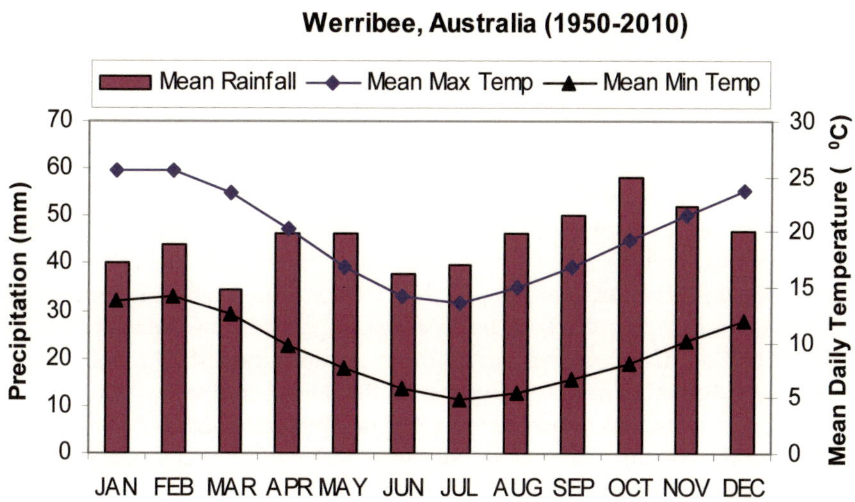

(b)

Fig. 4. (a) Cape Town mean monthly rainfall with maximum and minimum temperature. (b) Werribee mean monthly rainfall with maximum and minimum temperature is a gradual increase in rainfall and a reduction in temperatures moving from north to south. Mean annual rainfall (1950-2010) has a long-term average of 597 mm. There is a dry period with less than 20 mm rainfall per month from November to March (Figure 4a); the mean annual temperature is moderate, approximately 17 °C. In the Werribee Plains, rainfall variability throughout a typical year does not exhibit a clear seasonal bias like the Cape Flats but fairly distributed, with average monthly rainfall ranging from 36 mm/month (March) to 59 mm/month (October) (Figure 4b).

5. Groundwater level response

The results of the HARTT analysis for the selected bores are summarised into a Table in Appendix I. The groundwater trends determined in this analysis are comparable for both the Werribee Plain Delta aquifer and the Cape Flats aquifer. There is little or no delay in response to rainfall events. Although most of the bores showed a generally slight decline, few bores have a rising trend yet the rise is much less than 20 cm/yr. Most of the bores screened in the Werribee Delta aquifer have groundwater trend ranging -3 to 4 cm/yr (in exception of B112802 with more positive trend, 12 cm/yr). All of these bores showed no delayed response and a quick rise in response to the wet year 2010 (after lower than average rainfall from 2007-2009) (Appendix II a-d). The selected bores in the Werribee mostly showed the lowest groundwater levels (i.e. highest drawdown) in late 2003 and early 2004 except B59536 whose highest drawdown was in February 2007 (Appendix II d).

The groundwater trend of bores within the Cape Flats aquifer ranges from -8 to 8 cm/yr (except BA232 with more positive trend, 14 cm/yr; which is within the Philippi allotment portion). The bores on the Cape Flats showed marked seasonal fluctuations and a more slightly downward trend in comparison to the Werribee bores (see Table in Appendix I). The Cape Flats bores are examples of good data records with missing gaps (Appendix II e-j). Most of the bores selected along the south coast on the Cape Flats also showed no delayed response except BA002 (Appendix II e). Although there are no lithologic logs for most of these bores, there are reports of occurrence of thick lenses of clay within the Cape Flats sand aquifer (Adelana, 2011; Gerber, 1976) that may contribute to delayed response of bores to rainfall events.

There are no significant negative trends (groundwater trend all < -9 cm/yr) in both study area, even though a few bores were also selected from the intensively irrigated Atlantis area of Western Cape. Examples of bores from the Cape Flats sand in north-western Cape (Atlantis) showing influence of pumping in the 1990s are presented in Appendix II (k-n) with summary table in Appendix I). The Cape Flats aquifer in the Atlantis area has been under the Managed Aquifer Recharge (MAR) program since the last 20 years. Although the extent to which this has influenced the response of the bores is not known since the data are not accessible, it is expected to contribute to a more positive trend. Irrigation is intense in the Werribee area but the conjunctive groundwater use (with surface water, recycle water) may have been responsible for no significant negative trend.

However, the Philippi-Mitchells Plain bores are still more negative relative to both Atlantis and Werribee Irrigation districts even though both have longer history of groundwater use for irrigation. This may be in response to groundwater usage. It was estimated that approximately 13 million m^3 are abstracted from the Cape Flats aquifer by commercial farmers in the Philippi area of Cape Town (Colvin & Saayman, 2007), and an additional 5 million m^3 are abstracted by the City of Cape Town administration to irrigate sports fields at Strandfontein and Mitchell's Plain (Wright & Conrad, 1995). Moreover, another 20 million m^3 was abstracted from wellfields in the southern part of the aquifer during the Pilot Abstraction Scheme to understudy the Cape Flats aquifer response to stress conditions (Gerber, 1981; Vandoolaeghe, 1989).

The bores examined across the Werribee Plain showed declines in groundwater levels occurring from 1996 to 1999, 2003 to 2004 and in late 2006 to early 2007. This tends to follow the downward trend in the frequency and amount of rainfall and is consistent with the general groundwater trend observed across Victoria during this period (Hekmeijer et al., 2008; Reid, 2010). The groundwater level drawdown of Werribee Delta aquifer shows that during the early 1990s seasonal drawdown was less than 0.5 m but in 1996, the seasonal fall increased up to 2 m. This indicates more use of groundwater for irrigation due to the lack of supply from the Werribee River. Therefore, the seasonal fluctuations are mostly influenced by rainfall and usage; however, some observation bores show seasonal fluctuation that is believed to align with the pattern of channel deliveries (i.e. due to enhanced channel leakage) and groundwater pumping (SKM, 2009a, 2009b).

The observed groundwater trends and behaviour in the South African example (i.e. bores screened in the Cape Flats aquifer) are equally consistent with the fluctuations in rainfall pattern. It is obvious that the groundwater level falls due to less rain and possibly higher use from production bores, while rainfall recharge and recovery take place in wetter times when there is conversely less pumping. Some of the Cape Flats bores in Atlantis showed a marked response to pumping influences and have recorded higher groundwater level changes within short time. For example, WP167 with groundwater level decline of 3.5 m from August 1993-June 1995 and continuous decrease into the early 2000s. WP184 also show declines of 5.5m (September 1994-April 1995) and 4.8 m (October 1999-August 2000). Such high declines have influenced spring flows and base flow, and hence, the implications on groundwater management. Therefore, the groundwater declines are discussed in the context of groundwater resource sustainability and its implications on water security and resource management plans, including consideration of water conservation measures or conjunctive water use.

6. Salinity

The analysis of groundwater trends is critical in the study of salinity risk and the effectiveness of preventative measures. The majority of DPI bores were installed in response to reports of saline discharge outbreaks in the 1980s and 1990s (Clark & Harvey, 2008). Salinity has impacts on the social, economic and environmental values in any catchment. Therefore, groundwater monitoring co-ordinated by DPI and DSE provides an important tool in the understanding, measurement and management of salinity across the state of Victoria. Hence it is currently been reviewed and prioritised based on key assets in the state (Reid et al., 2011).

In both WID and the Cape Flats, salinity (as measured by electrical conductivity (EC) of groundwater or total dissolved solids (TDS)) revealed the varying quality of groundwater by comparing historic data with recent measurements. The groundwater salinity monitoring in the Werribee Plains commenced in 2002 while in the Cape Flats regular monitoring began in 1979. Groundwater salinity in the Werribee Plains varies from 1000 to 6000 mg/L TDS, and this (according to Leonard, 1979; SKM, 2002) represents the best quality water in the aquifer

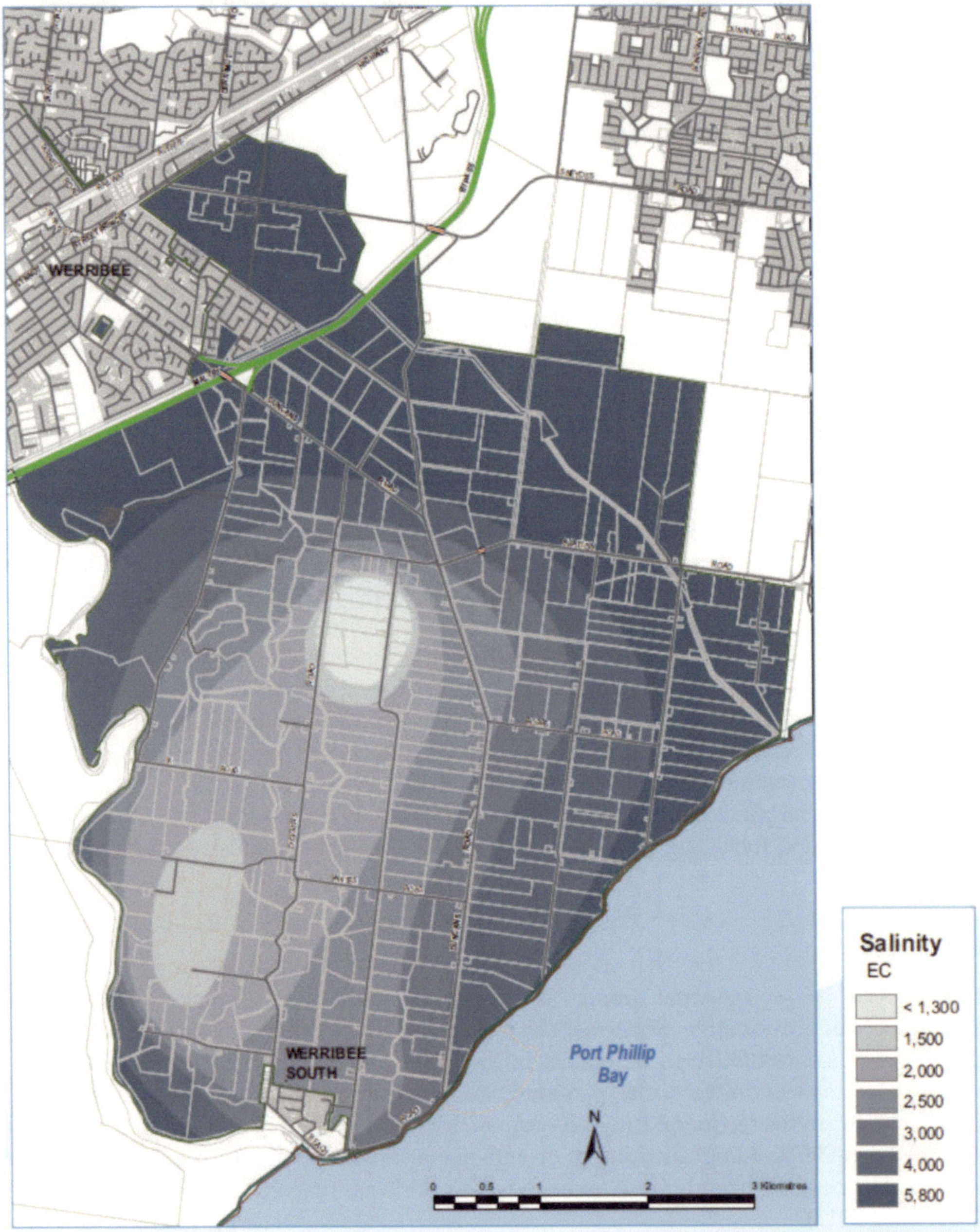

Fig. 5. The spatial distribution of salinity (i.e. variation of EC) across the WID (after SRWA, 2009)

systems in the area. The spatial distribution of salinity (i.e. variation of EC in µS/cm) across the WID is shown in Figure 5. This distribution showed the northern and eastern parts of the districts as relatively higher in salinity than the central-southern part. The highest groundwater salinity was recorded in bore B145273, located closest to the coastline (although the bore is not screened in the Werribee Delta aquifer). This is most probably primary salinity and does not coincide with any of the state's key asset areas (Reid et al., 2011). Nevertheless, under the Southern Rural Water plan on the WID, the key driver of groundwater management is to avoid drawing down the aquifer to the point where seawater intrusion takes place. The source of the salinity in the Werribee Delta has been traced to more saline adjacent aquifers or seawater intrusion (although studies to confirm this are on-going). Several studies in line with the management strategy have therefore been in place since the last 10 years (SRWA, 2004, 2006, 2009).

Total dissolved solids of the samples from bores screened in the Cape Flat sands are generally low compared to those of other aquifers in the area (Adelana, 2011). This salinity values varied from 67-4314 mg/L. The EC values of groundwater from the Cape Flats aquifer ranged from 9.2 to 4320 µS/cm. Field and monitoring data (Adelana, 2011) showed also that generally groundwater salinity increases following the groundwater flow direction, south-eastwards. Figure 6 illustrates the electrical conductivity areal distribution in the Cape Flats. The relations of chloride and EC with groundwater levels and well depths are not shown (in most cases) because the wells monitored for salinity are not necessarily used for groundwater level observations.

7. Resource management implications

A more appropriate and adequate dataset is essential for the planning and management of aquifers. Monitoring is, therefore, closely linked to the aquifer management, since the results of monitoring may require changes or modifications in the management practice. For example, the higher than average rainfall in the 2010/2011 season across Victoria (all reflected in the hydrograph analysis) may have influence on water use decisions and water restrictions. However, sustainable groundwater management decisions would require long-term monitoring and projections. Such long-term data covering all key elements of the hydrological cycle including groundwater fluctuations and water-level trends are essential as a basis for management and for evaluating the implications of changes in use.

Long-term monitoring using a number of observation bores has demonstrated that water levels have both declined and recovered over time and in the aquifer investigated. There has been full recovery of the aquifer over the past wet months (2010-2011), and this has been much more than what the recovery would have been over a normal wet year in Victoria. However, this is no cover against management measures except such higher than average rainfall becomes consistent over a longer period of time. Such monitoring data will be 'handy' information to support decision-making and demonstrate the impacts of climate on level changes in relation to resource management. Water level and quality data has been used a number of times in Victoria and (at least three occasions within the last 7 years) in Cape Town to change the extent of groundwater abstraction in order to support sustainable management.

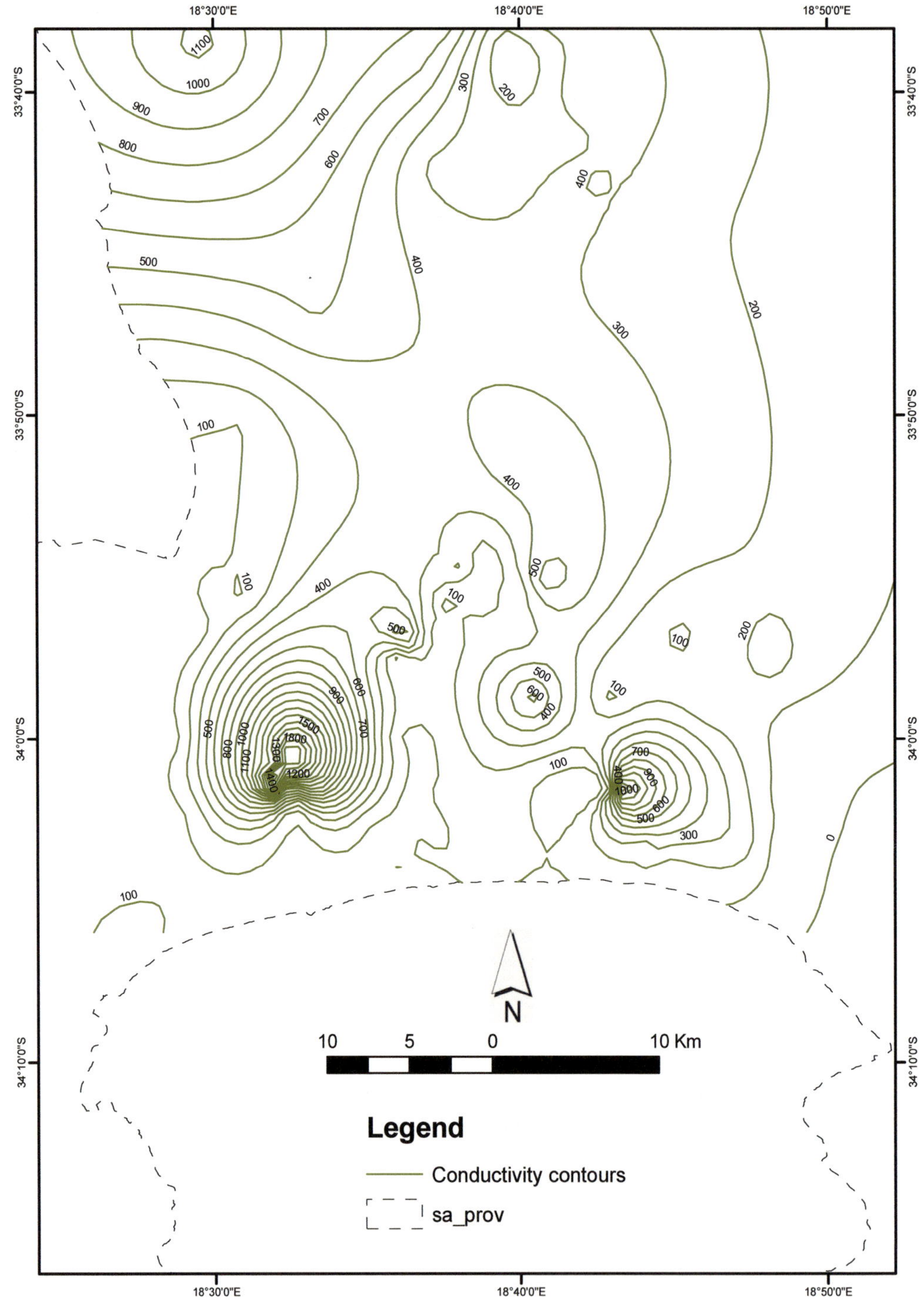

Fig. 6. Areal distribution map of electrical conductivity (in µS/cm) in the Cape Flats (Adelana et al., 2010)

In 1998, new groundwater management arrangements were put in place to maximise development opportunities in the Werribee Irrigation District, yet ensure that groundwater resources are managed in a sustainable way. Management arrangements include Groundwater Management Areas (GMAs); Groundwater Supply Protection Areas (GSPAs); and Groundwater Management Plans (SRWA, 2006). Water restrictions have since been in place and at different stages of restriction, they are periodically reviewed. For example, in March 2011, Southern Rural Water announced a substantial boost in groundwater allocation for landowners in the Deutgam Water Supply Protection Area based around Werribee. A full ban on groundwater use in Werribee was introduced in 2006 because of the threat of seawater intrusion into the groundwater through aquifer from Port Phillip Bay (SRWA, 2006). SRWA announced a partial lifting to 25% allocation in early January 2011, and recommended to the Minister a lifting to 75% after careful monitoring showed the aquifer is continuing to improve (SRWA, 2011). More than average mean rainfall over the last 12 months has seen groundwater levels rising and salinity levels improving. All bores can now be used for stock and domestic purposes.

Currently the Department of Water Affairs (Cape Town regional office) is capacity-constrained, which limits its ability to continue groundwater monitoring and the processing of licence applications (Colvin & Saayman, 2007). In such a situation, very little additional management of groundwater resources is possible. However, by the year 2012, DWA aims to complete institutional transformation with the establishment of Water User Associations (WUAs) and Catchment Management Agencies; and the licensing of all water use within another 5 years. Also, the City of Cape Town adopted an integrated approach to water management, which seeks a balance between water conservation and water demand management initiatives and conventional supply augmentation. But based on observations (Colvin & Saayman, 2007), formal government tend to focus on bulk water supply while household level bore use and development planning has not been fully integrated into water strategies.

Private (household) use of groundwater from the Cape Flats aquifer is widespread and increasing since the early 2000s when potable water tariffs increased. The immediate impact of such unregulated use was not feasible in this study due to prolonged missing gaps (mid-1990s to early 2000s) in water level data. The current gradual downward trend if projected would reflect in future bore responses as monitoring continues. To support this, the survey conducted by Colvin and Saayman (2007) revealed society's impacts on groundwater currently result from indirect drivers such as Water Demand Measure (WDM) introduced in the mid-1990s. This obviously occurs within the broader context of society supported by natural resources and a model which includes the resource base and its feedback.

Although the Department is aware of the increased private groundwater abstraction at a household level in Cape Town, this water use is covered under Schedule 1 of the National Water Act and therefore does not need to be registered with the Department. The cumulative impact of these small-scale abstractions generates concerns. Colvin and Saayman (2007) suggested that where the cumulative effect of these small-scale abstractions under Schedule 1 is too large and negative, by-laws or regulations can be promulgated—even by a municipality. Such a by-law or regulation would override the entitlements under

Schedule 1. As far as information available to date, no such by-laws or regulations have been promulgated either by DWA or the City of Cape Town.

Colvin and Saayman's (2007) survey further reveal there are concerns within government Departments (Department of Water Affairs and the Department of Agriculture) that the national land reform programme may be contributing to unsustainable resource exploitation in places. For example, some of the Cape Flats bores in Atlantis area (shown in the appendix) represent a marked response to pumping influences with a decline of 4-6 m within 1-2 years in mid-1990s and then continuous decline into the early 2000s. Such high declines may influence spring flows and base flows, and hence, have implications on groundwater resource management. Some of these monitoring bores are responding to pumping influences from the Atlantis wellfields. Bulk water supply wellfields at Atlantis have been in operation for over 20 years, supplying the satellite industrial town with water (Tredoux, 1982; Tredoux & Cave, 2002). This led to the establishment of a management scheme, known as the Atlantis Water Resource Management Scheme (AWRMS) to manage water resources in the area and to follow on the introduction of WDM in the South African Water Act 1997. The City of Cape Town also commissioned the Council for Scientific and Industrial Research (CSIR) to conduct intensive monitoring and numerical modelling of the Atlantis wellfields (Colvin & Saayman, 2007). Such information would help management of the groundwater resource.

Generally groundwater acts as the primary buffer against the impact of climate variability and spatial variability in drought. The buffering capacity of groundwater increases social resilience to drought in both urban and rural communities. However, as human development has become more susceptible to such variability, three major gaps in groundwater management were identified (FAO, 2003b): accelerated degradation of groundwater systems by over-abstraction, and effective resource depletion through quality changes (pollution, salinity), and the inability to resolve competition for groundwater between sectoral and environmental uses. Each of these has implications for sustainable development as demonstrated in this study.

Given the sensitivity of both aquifers to climate variability and pumping and the observed water quality changes noted above, it is considered necessary to uphold formal regulatory measures to avert further water level and quality decline. Effective institutional approaches need to be aware of the realities surrounding groundwater use and the inherent risks associated with development, the level of uncertainty (plus limitations in data quality) and the range of social pressures. The general lack of professional and public awareness about the sustainable use of groundwater resources will need to be continuously addressed. A more coherent planning framework should guide all scales of groundwater development and appropriate policy responses needed to prevent further degradation of the groundwater systems.

8. Conclusion

Climate (i.e. rainfall) is the primary factor influencing the fluctuation and trend of groundwater level although increased usage contributed to the drawdown especially during the dry years. The trend and seasonal fluctuation of groundwater level in the two study areas generally correlated with seasonal rainfall and linear trends were observed in a number hydrograph of bores in the area. The bore hydrographs of the Cape Flats aquifer showed marked seasonal fluctuations and a more slightly downward trend (-8 to 14 cm/yr) in comparison to the Werribee Delta aquifer bores (-3 to 4 cm/yr).

The resulting groundwater declines invariably affect groundwater resource sustainability and by implication water security. For example, the Werribee Delta aquifer groundwater level drawdown shows that during the early 1990s seasonal drawdown was less than 0.5 m but increased up to 2 m in 1996. The decline and general downward trend indicates increased reliance on groundwater. The fall in groundwater levels coincides with salinity increases from 2,500 EC to over 6,000 EC and, consequently yielding information that the source of salinity in the Werribee Delta could be more than saline adjacent aquifers or seawater intrusion (studies to confirm this are on-going).

Groundwater level responses and behaviour in observation bores, in response to climate and pumping in productive aquifers, is an indication of homogeneity and lateral hydraulic connection within the shallow coastal aquifers investigated in this study. However, the cases involving deeper aquifers and their responses were not considered. This is because the shallow aquifers are mostly used in both study area and the hydrogeological parameters have shown higher yield of these aquifers relative to the deeper ones. It is expected that vertical hydraulic conductivity will vary with the various underlying geological materials and only if aquifer connectivity exist that pumping from one productive aquifer can induce water level change in observation bores installed in other aquifers. Therefore, a more comprehensive study investigating the impacts of level changes in shallow aquifers on underlying deeper aquifer(s) would be necessary for effective resource management in these areas.

The groundwater trends and salinity increases are discussed in the context of groundwater resource sustainability and its implications on water security and resource management plans, including consideration of water conservation measures or conjunctive water use. However, aspects relating changes in groundwater level and zones of declining groundwater head to aquifer connectivity may be necessary to improve understanding of the system and, by implication, critical to the development of sustainable management frameworks for semi-arid regions.

In the face of the prolonged dry period (1995-2007) and a come-back of wet years (2010/2011), current irrigation and agricultural practices need to be reviewed in the catchments to ensure groundwater sustainability and secure future agricultural viability. Groundwater level responses in bores (consistent level records in WID, coupled with the data gaps in the Cape Flats farming districts), illustrate the importance of monitoring in relation to natural/environmental responsiveness and resilience. State-wide groundwater monitoring in Victoria (Australia) and the quarterly meter reading has continued to assist management decisions. There are realities surrounding groundwater use and inherent risks associated with development, the level of uncertainty and the range of social pressures. The social views of groundwater lag behind the formal policy of a public resource. Therefore, continued support for basic data collection and groundwater evaluation is justified on both scientific and social process grounds. The water authorities in the two case studies must adequately manage and maintain interactions with key stakeholders ensuring open and transparent relationships that are based on trust to promote good governance.

Apendix I: Summary table of groundwater trends for the selected bores from the two study areas

Bore	Location	Depth (m)		Best Fit Delay (months)	R^2 for selected one	C	Acc. Residual Rainfall (mm)		Time (month)		Monitoring period (years)	No. readings	Trend (cm/yr)
		Initial	Final	AARR			value	p	value	p			
G30944	Atlantis[1]	-7.0	-6.2	3	0.54	-6.23	0.0039	0.0000	-0.0061	0.0000	17	162	-7.3
PA20	Atlantis[1]	-3.6	-3.8	1	0.45	-3.70	0.0013	0.0000	-0.0014	0.0167	16	96	-1.7
WP167	Atlantis[1]	-10.6	-11.1	2	0.68	-11.08	0.0098	0.0000	0.0007	0.7462	13	111	0.9
WP184	Atlantis[1]	-8.2	-9.0	0	0.37	-10.42	0.0089	0.0000	0.0129	0.0003	12	105	15.5
DC182	Bellville[1]	-2.1	-4.1	0	0.33	-5.89	0.0062	0.0000	0.0068	0.0934	10	83	8.2
DC184	Bellville[1]	-1.8	-1.6	0	0.33	-2.14	0.0012	0.0000	0.0027	0.0001	10	73	3.3
BA002	Philippi[1]	-6.5	-6.8	2	0.38	-6.21	0.0008	0.0000	-0.0020	0.0000	28	156	-2.5
BA076	Philippi[1]	-4.7	-4.2	0	0.69	-3.67	0.0013	0.0000	-0.0015	0.0000	28	126	-1.8
BA083	Philippi[1]	-4.6	-3.8	0	0.38	-1.78	0.0047	0.0000	-0.0075	0.0000	32	172	-9.0
BA084	Philippi[1]	-4.5	-3.9	0	0.38	-2.48	0.0035	0.0000	-0.0050	0.0000	32	145	-6.0
BA232	Philippi[1]	-2.7	-2.3	0	0.57	-3.99	0.0030	0.0000	0.0121	0.0000	10	33	14.5
B59520	Werribee[2]	-2.6	-2.2	0	0.08	-2.80	0.0001	0.0039	0.0003	0.2578	26	250	0.4
B59521	Werribee[2]	-3.1	-2.6	0	0.05	-3.17	0.0001	0.0135	0.0003	0.2419	26	251	0.4
B59523	Werribee[2]	-2.0	-1.6	0	0.53	-2.41	0.0005	0.0000	-0.0008	0.1535	26	248	-1.0
B59525	Werribee[2]	-5.7	-3.8	0	0.51	-6.03	0.0013	0.0000	-0.0016	0.2216	26	263	-1.9
B59531	Werribee[2]	-10.1	-9.8	0	0.74	-10.23	0.0006	0.0000	-0.0008	0.0345	26	248	-0.9
B59533	Werribee[2]	-3.6	-3.2	0	0.16	-3.58	0.0001	0.0010	0.0001	0.5042	26	246	0.2
B59536	Werribee[2]	-5.9	-4.1	0	0.58	-6.69	0.0013	0.0000	-0.0028	0.0370	26	247	-3.4
B59537	Werribee[2]	-2.8	-2.5	0	0.05	-3.02	0.0001	0.1641	0.0000	0.8788	25	244	-0.1
B59539	Werribee[2]	-4.5	-2.5	0	0.32	-4.88	0.0009	0.0049	-0.0032	0.1043	25	247	-3.8
B112802	Werribee[2]	-7.7	-6.8	0	0.60	-11.68	0.0028	0.0000	0.0105	0.0046	19	176	12.7
B113018	Werribee[2]	-3.9	-3.6	0	0.50	-5.34	0.0009	0.0000	0.0036	0.0094	19	176	4.3

[1]Bore screened in the Cape Flats aquifer [2]Bore screened in the Werribee Delta aquifer

Appendix II: Some examples of HARTT analysis graphs from the selected bores in the study areas

Appendix II a: HARTT analysis graph for Bore 59539 (from Werribee Plains)

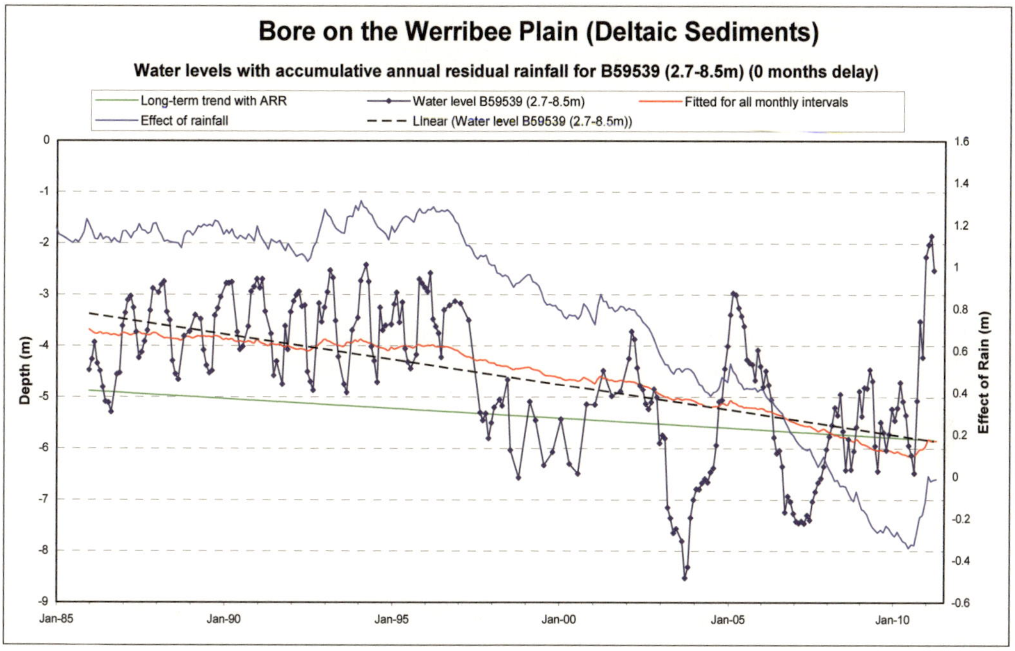

Appendix II c: HARTT analysis graph for Bore 59531 (from Werribee Plains)

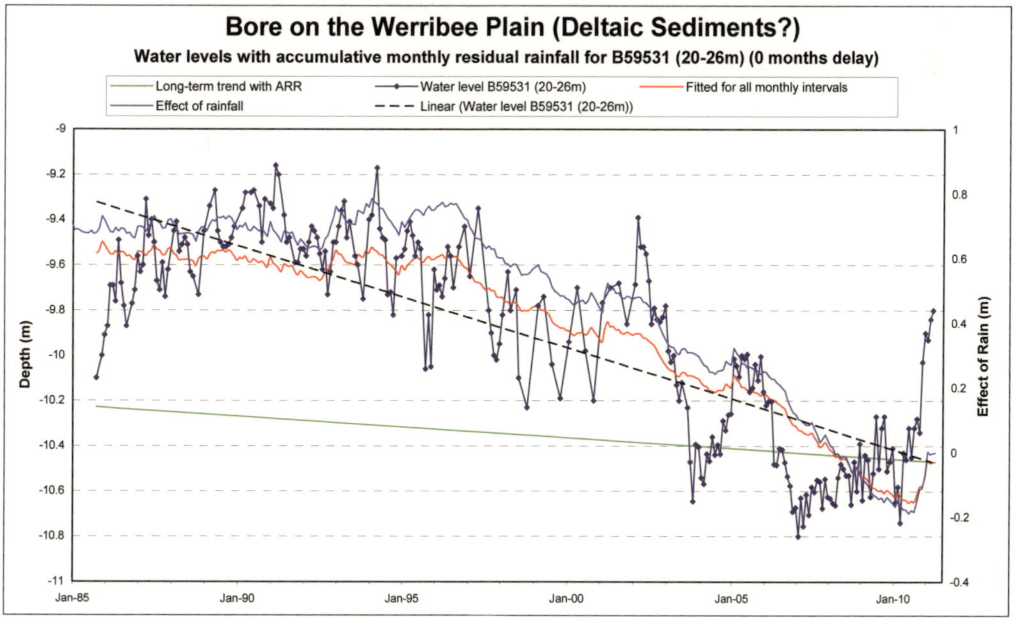

Appendix II d: HARTT analysis graph for Bore 112802 (from Werribee Plains)

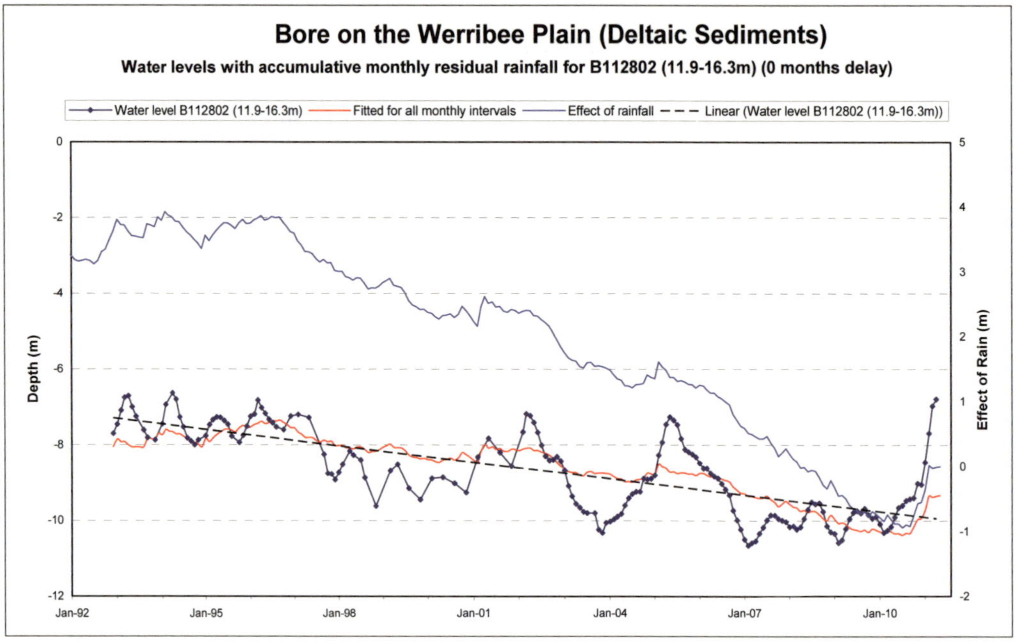

Appendix II e: HARTT analysis graph for BA002 (from Cape Flats, Philippi-Mitchells Plain)

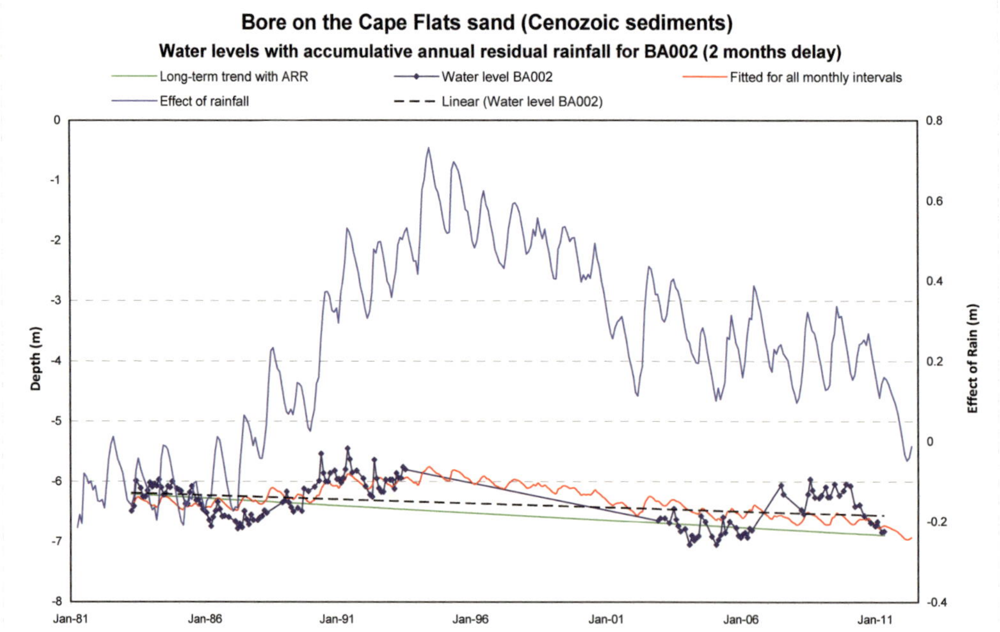

Appendix II f: HARTT analysis graph for BA076 (from Cape Flats, Philippi-Mitchells Plain)

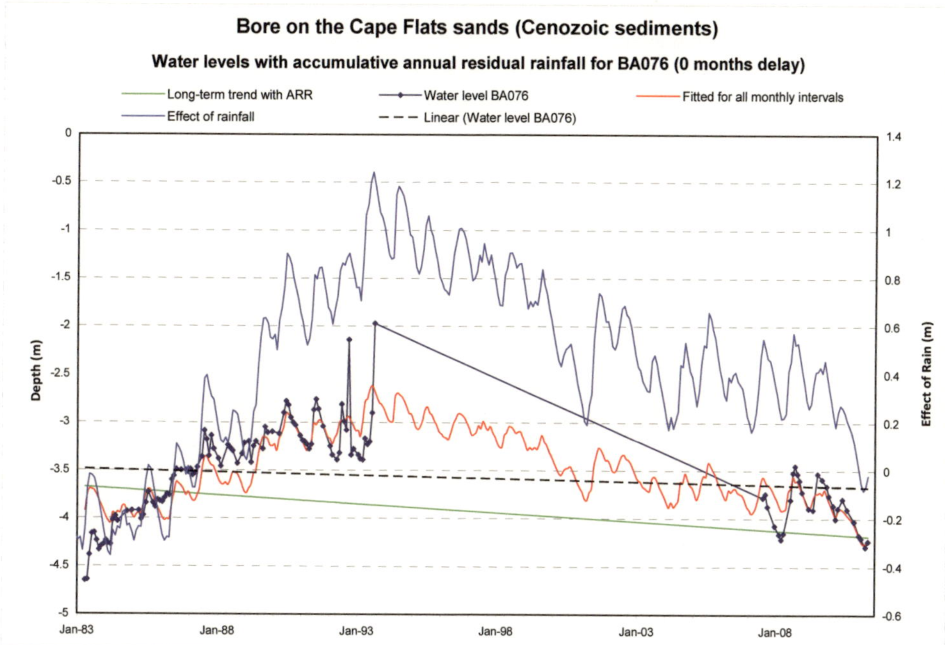

Appendix II g: HARTT analysis graph for BA083 (from Cape Flats, Philippi-Mitchells Plain)

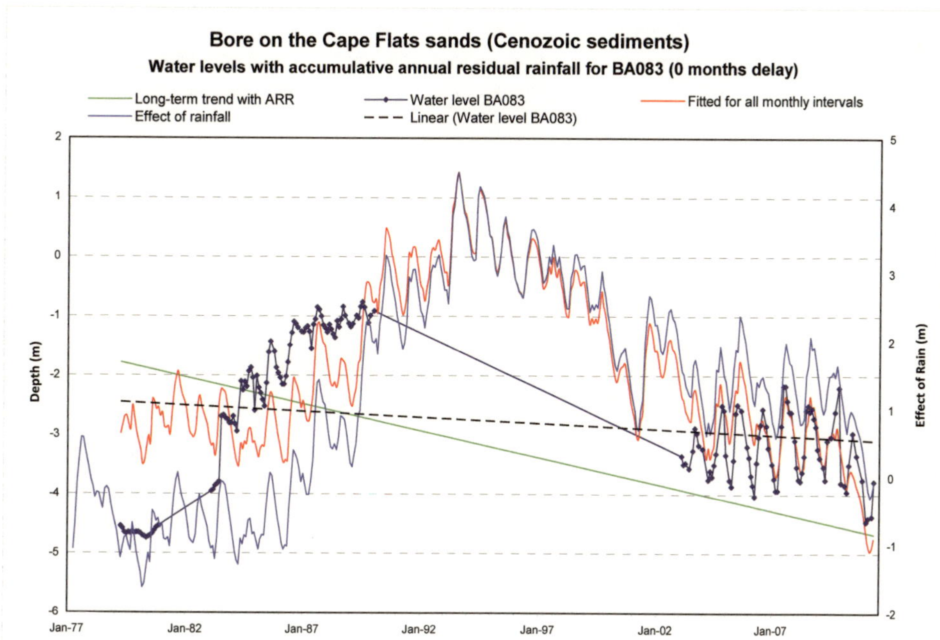

Appendix II h: HARTT analysis graph for BA084 (from Cape Flats, Philippi-Mitchells Plain)

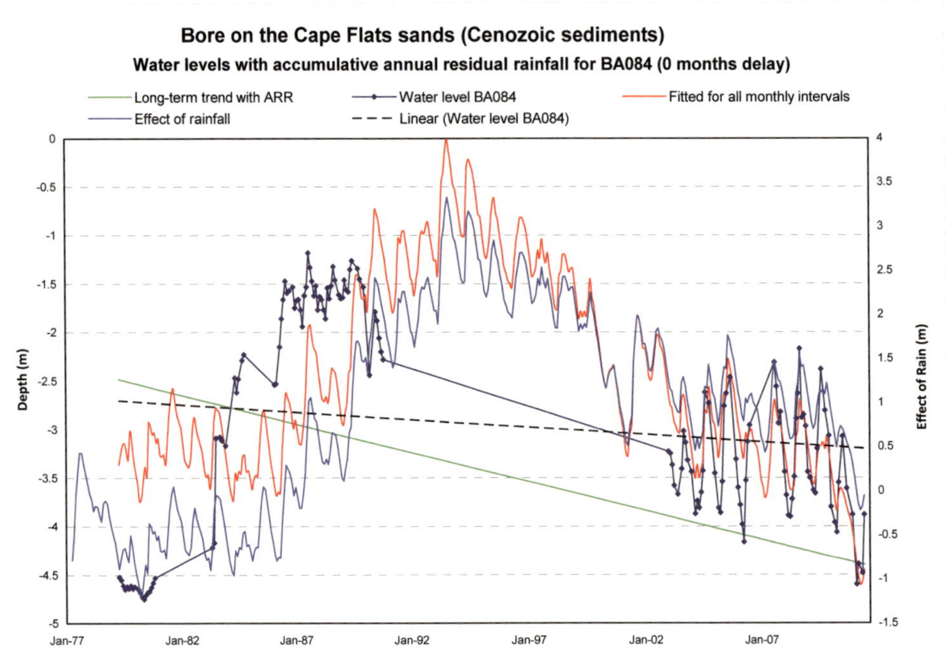

Appendix II i: HARTT analysis graph for BA232 (from Cape Flats, Philippi)

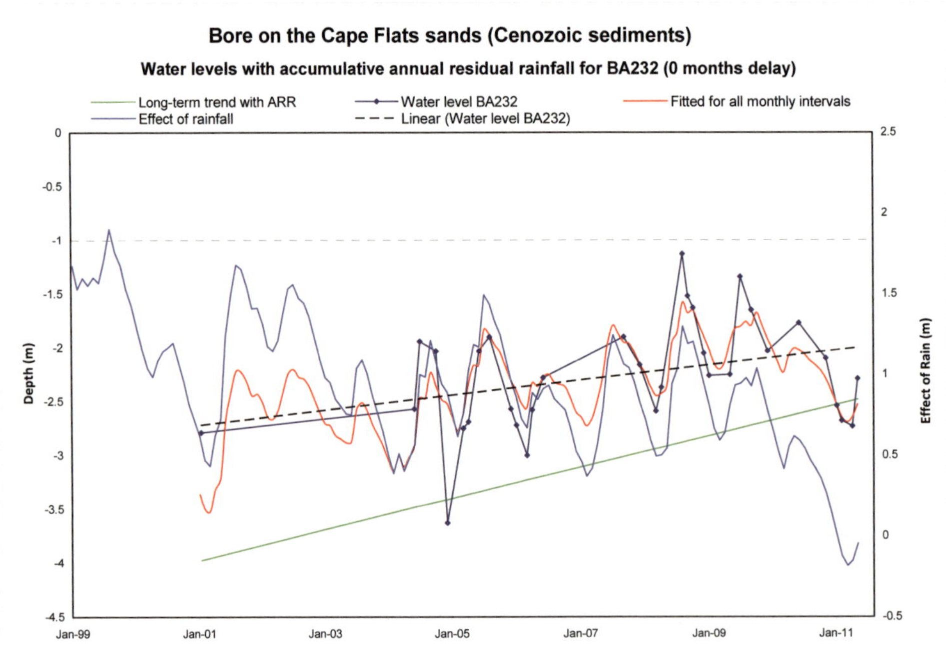

Appendix II j: HARTT analysis graph for DC182 (from Cape Flats, Bellville)

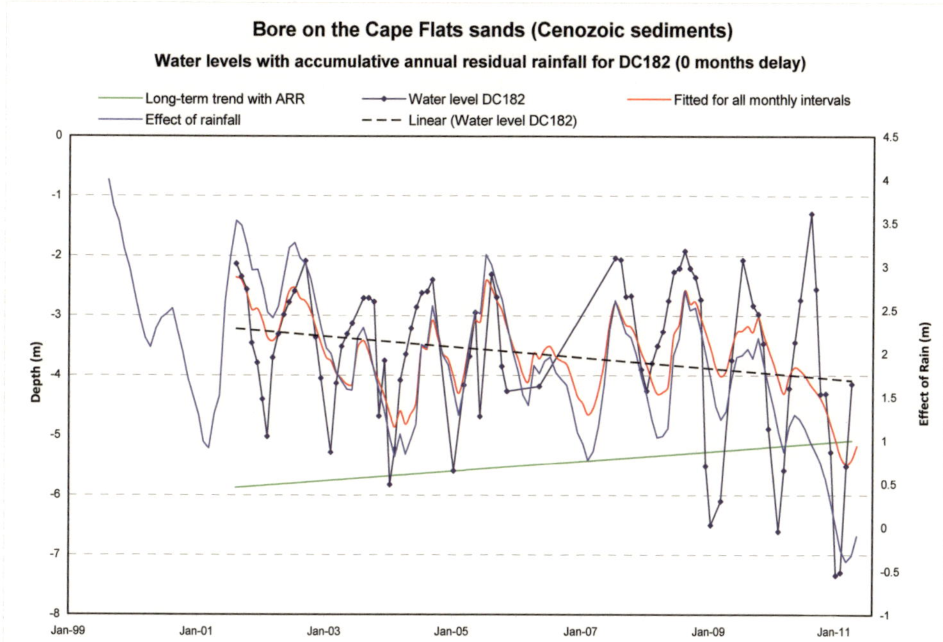

Appendix II k: HARTT analysis graph for G30944 (from Cape Flats, Atlantis)

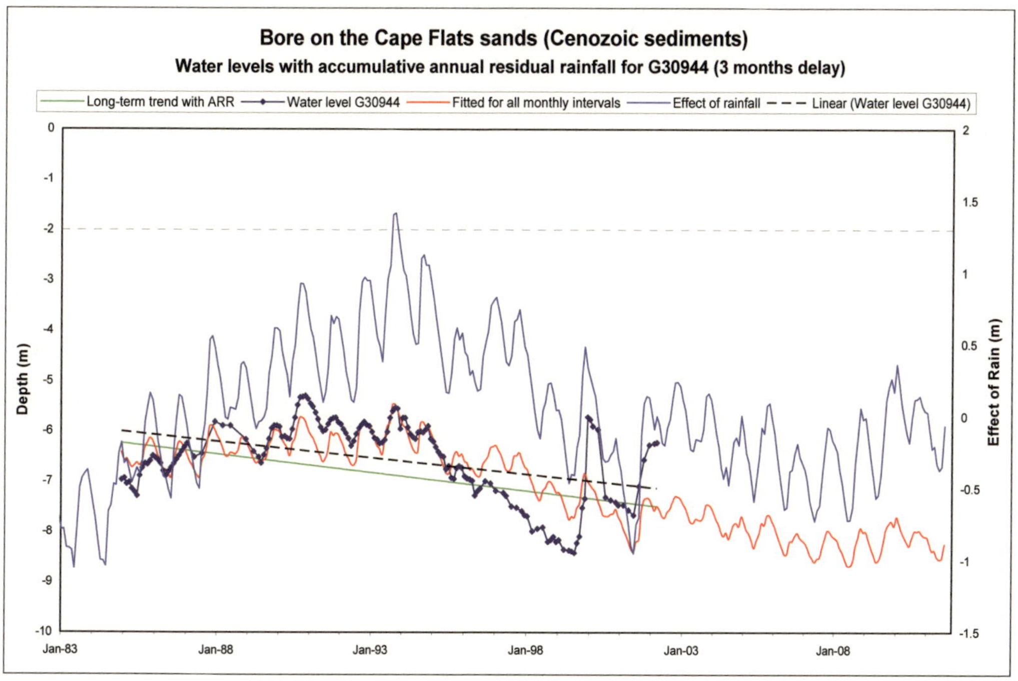

Appendix II (l): HARTT analysis graph for PA20 (from Cape Flats, Atlantis)

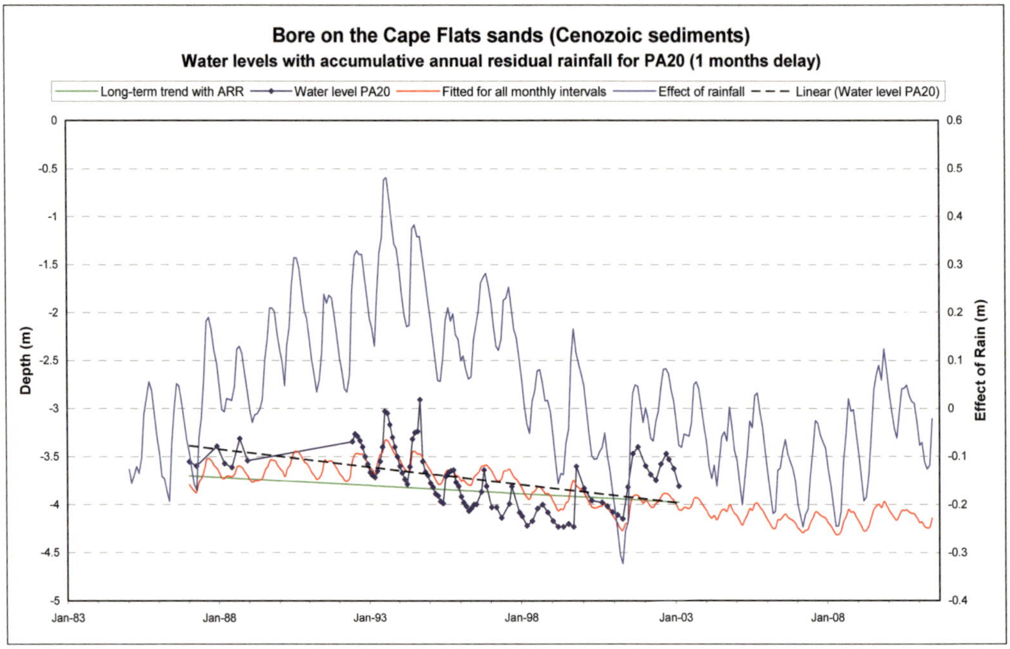

Appendix II (m): HARTT analysis graph for WP167 (from Cape Flats, Atlantis)

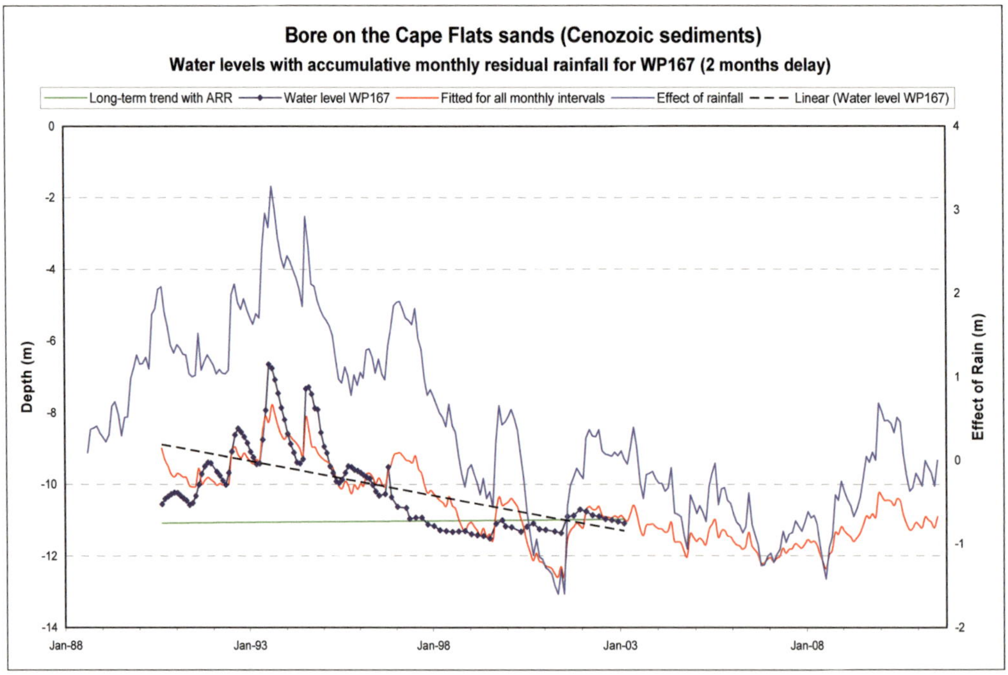

Appendix II (n): HARTT analysis graph for WP184 (from Cape Flats, Atlantis)

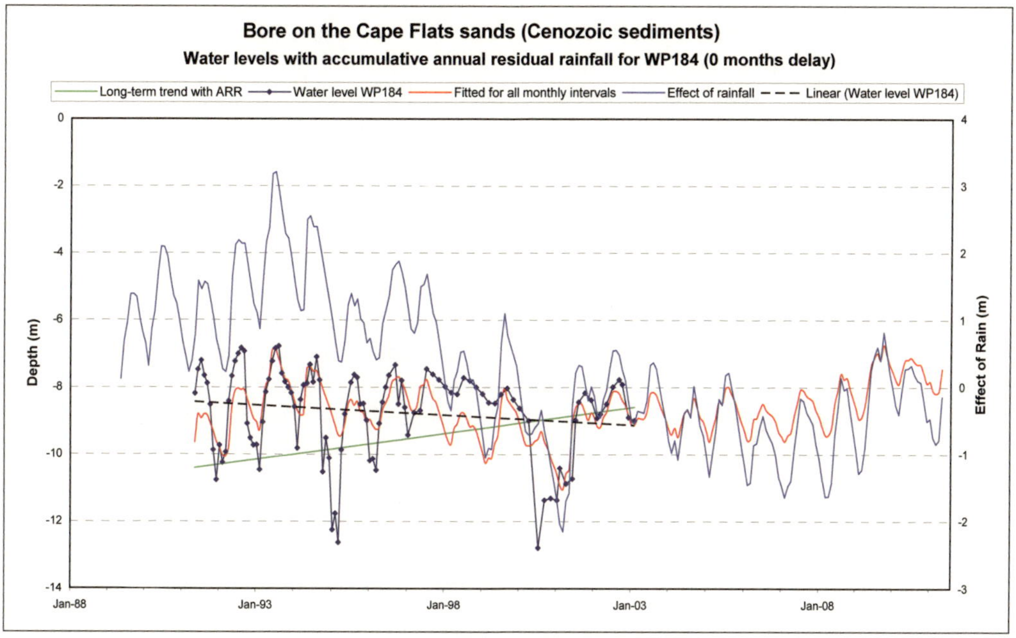

9. References

Abidin, H.Z. et al. (2001). Land Subsidence of Jakarta (Indonesia) and its Geodetic Monitoring System. *Natural Hazards* 23:365-387, Kluwer Academic Publishers, Netherlands

Adelana, S.M.A. (2009). Monitoring groundwater resources in Sub-Saharan Africa: issues and challenges, *IAHS Red Book Publ.* Vol. 334, pp. 103-113

Adelana, S.M.A. (2011). *Groundwater resource evaluation and protection*, LAP LAMBERT Academic Publishing, ISBN-13: 978-3-8443-2369-6, Saarbrücken, Germany

Adelana, S.M.A. & Xu, Y. (2006). Contamination and protection of the Cape Flats Aquifer, South Africa, In: Groundwater pollution in Africa Y. Xu & B. Usher (Ed.), 265-277, Taylor & Francis, ISBN 13 : 978-0-415-41167-7, London, UK

Adelana, S.M.A.; Xu, Y. & Adams, S. (2006a). Identifying sources and mechanism of groundwater recharge in the Cape Flats, South Africa: Implications for sustainable resource management. Proc. XXXIV Congress of the International Association of Hydrogeologists (IAH), Beijing, China, 9-13 October 2006

Adelana, S.M.A.; Olasehinde, P.I. & Vrbka, P. (2006b). A quantitative estimation of groundwater recharge in parts of Sokoto Basin, Nigeria, International Journal Environmental Hydrology, Vol. 14, paper 5, (May 2006), pp.1-17, ISSN 1058-3912

Adelana, S.M.A.; Xu, Y. & Vrbka, P. (2010). An integrated conceptual model for the development and management of the Cape Flats aquifer, South Africa, Water SA, Vol. 36, No. 4, (July 2010), pp. 461-474, ISSN 0378-4738

Burke, J.J. (2002). Groundwater for irrigation: productivity gains and the need to manage hydro-environmental risk, In: *Intensive use of groundwater challenges and opportunities*, R. Llamas & E. Custodio, (Ed.), 478 pp, Balkema, Abingdon, U.K

Cape Water Solutions (August 2010). Cape Town Rainfall is drying up, 11.08.2010, Available from: http://www.capewatersolutions.co.za/2010/08/11/cape-town-rainfall-is-drying-up/

Chai, J.C.; Shen, S.L.; Zhu, H.H. & Zhang, X.L. (2004). Land subsidence due to groundwater drawdown in Shanghai, *Geotechnique*, 54, 143–147.

Cheng, X.; Adelana, M. & Reid, M. (2011). Groundwater trend and behaviour in the Campaspe dryland catchments, *Technical Report*, Department of Primary Industries, Victoria, Australia

Clark, R. & Harvey, W. (2008). Dryland salinity in Victoria in 2007 - An analysis of data from the soil salinity database and Victorian discharge monitoring network. *Technical Report*, Department of Primary Industries, Victoria, Australia

City of Cape Town, CCT (2010). Cape Flats District Plan - Spatial development plan and environmental management framework. *Technical Report*, City of Cape Town

Colvin, C. & Saayman, I. (2007). Challenges to groundwater governance: a case study of groundwater governance in Cape Town, South Africa. *Water Policy* 9, Supplement 2, (September 2007), pp. 127–148

Condon, M.A. (1951). The Geology of The Lower Werribee River, Victoria. Proceedings Royal Society Victoria, pp. 63:1-24

Deutgam Water Supply Protection Area Consultative Committee, DWSPACC (2002). Explanatory notes to the Deutgam groundwater management plan. Draft for consultation, (October 2002), 35p

FAO (2003a). Re-thinking the Approach to Groundwater and Food Security, *Water Reports* No. 24, Food and Agriculture Organisation of the United Nations, Rome

FAO (2003b). Groundwater Management - The Search for Practical Approaches. *Water Reports* No. 25, Food and Agriculture Organisation of the United Nations, Rome

Ferdowsian, R.; Pannell, D.J.; McCarron, C.; Ryder, A.T. & Crossing, L. (2001). Explaining groundwater hydrographs: separating atypical rainfall events from time trends, *Australian Journal Soil Research*, Vol.39, No.4, (January 2001), 861 - 876

Gerber, A. (1976). An investigation into the hydraulic characteristics of the groundwater source in the Cape Flats, Unpublished M.Sc. thesis, University of the Orange Free State, Bloemfontein

Gerber, A. (1981). A digital model of groundwater flow in the Cape Flats, *CSIR Contract Report* C WAT 46, Pretoria

Giambastiani, B.M.S.; Kelly, B.F.J. & McCallum, A. (2009). Three dimensional temporal analysis of surface and groundwater interactions. Paper presented at the First Australian 3D hydrogeology Workshop, GeoScience Australia, Canberra.

Hekmeijer, P.; Gll, B.; Reid, M.; Fawcett, J. & Cheng, X. (2008). Dryland groundwater monitoring review, 2006-07. Department of Primary Industries, *Technical Report*, Victoria, Australia

Holdgate, G.R.; Gallagher, S. J. & Wallace, M. W. (2002). Tertiary coal geology and stratigraphy of the Port Phillip Basin, Victoria, *Australian Journal of Earth Sciences*, Vol. 49, Issue 3, (June 2002), pp. 437–453, ISBN1440-0952

Holdgate, G.R.; Geurin, B.; Wallace, M.W. & Gallagher, S.J. (2001). Marine geology of Port Phillip, Victoria, Vol. 48, Issue 3, (June 2001), pp: 439–455, ISSN 1440-0952

Holdgate, G.R. & Gallagher, S.J. (2003). Tertiary, In: *Geology of Victoria*, Birch W.D. (Ed), 289-335, Geological Society of Australia Special Publication 23, Geological Society of Australia (Victorian Division)

Hotta, N.; Tanaka, N.; Sawano, S.; Kuraji, K.; Shiraki, K. & Suzuki, M. (2010). Changes in groundwater level dynamics after low-impact forest harvesting in steep, small watersheds. *Journal of Hydrology,* Vol. 385 (June 2010), 120–131

Kelly, B.F.J.; Giambastiani, B.M.S. & Timmis, W. (2009). 3D hydrograph analysis for constraining the construction of hydrogeological models. Paper presented at the First Australian 3D hydrogeology Workshop, GeoScience Australia, Canberra

Leonard, J.G. (1979). Preliminary assessment of the groundwater resources in the Port Phillip Region. *Geologial Survey Report,* 1980-66

Lopez-Quiroz, P.; Doin, M.P; Tupin, F.; Briole, P. & Nicolas, J.M. (2009). Time series analysis of Mexico City subsidence constrained by radar interferometry, *Journal Applied Geophysics,* Vol. 69, No. 1, 1–15, doi:10.1016/j.jappgeo.2009.02.006

McLean, W. & Jankowski, J. (2002). Changes in groundwater chemistry and salinity in alluvial aquifer system induced by irrigation, In: *Groundwater and Human Development,* E. Bocanegra; Martinez, D. & Massone, H. (Ed.), pp.400-410, Mar de Plata, Argentina

McLean, W.; Jankowski, J. & Lavitt, N. (2000). Groundwater quality and sustainability in an alluvial aquifer Australia, *Proceedings of IAH 2000 30th International Congress,* pp. 567-57326, Cape Town, South Africa, 28 Nov - 1 Dec, 2000

Meyer, P.S. (2001). An explanation of the 1:500 000 hydrogeological map of Cape Town 3317. Department of Water Affairs & Forestry, 59pp

Rabe, L. (1992). The Germans of Philippi. Lantern, Special Edition, Available from http://www.safrika.org/phipi_en.html

Reid, M. (2010). Impacts of prolonged dry climate conditions on groundwater levels and recharge in North-Central Victoria. *Transactions of the Royal Society of Victoria,* Vol. 122, No.2, (November 2010), 1xii-1xxv, ISSN 0035-9211

Reid, M.; Cheng, X. & Huggins, C. (2006). Using ground water responses to improve understanding of climate variation impacts and salinity risk, *Proceedings of 10th Murray-Darling Basin Groundwater Workshop,* pp. 12-16, Canberra, ACT, Australia, September 2006

Reid, M.; Cheng, X; Adelana, M.; Hekmeijer, P. & Zydor, H. (2011). Salinity risk assessment of environmental asset regions in Victoria - 2011. *Technical Report,* Department of Primary Industries, Victoria, Australia

Rodda, C. & Kent, M. (2004) Werribee Irrigation District recycling scheme: First years, *Southern Rural Water Report,* 14p

Schaffer, D. & Pigois, J.P. (2009). 3D hydrogeology in Western Australia - aquifer mapping, regional models and stratigraphic analysis. Paper presented at the First Australian 3D hydrogeology Workshop, GeoScience Australia, Canberra

Schalke, H.J.W. (1973). The upper Quaternary of the Cape Flats area, Cape Province, *South Africa. Scripta Geol.* 15, pp1-37

Sinclair Knight Merz, SKM (1998). Permissible Annual Volume Project - The Deutgam Groundwater Management Area. Department of Natural Resources and Environment Victoria, (January 1998), 25p

Sinclair Knight Merz, SKM (2002). Second Generation Dryland Salinity Management Plan for the North Central Region. Draft B, Unpublished, Maffra 3860, Victoria

Sinclair Knight Merz, SKM (2004). Werribee Irrigation District Groundwater Investigations, *Consultancy Report* - Sinclair Knight Merz Ltd., Maffra 3860, Victoria

Sinclair Knight Merz, SKM (2009a). Melbourne Groundwater Directory Technical Report. Unpublished Report no VW03963, Maffra 3860, Victoria

Sinclair Knight Merz, SKM (2009b). Hydrogeological Mapping of Southern Victoria, Project Report, 14th July 2009, Sinclair Knight Merz Ltd., Maffra 3860, Victoria

Southern Rural Water Authority, SRWA (2006) Annual Report http://www.srw.com.au/Files/Annual_reports/Annual_Report_2006.pdf

Southern Rural Water Authority, SRWA (2004) Annual Report http://www.srw.com.au/Files/Annual_reports/Annual_Report_2004.pdf

Southern Rural Water Authority, SRWA (2009). Western Irrigation Futures: A Southern Rural Water 2009 Atlas, Southern Rural Water Authority, Melbourne.

SRWA (March 2011) Werribee farmers can now use 75% groundwater - 11.03.2011, *SRW News*. Available for download http://www.srw.com.au/Page/NoticeBoard.asp

Theron, J.N.; Gresse, P.G.; Siegfried, H.P. & Rogers, J. (1992). The geology of the Cape Town area. Explanation on Sheet 3318, Geological Survey, South Africa, 140p

Tredoux, G. (1982). Artificial recharge of the Cape Flats aquifer with reclaimed effluents. Paper presented at a meeting of the Western Cape Group of the IWPC, (February 1982)

Tredoux, G. & Cave´, L. (2002). Atlantis Aquifer: A Status Report on 20 Years of Groundwater Management at Atlantis. *CSIR Contract Report*, Stellenbosch.

The Water Act 1989, 2002 Reprint No.6 - 4 April 2002

Usher, B.H.; Pretorius, J.A.; Dennis, I.; Jovanovic, N.; Clarke, S.; Titus, R. & Xu, Y. (2004). Identification and prioritisation of groundwater contaminants and sources in South Africa's urban catchments, *WRC Report* No. 1326/1/04

Vandoolaeghe, M.A.C. (1989). The Cape Flats groundwater development pilot abstraction scheme. *Technical Report*, No. GH3655, Directorate Geohydrology, DWA, Cape Town

Water Services Act (1997). Act No. 107 of 1997, Republic of South Africa, Pretoria.

Wright, A. W. & Conrad, J. (1995). The Cape Flats Aquifer - Current Status. *Report* No. 11/95. CSIR, Stellenbosch.

Part 2

Water Quality Management

5

Modelling of Surface Water Quality by Catchment Model SWAT

Matjaž Glavan and Marina Pintar
University of Ljubljana, Biotechnical Faculty, Agronomy Department, Chair for Agrometeorology, Agricultural Land Management, Economics and Rural Development
Slovenia

1. Introduction

Catchment represents a logical administrative unit of governance as a biological, physical, economic and social system, which is affected by natural (rain, sun) and human influences (industry, agriculture, population). The effective implementation of the river basin management plans are necessary and should include clear and strong objectives and instructions for maintaining the quality of surface water, even if needs of the society are changed in the future (Wagner et al., 2002).

The European Union Water Framework Directive (WFD) (2000/60/EC) set new rules for the catchments water management. The main objectives of the WFD are to improve, protect and prevent a further decreasing of water quality and to achieve good quality status of water bodies in Europe by 2015. The lack of studies and data put doubts on ambitious goals as it is difficult to examine the environmental changes associated with nutrients from biology to ecology (Neal & Heathwait, 2005). Volk et al. (2009) showed that to reach the WFD target water quality in German study catchments, dramatically unrealistic socio-economic measures would be needed (reduction of cultivated land from 77% to 46%, 13% of organic farming, increasing pastures from 4% to 15% of the forest from 10% to 21% and wetlands from 0% to 9%. Clean Water Act implemented in 1972 in the USA still did not achieve all objectives for drinking and bathing waters even after more than 30 years (Randhir & Hawes, 2009). Single or uniform integrated catchment management does not meet all the goals in soil and water protection due to usually very heterogeneous catchment characteristics (precipitation, geomorphology, slope, soils, agricultural crops) (Hatch et al., 2001).

Agricultural intensification since 1940 resulted in higher nutrients leaching to water and increased rate of soil erosion. The soil loss with surface migration of soil particles, which exceeds more than 1 t ha^{-1} $year^{-1}$ is regarded as irreversible within a time span of 50-100 years (EUSOILS, 2004). In Europe over 54 million km^2 of land is suffering similar or a higher rate of loss (Čarman et al., 2007). Erosion can cause significant reduction of the fertile soil depth, a significant loss of nutrients (Ramos & Martinez-Casasnovas, 2006) and depositions of the fine sediment in rivers, affecting fish spawning and egg development (Lohse, 2008).

Nitrogen (N) is an easily available nutrient and to the most crops is the limiting factor in production. Majority of the loss is associated with leaching in to groundwater and minority with surface runoff, depending on the geology and soil type. N leaching occur during wet periods of the year, after crops are harvested, fertilizers and mineralized crop biomass residues are exposed to leaching (Glavan & Pintar, 2010), and when N is not actively absorbed by plants and precipitation exceeds evapotranspiration (Rusjan, 2008).

Phosphorus (P) is known as the limiting factor in eutrophication of freshwater ecosystems (Khan & Ansari, 2005). P is a macronutrient required for the life of all living cells that plants absorb directly in the form of ortho-phosphorus (PO_4^{3-}) (Khan & Ansari, 2005). Excessive use of P fertilizers may lead to P soil saturation, causing P transport with runoff bound to soil particles or through drainage (Bowatte et al., 2006). Most P in inland waters is contributed by point sources (wastewater treatment plants). Due to advances in wastewater, P stripping has put more emphasis on P from agriculture (Buda et al., 2009).

Computer models in modern integrated catchment management are indispensable for studying the levels of pollutants from diffused sources, as they are capable of merging different spatial and environmental data (Dymond et al., 2003; Kummu et al., 2006). Catchment models can be divided into empirical-statistical (GLEAMS, MONERIS, N-LES), physical (WEPP, SA) and conceptual (distributed or partially distributed - SWAT, NL-CAT, TRK, EveNFlow, NOPOLU, REALTA) (Hejzlar et al., 2009; Kronvang et al., 2009a). Models connected with the Geographic Information System (GIS) has gained new values, as they are more accessible and understandable to different target groups.

Agricultural Research Service (ARS) of the U.S. Department of Agriculture is very active in developing models for agricultural hydrology, erosion and water quality. The Soil and Water Assessment Tool (SWAT) model was developed to assist the water managers in examining the impacts of agricultural activities in catchments (Arnold et al., 1998). The SWAT model is widely used for modelling the hydrology in terms of quantity of water (discharge, soil water, snow and water management), quality of water (land use, production technologies, good agricultural practices, agri-environmental measures) and the effects of climate changes (Gassman et al., 2007; Krysanova & Arnold, 2008). This model enables the modelling of long-term (more than 25 years) effects of agri-environmental measures (Bracmort et al., 2006). SWAT model has undergone several refinement and upgrades resulting in different model versions (SWAT2000, SWAT2005 and SWAT2009). The overall desire to adapt the model for the local conditions has resulted in many adaptations like G-SWAT, SWIM, E-SWAT, K-SWAT (Gassman et al., 2007).

The European Commission has, for the purposes of ensuring adequate tools, for the end user, that could meet the current European needs for harmonization and transparency in the quantitative assessment of diffused sources of nutrient losses, financially supported EUROHARP project (Kronvang et al., 2009b). This project compared nine different catchment models for simulation of the non-point sources of pollution from agriculture on numerous catchments in Europe. The results of the project ranked SWAT, along with NL-CAT and TRK models, in the top three of the best (Schoumans et al., 2009). EUROHARP study showed that the modellers are not yet able to propose only on the best and the most appropriate model for all river basins in Europe, because the quality of the models is based on the input data quality along with quality of the modellers (Kronvang et al., 2009a).

The aim of this chapter is to examine modelling of surface water quality by the catchment model Soil And Water Assessment Tool (SWAT). The capabilities of the model were tested through agri-environmental measures and their impacts on quantity and quality of the surface waters.

2. Materials and methods

2.1 Descriptions of the study areas

The river Reka catchment spreads over 30 km^2 and is located in the northwestern part of the country (Goriška Brda) (Fig. 1). Altitude ranges between 75 m and 789 m a.s.l. Very steep ridges of numerous hills, which are directed towards the southwest, characterizes the area. The catchment landscape is very agricultural with higher percentages of forest (56 %) and vineyards (23 %). The river Dragonja catchment area spreads over 100 km^2 and is located in the far southwestern part of the country (Istria) (Fig. 1). This is a coastal catchment (Adriatic Sea), with an altitude ranging between 0 and 487 m a.s.l. The ridges of the hills are designed as a plateau with flat tops and steep slopes. The landscape is largely overgrown with forest (63 %) and grassland (18 %). Steep slopes allow cultivation only on the terraces.

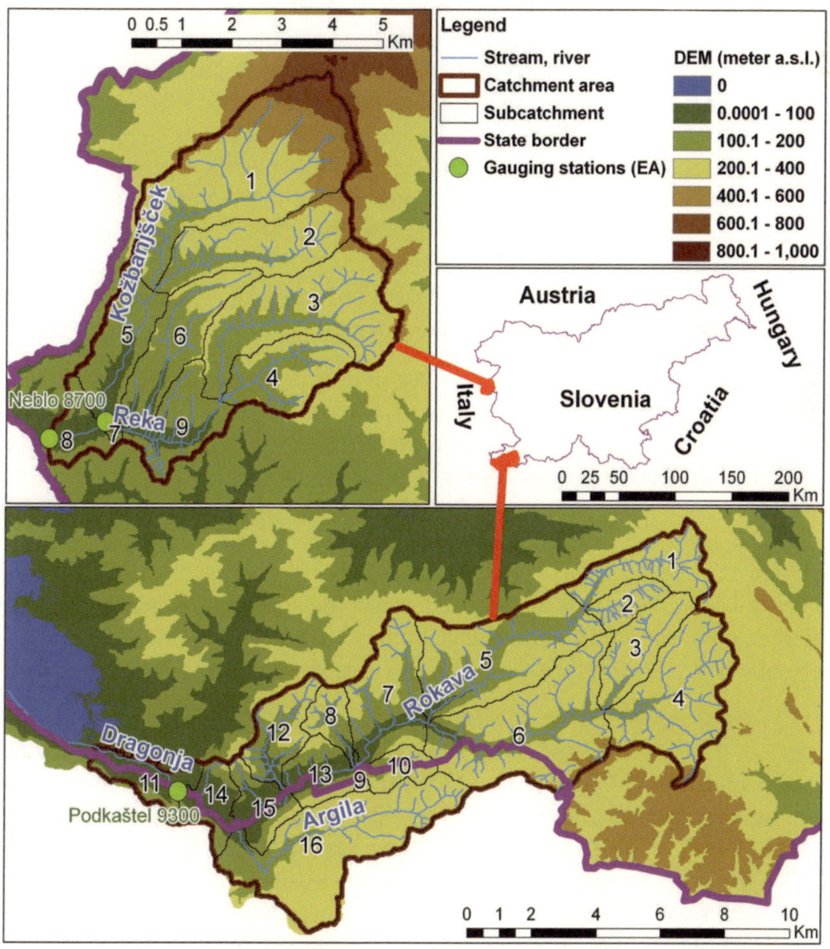

Fig. 1. The river Reka and Dragonja catchment case areas divided in sub-catchments

Flysch bedrock of the case areas was formed in Eocene as a product of the sea sediments and undersea landslides. Flysch consists of repeated sedimentary layers of sandstones, marl, slate and limestone, which can quickly crumble under the influence of precipitation and temperature changes. Brown eutric soils are shallow and due to silt-loam-clay texture difficult for tillage, with appropriate agro-technical measures (deep ploughing, organic fertilisers) they obtain properties for vine or olive production. In case of inappropriate agricultural activities and land management, we can witness very strong erosion processes.

Both areas are characterized by sub-Mediterranean climate (NE Mediterranean) with southwestern winds and warm and moist air. Average annual temperature at the station Bilje (the Reka catchment), for the period 1991–2009, was 13.3 °C, with the highest and lowest monthly average in August (22.8 °C) and January (4 °C). Average annual rainfall in the period 1992 – 2008, was 1397 mm, with peaks between September and November (max. in September 184 mm). Average annual temperature at the Portorož station (the Dragonja catchment), for the period 1991 – 2009, was 14.1 °C, with the highest and lowest monthly average in August (23.4 °C) and January (5.2 °C). Average annual rainfall in the period 1993 – 2008, was 930 mm, with peaks between September and November (max. in September 130 mm). Both catchments are characterized by fractured aquifer, where water trapped between flysch layers forms surface springs. Alluvial aquifer in the valley bottom overlays impermeable flysch. River network, of the two areas is very extensive. Rivers character is torrential and mediterranean. The river Reka hydrograph recorded (1993-2008) the highest flow rates between October and January with the average flow of 0.98 m^3s^{-1} in November and the maximum 24.50 m^3s^{-1} in October 1998, and extremely low in the summers. Hydrograph of the river Dragonja recorded (1993-2008) the highest flows between November and April with the average flow of 1.41 m^3s^{-1} in January and the maximum 64.70 m^3s^{-1} in October 1993, however in the summer the river dries up every year.

The favourable climate and terrain influences at the higher average temperature, better lighting, soil temperatures, minimal risk of frost, wind prevents diseases development. Viticulture is economically most important agricultural sector in both areas, with important share of olive and vegetable productions in the Dragonja area, and fruit production in the Goriška Brda. Terracing is typical for both areas and depends on natural conditions, steepness of the slopes (erosion), geological structure (sliding) and climatic conditions (rainfall). In Goriška Brda (Reka) is 78% of vineyards terraced while in the Slovenian Istria (Dragonja) about 18%. Vine and olive growing are the sole agricultural sectors, which can withstand the cost of the terraces installation. Terraces in the Dragonja area are characterized by overgrowing, which results in a disordered ownership structure.

The annual average concentration of sediment in the river Reka catchment for one year of research period (July 2008 - June 2009) was 32.6 mg l^{-1}, nitrate (NO_3^-) 2.7 mg l^{-1} and TP concentration of 0.109 mg l^{-1}. In the river Dragonja catchment average annual concentration of sediment in the research period (August 1989 - December 2008), was 29.1 mg l^{-1} (107 samples), NO_3^- 2.7 mg l^{-1} (87 samples) and TP concentration of 0.043 mg l^{-1} (92 samples). In January 2007, the highest sediment concentration measured so far, was 1362 mg l^{-1}. The water quality with exception of sediments does not cause any serious problems in these two study areas. Data shows that sediment concentrations are well in excess of Environment Agency guide level (25 mg l^{-1}).

2.2 Database development for the model build

Before the modelling a field tour to the research areas and review of available data was carried out (Table 1). Since the available data was insufficient for modelling, we perform additional monitoring of surface water quality at the Reka tributary Kožbanjšček hydrological station Neblo, excavation of soil profiles, laboratory measurements and using established model standards (texture, albedo, organic carbon etc) and water-physical soil

Data type	Scale	Source	Description/properties
Topography	25m×25m	The Surveying and Mapping Authority of the Republic of Slovenia	Elevation, overland and channels slopes, lengths
Soils	Slovenia: 1:25,000 Croatia: 1:50,000	Ministry of Agriculture, Forestry and Food of the Republic of Slovenia; Biotechnical Faculty (University of Ljubljana); Faculty of Agriculture (University of Zagreb)	Spatial soil variability. Soil types and properties.
Land use	Slovenia: 1m×1m (Graphical Units of Agricultural Land) Croatia: 100m×100m (CORINE)	Ministry of Agriculture, Forestry and Food of the Republic of Slovenia; European Environment Agency	Land cover classification and spatial representation
Land management	/	Chamber of Agriculture and Forestry of Slovenia; Guidelines for expert justified fertilization (Mihelič et al., 2009); Interviews with farmers	Crop rotations: planting, management, harvesting. Fertiliser application (rates and time)
Weather stations	Reka: 2 precipitation, 1 meteo (wind, temp., rain, humidity, solar) Dragonja: 3 precipitation, 1 meteo	Environment Agency of the Republic of Slovenia	Daily precipitation, temperature (max., min.), relative humidity, wind, solar radiation.
Water abstraction	46 permits (136 points)	Environment Agency of the Republic of Slovenia	From surface and groundwater.
Waste water discharges	Reka: 2 points Dragonja: 1 point	Environment Agency of the Republic of Slovenia	Registered domestic, Industrial discharge
River discharge	Reka: 2 stations Dragonja 1 station	Environment Agency of the Republic of Slovenia	Daily flow data (m^3 day^{-1})
River quality	Reka: 0 monitoring station; Dragonja: 1 monitoring station	Environment Agency of the Republic of Slovenia	Water quality (mg l^{-1}): sediment, NO_3^-, ortho-P, TP

Table 1. Model input data sources for the Reka and Dragonja catchments

properties (hydraulic conductivity, water-retention properties etc) (Saxton et al., 1986; Neisch et al., 2005; Pedosphere, 2009). For certain input data an expert assessment was performed, as required measured data was not available. For the purpose of this study we used the SWAT 2005 model and Geographic Information System (GIS) 9.1 software and ArcSWAT interface. Extensions necessary for SWAT functioning in GIS environment are Spatial Analyst, Project Manager and SWAT Watershed Delineator, which enables visualisation of the results.

SWAT is capable of simulating a single catchment or a system of hydrological linked subcatchments. The model of GIS based interface ArcSWAT defines the river network, the main point of outflow from the catchment and the distribution of subcatchments and Hydrological Response Units (HRU). HRUs are basically parts of each subcatchment with a unique combination of land use, soil, slope and land management. This allows the model modelling different ET, erosion, plant growth, surface flow, water balance, etc for each subcatchment or HRU, thus increases accuracy of the simulations (Di Luzio et al., 2005). The river Reka catchment was delineated on 9 subcatchment and 291 HRUs and the river Dragonja catchment on 16 subcatchments and 602 HRUs.

2.3 Model performance objective functions

The Pearson coefficient of correlation (R^2) (unit less) for n time steps (1) describes the portion of total variance in the measured data that can be explained by the model. The range is from 0.0 (poor model) to 1.0 (perfect model). A value of 0 for R^2 means that none of the variance in the measured data is replicated by the model, and value 1 means that all of the variance in the measured data is replicated by the model predictions. The fact that only the spread of data is quantified is a major drawback if R^2 is considered alone. A model which systematically over or under predicts all the time will still result in good values close to 1.0 even if all predictions were wrong.

$$R^2 = \left(\frac{\sum_{i=1}^{n} (simulated_i - simulated_{average})(measured_i - measured_{average})}{\sqrt{\sum_{i=1}^{n} (simulated_i - simulated_{average})^2} \sqrt{\sum_{i=1}^{n} (measured_i - mesured_{average})^2}} \right)^2 \quad (1)$$

The Nash-Sutcliffe simulation efficiency index (E_{NS}) (unit less) for n time steps (2) is widely used to evaluate the performance of hydrological model. It measures how well the simulated results predict the measured data. Values for E_{NS} range from negative infinity (poor model) to 1.0 (perfect model). A value of 0.0 means, that the model predictions are just as accurate as using the measured data average. A value greater than 0.0 means, that the model is a better predictor of the measured data than the measured data average. The E_{NS} index is an improvement over R^2 for model evaluation purposes because it is sensitive to differences in the measured and model-estimated means and variance (Nash & Sutcliffe, 1970). A major disadvantage of Nash-Sutcliffe is the fact that the differences between the measured and simulated values are calculated as squared values and this places emphasis on peak flows. As a result the impact of larger values in a time series is strongly overestimated whereas lower values are neglected. Values should be above zero to indicate minimally acceptable performance.

$$E_{NS} = 1 - \left(\frac{\sum_{i=1}^{n} (measured_i - simulated_i)^2}{\sum_{i=1}^{n} (measured_i - measured_{average})^2} \right) \quad (2)$$

Root Mean Square Error - RMSE (3) is determined by calculating the standard deviation of the points from their true position, summing up the measurements, and then taking the square root of the sum. RMSE is used to measure the difference between flow (q) values simulated by a model and actual measured flow (q) values. Smaller values indicate a better model performance. The range is between 0 (optimal) and infinity.

$$RMSE = \sqrt{\frac{\sum_{i=1}^{n} (q_t^{simulated} - q_t^{measured})^2}{n}} \quad (3)$$

Percentage bias - PBIAS (%) (4) measures the average tendency of the simulated flows (q) to be larger or smaller than their observed counter parts (Moriasi et al., 2007). The optimal value is 0, and positive values indicate a model bias toward underestimation and vice versa.

$$PBIAS = \left(\frac{\sum_{i=1}^{n} (q_t^{measured} - q_t^{simulated})}{\sum_{i=1}^{n} (q_t^{measured})} \right) \cdot 100\% \quad (4)$$

Model calibration criteria can be further based on recommended percentages of error for annual water yields suggested from the Montana Department of Environment Quality (2005) who generalised information related to model calibration criteria (Table 2) based on a number of research papers.

Errors (Simulated-Measured)	Recommended Criteria
Error in total volume	10%
Error in 50% of lowest flows	10%
Error in 10% of highest flows	15%
Seasonal volume error (summer)	30%
Seasonal volume error (autumn)	30%
Seasonal volume error (winter)	30%
Seasonal volume error (spring)	30%

Table 2. Model calibration hydrology criteria by Montana Department of Environment Quality (2005)

For the detection of statistical differences between the two base scenarios and alternative scenarios Student t-test statistics should be used ($\alpha = 0.025$, degrees of freedom (SP = n-1)), for comparing average annual value of two dependent samples at level of significance 0.05 (5). Variable, which has approximately symmetrical frequency distribution with one modus class, is in the interval $\bar{x} \pm s$ expected 2/3 of the variables and in $\bar{x} \pm 2s$ approximately 95% of the variables and in $\bar{x} \pm 3s$ almost all variables. Confidence interval ($l_{1,2}$) (6) for Student distribution for all sample arithmetic means ($\bar{x}$) can be calculated (6).

$$t = \frac{\bar{x} - \mu}{s / \sqrt{n}} \quad (5)$$

$$t = \overline{x} \pm t_{\frac{\alpha}{2}}(n-1) \cdot \frac{s}{\sqrt{n}} \tag{6}$$

$\overline{x}$	sample arithmetic mean (alternative scenario)
μ	average of the corresponding random variables (base scenario)
s	sample standard deviation (alternative scenario)
n	number of pairs (alternative scenario)
$t_{\frac{\alpha}{2}}$	Student distribution

3. Sensitivity analysis

If the model in certain areas has not been used, then it is necessary to carry out sensitivity analysis. Sensitivity analysis limits the number of parameters that need optimization to achieve good correlation between simulated and measured data. The method of analysis in the SWAT model called PARASOL is based on the method of Latin Hypercube One-factor-at-a-Time (LH-OAT). LH-OAT combines the advantages of global and local sensitivity analysis (van Griensven et al., 2006). This method performs LH sampling of data at first, followed by OAT sampling. The new scheme allows the LH-OAT to unmistakably link the changes in the output data of each model to the modified parameter (van Griensven et al., 2006). For the sensitivity analysis and calibration a special tool called SWAT-CUP is available which includes all important algorithms (GLUE, PSO, MCMC, PARASOL and SUFI2) of which Sequential Uncertainty Method (SUFI2) was shown to be very effective in identifying sensitive parameters (Abbaspour et al., 2007).

Tool within the model can automatically carry out the sensitivity analysis without the measured data or with the measured data. The tool varies values of each model parameter within a range of (MIN, MAX). Parameters can be multiplied by a value (%), part of the value can be added to the base value, or the parameter value can be replaced by a new value. The final result of the sensitivity analysis are parameters arranged in the ranks, where the parameter with a maximum effect obtains rank 1, and parameter with a minimum effect obtains rank which corresponds to the number of all analyzed parameters. Parameter that has a global rank 1, is categorized as "very important", rank 2 – 6 as "important", rank 7 – 41 (i.e. the number of parameters in the analysis – i.e. flow 7 - 26) as "slightly important" and rank 42 (i.e. flow 27) as "not important" because the model is not sensitive to change in parameter (van Griensven et al., 2006).

Sensitivity analysis was performed using the measured data of the river Reka tributary Kožbanjšček (subcatchment 5) and the river Dragonja (subcatchment 14). The analysis was performed for an average daily flow, sediments, TP and NO_3^-. Table 3 represents for each model the first 10 parameters that have the greatest impact on the model when they are changed. The sensitivity analyses demonstrated great importance of the hydrological parameters that are associated with surface and subsurface runoff.

Alpha_Bf factor determines the share between the base and surface flow contribution to the total river flow. Cn2 curve runoff determines the ratio between the water drained by the surface and subsurface runoff in moist conditions. Ch_K2 describes the effective hydraulic conductivity of the alluvial river bottom (water losing and gaining). Surlag represents the surface runoff velocity of the river and Esco describes evaporation from the soil. For the sediment modelling the most important parameters are Spcon and Spexp that affect the movement and separation of the sediment fractions in the channel. Ch_N – Manning coefficient for channel, determines the sediment transport based on the shape of the channel and type of the river bed material. Ch_Cov – Channel cover factor and Ch_Erod – Channel erodibillity factor proved to be important for the Dragonja catchment. Soil erosion is closely related to the surface runoff hydrological processes (Surlag, Cn2). The analysis showed importance of the hydrological parameters that are associated with surface and subsurface runoff (Cn2, Canmx, Sol_Awc), evaporation (Revapmin, Esco, Blai), base flow (Alpha_Bf) and groundwater (Rchrg_Dp, Gwqmn), suggesting numerous routes by which sediment nitrate nitrogen (NO_3-N) and TP are transported (Table 3). We noticed that the amount of N is also influenced by other parameters that are not included in the sensitivity analysis tool like Rate factor for humus mineralization of organic nutrients active N and P (CMN.bsn), half-life of nitrates and the shallow aquifer (HLIFE_NGW.gw), fraction of algal biomass that is N (Al1.wwq). TP results are significantly affected by the parameters that control surface runoff (Cn2, Canmx, Usle_P). Usle_P factor adjusts the USLE value for a particular land management. This means that the soil loss from the terraced land is different, from non terraced slopes. Parameters which have a significant impact on P, but not included in the sensitivity analysis tool are: fraction of algal biomass that is P (Al2.wwq), P availability index (PSP.bsn), P enrichment ratio for loading with sediment (ERORGP.hru), BC4.swq, benthic sediment source rate for dissolved P in the reach (RS2.swq), organic P settling rate (RS5.swq).

Base model	Sensitivity Analysis Objective function (SSQR)				Category
	Flow	Sediment	NO_3-N	TP	
Brda	Surlag	Spcon	Cn2	Usle_P	Very important
	AlphaBf	Ch_N	Revapmin	Cn2	Important (2-6)
	Cn2	Surlag	Alpha_Bf	AlphaBf	
	Ch_K2	Spexp	Esco	Surlag	
	Esco	Cn2	RchrgDp	Ch_K2	
	Ch_N	Alpha_Bf	Sol_Awc	Slope	
Dragonja	Cn2	Spcon	Blai	Canmx	Very important
	AlphaBf	Ch_Erod	Sol_Awc	AlphaBf	Important (2-6)
	Ch_K2	Ch_Cov	Cn2	Blai	
	RchrgDp	Ch_N	Revapmin	Surlag	
	Esco	Spexp	RchrgDp	Cn2	
	Surlag	Surlag	Sol_Z	Sol_Z	

Table 3. SWAT parameters ranked by the sensitivity analysis for the Reka subcatchment 5 and Dragonja subcatchment 14 (1998 - 2005)

4. Calibration and validation

During the model calibration parameters are varied within an acceptable range, until a satisfactory correlation is achieved between measured and simulated data. Usually, the parameters values are changed uniformly on the catchment level. However, certain parameters (Sol_Awc, Cn2, Canmx) are exceptions, because of the spatial heterogeneity. Firstly manual calibration, parameter by parameter, should be carried out with gradual adjustments of the parameter values until a satisfactory output results (E_{NS} and R^2> 0.5) (Moriasi et al., 2007, Henriksen et al., 2003). This procedure may be time consuming for inexperienced modellers. In the process of autocalibration only the most sensitive parameters are listed that showed the greatest effect on the model outputs. For each of the parameter a limit range (max, min) has to be assigned.

Validation is performed with parameter values from the calibrated model (Table 4) and with the measured data from another time period. Due to the data scarcity, the model was validated only for the hydrological part (flow). The river Reka water quality data covers only one year of daily observations, which was only enough for the calibration. For the river Dragonja a 14 years long data series of water quality was available, but the data was scarce in the number of observations (for sediment, NO_3^- and TP only 92, 73, 75, 78 measurements). It should be pointed out that samples taken during monitoring represents only the current condition of the river in a certain part of the day (concentration in mg l^{-1}), while the simulated value is a total daily transported load (kg day^{-1}) in a river.

Calibration of the daily flow for the rivers Reka and Dragonja catchments was performed for the period from 1998 to 2005. According to the availability of data we selected different periods for the daily flow validation of the Reka (1993–1997, 2006–2008) and Dragonja (1994–1996, 2006–2008). Due to the lack of data, on sediment, NO_3^- and TP, we performed only the calibration for Kožbanjšček (1. 7. 2008 - 30. 6. 2009) and Dragonja (1994–2008).

Parameter		Default	Range	Calibrated values	
				Reka	Dragonja
1	Alpha_Bf	0.048	0–1	0.30058	0.45923
2	Canmx[1]	0	0–20	8, 4, 2	8, 4, 2
3	Ch_K2	D	0–150	7.0653	3.7212
4	Ch_N	D	0–1	0.038981	0.04363
5	Cn2	D	-25/+25%	-8, -15 [2]	+14
6	Esco	0.95	0–1	0.8	0.75
7	Gw_Delay	31	0–160	131.1	60.684
8	Gw_Revap	0.02	0–0.2	0.19876	0.069222
9	Gwqmn	0	0–100	100	0.79193
10	Sol_Awc	D	+50%	no change	no change
11	Surlag	4	0.01–4	0.28814	0.13984
E_{NS}				0.61	0.57

Legend: [1] - forest, permanent crops, grassland, arable; [2] - subcatchment 1-2-5, subcatchment 3-4-6-7-8-9; D - default value - depends on soil type, land use and modeller set up

Table 4. Hydrological parameters, ranges and final values selected for the calibration of models (SWAT) for the rivers Reka and Dragonja catchments

4.1 Hydrology calibration and validation

Objective functions show that the simulated total flows are within the acceptable range (Table 5, Fig. 2). Correlation coefficient (R2) for a daily flow is influenced by low flows.

Official measurements of a flow showed that on certain days the flow was not present or it was negligible. Model does not neglect extremely low flows, as is evident from the cumulative distribution of the flow (Fig. 2). Errors in flow measurements, in the worst case may be upto 42 % and in best case upto 3 % of the total flow (Harmel et al., 2006).

The E_{NS} values for total flow fall into the category of satisfactory results (Moriasi et al., 2007, Henriksen et al., 2003), R^2 values fall into the category of good results, RMSE into the category of very good results (Henriksen et al., 2003) and PBIAS into the category of very good and good results (Moriasi et al., 2007). The reasons for lower results of the objective functions in the validation lie in the representation of the soil, rainfall and in the river flow data uncertainty.

	Reka				**Dragonja**			
	Calibration		Validation (Total Flow)		Calibration		Validation (Total Flow)	
Objective function	Base Flow	Total Flow	1993 - 1997	2006 - 2008	Base Flow	Total Flow	1994 - 1996	2006 - 2008
E_{NS}	0.61	0.61	0.39	0.69	0.55	0.57	0.45	0.42
R^2	0.72	0.64	0.57	0.70	0.66	0.59	0.49	0.49
RMSE	0.13	0.82	1.21	0.74	0.35	1.06	1.98	1.50
PBIAS	-12.79	7.04	-14.19	19.40	1.49	4.69	23.15	-3.31

Table 5. Daily time step river flow performance statistics for the rivers Dragonja and Reka for the calibration (2001-2005) and validation periods

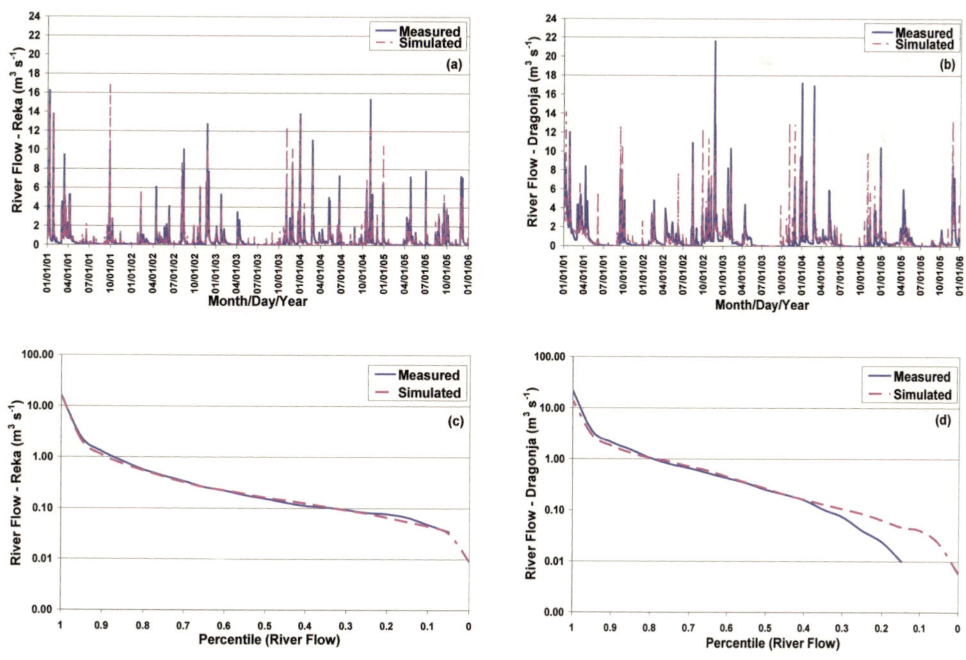

Fig. 2. Comparison between simulated (SWAT) and measured daily flows ($m^3 s^{-1}$) (a, b) and cumulative distribution (c, d) of daily river flows for the calibration period (2001-2005)

4.2 Sediment, nitrogen and phosphorus calibration

Sediment calibration is essential for the proper P calibration, as P is preferentially transported adsorbed on the sediment particles. Parameters used for the calibration were USLE_P, SPCON, SPEXP, CH_EROD, CH_COV. Simulation results for the river Reka show lower E_{NS} = 0.23 and a good result in predicting the variability of $E_{NSpercentile}$ = 0.83 (Table 6). In the case of Dragonja, model achieved good results for E_{NS} = 0.70 and $E_{NSpercentile}$ = 0.73. PBIAS values fall within the category of very good results as deviation is less than 15% (Moriasi et al., 2007).

Parameters with impact on the N calibration results were FRT_SURFACE, NPERCO, AL1, CMN, HLIFE_NGW. The river Dragonja statistic is lower (E_{NS} = 0.10, $E_{Nspercentile}$ = 0.78) and for the river Reka is in satisfactory range with E_{NS} = 0.40 and $E_{Nspercentile}$ = 0.72 (Table 6). The PBIAS results fall into the very good (Dragonja) and satisfactory (Reka) category (Moriasi et al., 2007). The lower performance of the objective functions is connected to data scarcity in the Dragonja catchment with only 73 measurements in 14 years and in river Reka with only one year of daily data. Therefore, it is difficult to say whether the model is a good predictor of nitrate nitrogen (NO_3-N) loads and dynamics. Monthly sampling rate leads to inaccurate estimates of the transported loads of nutrients in rivers (Johnes, 2007); especially NO_3^- (Harmel et al, 2006).

4.3 Model performance indicators

An important step before calibrating sediment and water quality parameters is to look at other model performance indicators. Three main parameters are crop growth, evapotranspiration (ET) and Soil Water Content (SWC), as all of them have a great effect on the water balance. Evapotranspiration is a primary mechanism by which water is removed from the catchment. It depends on air temperature and soil water content. The higher the temperature, the higher is potential evapotranspiration (PET) and consequently ET, if there is enough of water in the soil. A simple monthly water balance between monthly precipitation and PET showed that average monthly water balance in the Reka catchment (station Bilje) is negative between May and August (Fig. 3). In the Dragonja catchment (station Portorož) water balance is negative from April to August (growing season) (Fig. 3).

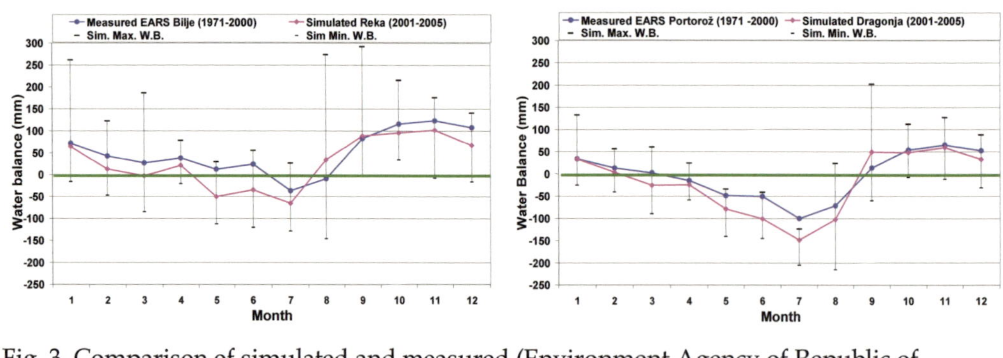

Fig. 3. Comparison of simulated and measured (Environment Agency of Republic of Slovenia - EARS) water balance (mm) for the Reka subcatchments 8 and Dragonja subcatchment 14

Parameter		Default	Range	Calibrated values	
Sediment				*Reka – Kožbanjšček*	*Dragonja*
1	SpCon	0.0001	0.0001-0.01	0.002	0,002
2	SpExp	1	1-1.5	1.3	1
3	Ch_Erod	0	0-1	0.092	0,06
4	Ch_Cov	0	0.05-0.6	0.1	0,1
5	USLE_P	1	0-1	slope dependent	slope dependent
E_{NS}				**0.23**	**0.70**
E_{NS} percentile				**0.83**	**0.73**
R^2				**0.24**	**0.80**
RMSE				**10.35**	**19.81**
PBIAS				**-0.15**	**-6.33**
Nitrate nitrogen (NO_3-N)					
1	Nperco	0.2	0.01-1	1	0,2
2	Al1	0.08	0.07-0.09	0.071	0,08
3	CMN	0.0003	0.0001-0.001	-	0,0001
4	HLIFE_NGW	0	0-200	-	0,02
5	FRT_surface	0.2	0-1	management dependent	management dependent
E_{NS}				**0.40**	**0.10**
E_{NS} percentile				**0.72**	**0.78**
R^2				**0.46**	**0.17**
RMSE				**79.89**	**5.11**
PBIAS				**21.24**	**-3.43**
Total phosphorus (TP)					
1	Pperco	10	10-17.5	15	10
2	Phoskd	175	100-200	175	200
3	Al2	0.015	0.01-0.02	0.003	0,001
4	PSP	0.4	0.01-0.7	0.22	0,04
5	ERORGP	0	0.001-5	0	0,003
6	BC4	0.35	0.01-0.7	0.1	0,1
7	RS2	0.05	0.001-0.1	0.1	0,1
8	RS5	0.05	0.001-0.9	0.08	0,001
9	FRT_surface	0.2	0-1	management dependent	management dependent
E_{NS}				**-0.05**	**0.36**
E_{NS} percentile				**0.95**	**0.85**
R^2				**0.11**	**0.46**
RMSE				**48.17**	**0.18**
PBIAS				**3.43**	**49.21**

Table 6. SWAT water quality parameters, their ranges and the final values chosen for the models calibration periods (Reka 2008 - 2009; Dragonja 1994 - 2008)

Water that enters the soil may move along one of the several different pathways. It may be removed by plant uptake or evaporation; it may percolate past the bottom of the soil profile or may move laterally in the profile. However, plant uptake removes the majority of water that enters the soil profile (Neitsch et al., 2005). The soil water content will be represented correctly if crops are growing at the expected rate and soils have been correctly parameterized. Figure 4 shows the average of HRU for both catchments, with a silt clay soils, with the prevailing surface runoff and slow lateral subsurface flow. Soils exit the field capacity in the spring and return to that state in the autumn (Fig. 4). Soils in the summer are often completely dry with occasional increasing induced by storms.

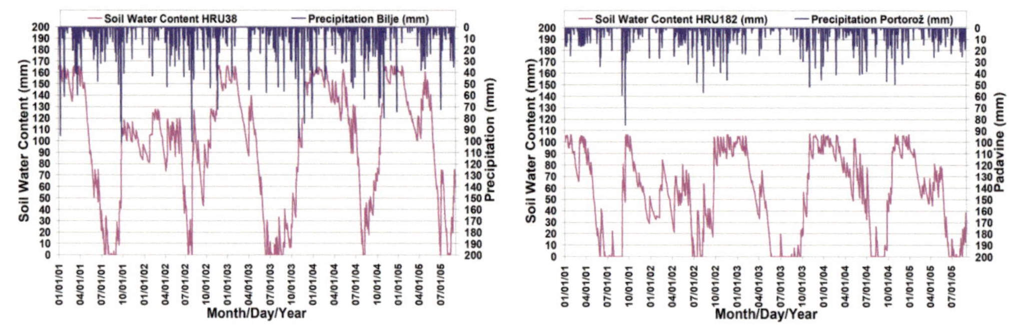

Fig. 4. Comparison of simulated soil water content (mm) for the HRU No. 38 (Reka) and HRU No. 182 (Dragonja) and observed precipitation (mm) in the calibration period (2001–2005)

The plant growth component of SWAT is a simplified version of the plant growth model. Phenological plant development is based on daily accumulated heat units, leaf area development, potential biomass is based on a method developed by Monteith, a harvest index is used to calculate yield, and plant growth can be inhibited by temperature, water, N or P stress. (Neitsch et al., 2005). In the crop database a range of parameters can be changed to meet the requirements for optimal plant growth. We used default SWAT database parameters that were additionally modified (Frame, 1992). An example crop growth profile for development of leaf area index (LAI) and plant biomass (BIOM) for vineyard is presented on figure 5.

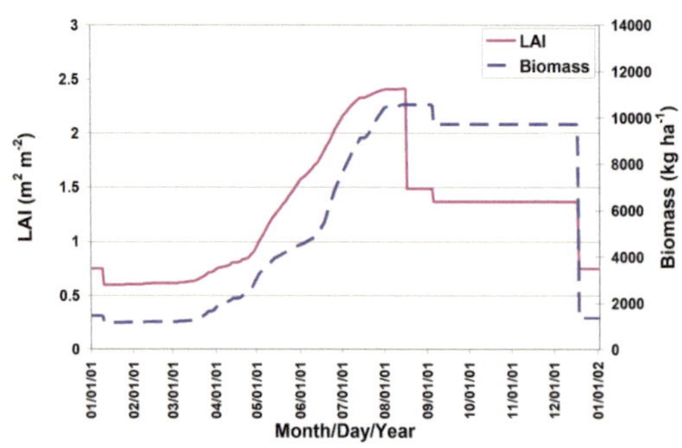

Fig. 5. Simulated vineyard biomass growth (kg ha^{-1}) and leaf area index (m^2 m^{-2}) for the HRU No. 38 in the river Reka catchment

5. Agri-environmental scenarios

The aim of this scenario was to investigate possible effects of the agri-environmental measures on the river water quality. To achieve the aim seven different scenarios were applied to the study area EVP, EKO20, EKO100, S35, S50, STV35, ETA.

The field erosion buffer strips scenario (EVP) is a function of how to minimize influences of diffuse pollution resulting from agricultural activities without drastic management changes. They are planted or indigenous bands of vegetation that are situated between source areas and receiving waters to reduce surface runoff velocities and to remove pollutants from surface and subsurface runoff. The effectiveness of strips is closely correlated with their slope and width (Dillaha et al., 1989). An option of 3 m wide strips was modelled on all arable (AGRC, AGRR), vineyard (VINE), orchard (ORCI, ORCE) in olive grove (OLEA) HRUs.

Organic farming scenarios on 20 % of the area (EKO20) and on the 100 % area (EKO100) aim to reduce the use of mineral fertilizers and to reduce the intensity of production. Special organic rotations with green manure and composted farmyard manure were created. The lack of P was compensated with the use of triple-superphosphate that is allowed in organic production. Both organic scenarios were designed to ensure normal production for the market.

Steep meadows, being an agricultural landscape, should be cut regularly, but due to the steep slopes and the associated costs and risks, are abandoned and overgrown. Scenarios having steep meadows with slope inclination above 35 % (S35) and 50 % (S50) should prevent overgrowth. To verify the effects of scenarios on water quantity and nutrients transport, meadows (TRAV) of both case studies located on slopes greater than 35 % and 50 % were changed into the forest (FRSD) (Fig. 6). In the S35 scenario 18 % (Reka) and 3.6 % (Dragonja) of grassland was changed into forest, which is equivalent to 1.43 % (Reka) and 0.67 % (Dragonja) of the total catchments. In the S50 scenario only 2 % (Reka) and 0.3 % (Dragonja) of grassland was changed into forest, which is equivalent to 0.16% (Reka) and 0.06% (Dragonja) of the total catchments.

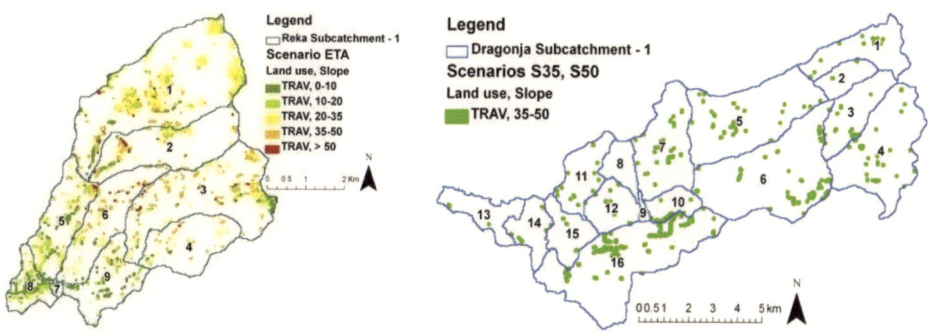

Fig. 6. Hydrological response units with the grassland land use (TRAV) and slopes greater than 35 % and 50 % for the Reka and Dragonja catchment

Conservation of vineyards on steep slopes has proved to be difficult because of unprofitable production. Economic reasons were followed by a trend of wine production abandonment.

In the steep vineyards scenario (STV35), all vineyards on the slopes greater than 35 % were changed into forest, to verify the environmental impact of abandonment of vineyards on steep slopes (Fig. 7). In the STV35 scenario, 17 % (Reka) and 1.4 % (Dragonja) of grassland is changed into forest, which is equivalent to 3.93% (River) and 0.06% (Dragonja) of the total catchments.

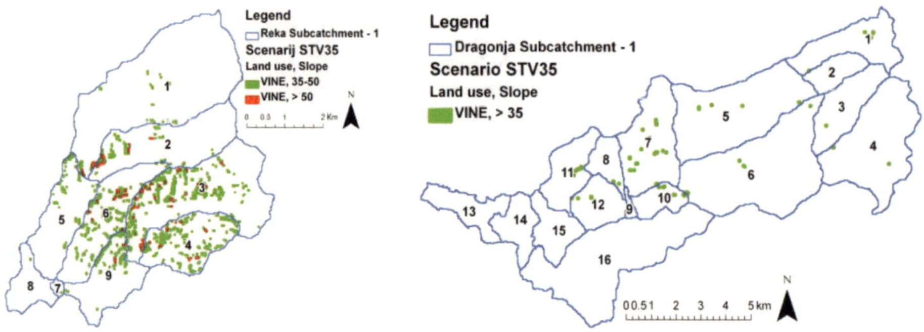

Fig. 7. Hydrological response units with the vineyard land use (VINE) and slopes greater than 35 % for the Reka and Dragonja catchment

Extensive grassland scenario (ETA) objective was to determine what would be the impact on water quantity and quality, if the whole grassland would be overgrown with forest. Extensive grassland use with one cutting is widespread in both areas. Whole grassland in the Reka (8 %) and Dragonja (18%) catchments area was turned into a forest (Fig. 8).

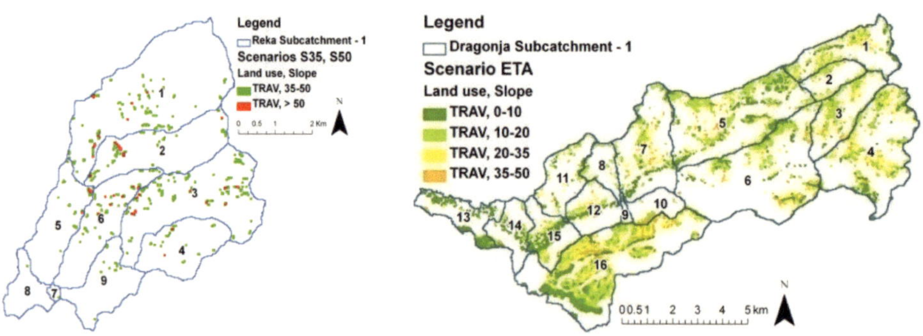

Fig. 8. Hydrological response units with the grassland land use (TRAV) and slope classes for the Reka and Dragonja catchment

6. Results and discussion

The base scenario indicates a high average annual variability in the transport of the sediment, total nitrogen (TN) and total phosphorus (TP) in the river flow (Table 7). The standard deviations for the Reka subcatchment 8 reveal that the sediment, TN and TP 2/3 of transported quantities are expected in the interval 1,844 ± 1,075 t sediment year^{-1}, 88,728 ± 63,255 kg TN year^{-1} and 3,489 ± 2,993 kg TP year^{-1} and for the Dragonja subcatchment 14 in the interval 4,804 ± 1,576 t sediment year t^{-1}, 163,763 ± 98,949 kg TN year^{-1} and 3,489 ± 11,742 kg TP year^{-1}.

Catchment/subcatchment	Average	Median	Standard deviation	Min.	Max.
Flow (m^3 s^{-1})					
Reka/8	0.57	0.56	0.21	0.27	1.00
Dragonja/14	0.80	0.78	0.21	0.42	1.11
Sediment (t $year^{-1}$)					
Reka/8	1,844	1,576	1,075	571	4,185
Dragonja/14	4,804	4,934	1,576	1,917	7,734
Total nitrogen (kg $year^{-1}$)					
Reka/8	88,728	74,260	63,255	33,376	278,227
Dragonja/14	163,763	134,801	98,949	59,922	406,330
Total phosphorus (kg $year^{-1}$)					
Reka/8	3,489	2,729	2,993	947	11,742
Dragonja/14	2,420	1,950	1,447	896	6,009

Table 7. Average annual flow (m^3 s^{-1}) and river load of sediment (t $year^{-1}$), total nitrogen and total phosphorus (kg $year^{-1}$) for the Reka subcatchment 8 and Dragonja subcatchment 14 (1994–2008)

6.1 River flow

Changes in average annual flow between base and agri-environmental scenarios are minimal for both catchments for the research period. Maximum changes on an annual basis are less than 0.5 % (Table 8) and on a monthly basis close to 1% (Reka) and 5% (Dragonja) (Fig. 9). Student t-statistics for average annual flows reveal that the results of the agri-environmental scenarios are not statistically different from the base scenario (Table 9).

Catchment/ subcatchment	Average annual percentage change (%)						
	EVP	EKO20	EKO100	S35	S50	STV35	ETA
River Flow							
Reka/8	0.00	0.09	0.04	0.02	0.00	0.17	0.16
Dragonja/14	0.00	-0.32	0.36	0.00	0.00	0.00	0.09
Sediment							
Reka/8	-14.93	-4.95	-25.42	-0.85	-0.05	-2.28	-3.12
Dragonja/14	-31.95	-20.82	-20.92	-2.26	-0.05	-0.01	-52.96
Total nitrogen							
Reka/8	-2.67	9.00	-1.91	-0.43	-0.02	-5.15	-2.32
Dragonja/14	-1.46	12.51	3.71	-0.22	-0.01	0.00	-6.63
Total phosphorus							
Reka/8	-14.15	9.28	-26.15	-0.58	-0.04	-2.44	-3.45
Dragonja/14	-3.28	9.90	1.39	-0.29	0.00	0.00	-8.58

Table 8. Impacts (change in %) of agri-environmental scenarios on the river flow, sediment load, total nitrogen and total phosphorus load in the watercourse; compared to the baseline scenario

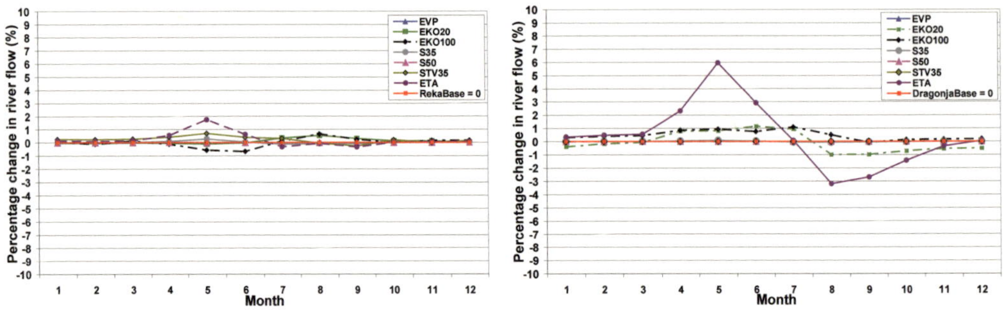

Fig. 9. Change in average monthly flow (%) between the base (Base = 0) and agri-environmental scenarios for the Reka subcatchment 8 and Dragonja subcatchment 14 (1994–2008)

Scenario	**Student t-test** (Significance level 0.05) Student distribution of the sample with n-1 degrees of freedom α=0.025, SP=14, t_α =2.145							
	Reka - subcatchment 8				**Dragonja - subcatchment 14**			
	Flow	**Sediment**	**TN**	**TP**	**Flow**	**Sediment**	**TN**	**TP**
EVP	0.000	-1.214	-0.148	-0.712	0.000	**-5.630**	-0.122	-0.215
EKO20	0.009	-0.348	0.448	0.389	-0.047	**-3.056**	0.750	0.603
EKO100	0.005	**-2.435**	-0.105	-1.439	0.053	**-3.023**	0.209	0.080
S35	0.002	-0.057	-0.023	-0.027	0.000	-0.274	-0.014	-0.019
S50	0.000	-0.004	-0.001	-0.002	0.000	-0.006	-0.001	0.000
STV35	0.018	-0.157	-0.281	-0.112	0.000	-0.001	0.000	0.000
ETA	0.018	-0.216	-0.127	-0.159	0.013	**-14.386**	-0.450	-0.594

Note: The results of the scenarios are statistically significantly different from the base scenario, if the value of Student t-test exceeds t_α = 2.145. If the value is negative, scenario is reducing the quantities in the river flow, and vice versa.

Table 9. Review of statistically significant results of Student t-statistics for average annual flow and average annual load of sediment, total nitrogen and total phosphorus

6.2 Sediment

Impacts of agri-environmental scenarios EVP, EKO20, EKO100, S35, S50, STV35, ETA on an average annual load of sediment transported with the flow are evident for certain scenarios (Table 8). Statistically significant changes in the Reka catchment have been calculated for the EKO100 scenario, while the EVP scenario result is slightly lower to be statistically significantly different (Table 9). The river Dragonja results show that changes in the scenarios EVP, EKO20, EKO100 and ETA are statistically significantly different from the base scenario (Table 9). The biggest differences between scenarios in transported sediment load are in autumn and winter months, when the loads for scenarios EKO100 (Reka) and ETA (Dragonja) get considerably reduced (Fig. 10).

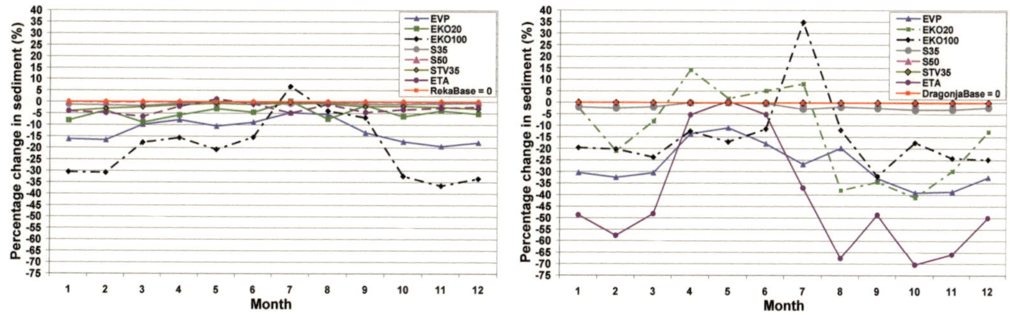

Fig. 10. Change in average monthly river loads of sediment (%) between the base (Base = 0) and agri-environmental scenarios for the Reka subcatchment 8 and Dragonja subcatchment 14 (1994–2008)

6.3 Total nitrogen

The effect of agri-environmental scenarios on the annual TP transport in the river flow has proved to be negligibly small, due to the small proportion of land on which the scenarios were set up (Table 8). The results of the agri-environmental scenarios for the TN transport in both catchments are not statistically significantly different from the base scenario (Table 9). Large monthly variations in the loads of TP transported were typical for the scenarios with higher levels of organic matter (EKO20, EKO100, ETA) (Fig. 11). The decomposition of the organic matter is difficult to control, monitor and predict. However, on an annual basis, the variation between months are equalized.

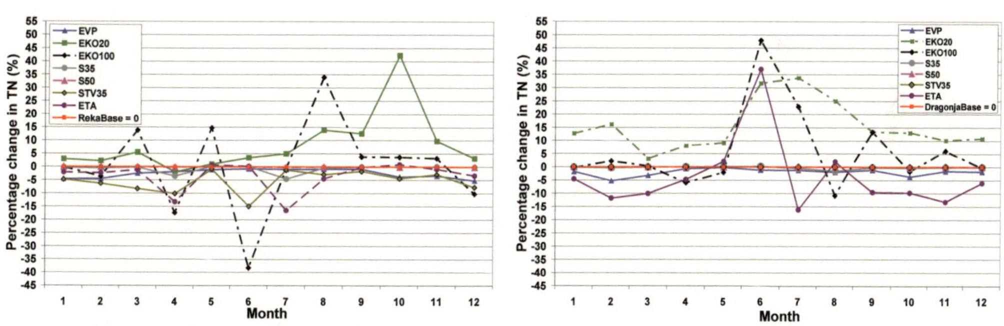

Fig. 11. Change in average monthly river loads of total nitrogen (%) between the base (Base = 0) and agri-environmental scenarios for the Reka subcatchment 8 and Dragonja subcatchment 14 (1994–2008)

6.4 Total phosphorus

The effects of agri-environmental scenarios on the TP transport in the stream are low (Table 8) and may be observed in scenarios EKO 100 and EVP (Reka) and ETA (Dragonja) (Fig. 12). Student t-statistics for average annual TP load in both catchments are not statistically significantly different (Table 9). In case of Rivers, maximum difference between the scenarios resulting in cooler and wetter period of the year, and in the Dragonja catchment, in the warmer and more stormy period.

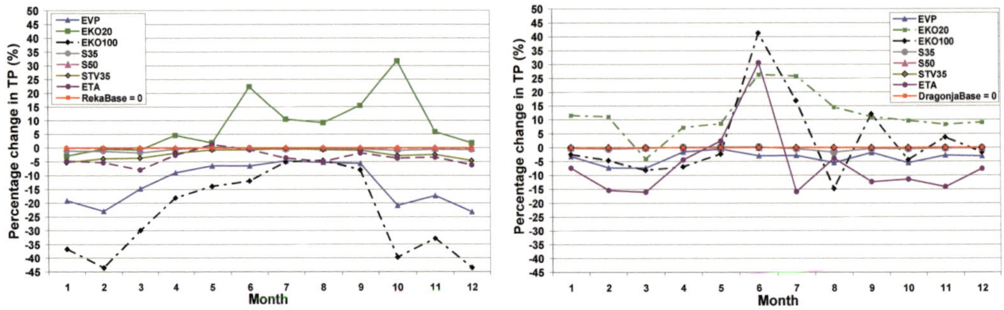

Fig. 12. Change in average monthly river loads of total phosphorus (%) between the base (Base = 0) and agri-environmental scenarios for the Reka subcatchment 8 and Dragonja subcatchment 14 (1994–2008)

6.5 Scenario evaluation

The evaluation of impacts of the agri-environmental scenarios on the sediment and nutrients transport processes on the catchment level was performed in the light of the EU Water Framework Directive (WFD 2000/60/ES) and Republic of Slovenia legislation. Both set guide concentrations with the purpose of limiting impacts of excessive levels on flora and fauna in the rivers. When interpreting the concentrations we need to have in mind the geological and pedological characteristics of the catchment. There is also the question of whether to consider set guide levels for the rivers that do not represent an economic interest (Lohse, 2008), however rivers are not only economic asset. When recommending possible agri-environmental mitigation measures to deliver water quality improvements, careful evaluation and prioritization of each measure has to be performed according to its positive and negative issues on the environment, agriculture, social life and economy (Bockstaller et al., 2009; Everard, 2004; Glavan et al., 2011).

The results of the scenarios demonstrate that in the Reka and Dragonja catchments major problems with the concentrations of NO_3^- and TP are excluded, as both are lower than the limit values (Table 10). Nevertheless, the results reveal the difficult path to achieve the recommended value for sediment in both catchments, especially in the case of the river Reka catchment. With the realization of agri-environmental scenarios for the Dragonja catchment, particularly the EVP and ETA, we could expect reduction of the sediment concentration below the recommended level and consequently water quality improvements. In the Dragonja catchment, the guide concentration of 25 mg l^{-1} was reached with the scenarios EVP, EKO20, EKO10 and ETA. However, in the Reka catchment, scenarios sediment reductions are not sufficient to reduce the concentration below the guide level. This leads us to thinking, that catchment is dominated by certain land use (vineyard) and soils, which have a negative impact on the river concentrations (Komac & Zorn, 2007; Petek, 2007; Volk et al., 2009).

The EKO100 scenario is considering the low proportion of land involved in organic production in research areas almost impracticably, since it would require too much labour-intensive work, which results in a higher final price of the crop. Organic production is advised in the areas with long-term organic fertilization where soils were sufficiently enriched with organic matter and nutrients to supply plants for a several decades (Mihelič et al., 2009). In the Dragonja catchment, which is subject to a high degree of afforestation, the scenario EVP reflected in the significant concentration reduction below the recommended value. We used 3 meters wide vegetation bands that have reflected a 14 % (Reka) and 31 % (Dragonja) reduction of sediment in the watercourse, but with broader bands, an even greater impact could be achieved. For the effectiveness of the bands, the identification of critical points is important (Garen & Moore, 2005; Wolfe, 2000). A small proportion of the area can have a significant impact on the sediment, N and P loads in the watercourses.

		Average annual concentration (mg l^{-1})					
		DRAGONJA - subcatchment 14 (Podkaštel 9300) - cyprinid river			REKA - subcatchment 5 (Neblo 8700) - salmonid river		
		Sediment	**Nitrate**	**TP**	**Sediment**	**Nitrate**	**TP**
Measured		**29.3**	**2.7**	**0.043**	**32.6**	**2.7**	**0.109**
Scenarios	EVP	19.9	2.6	0.042	28.8	2.7	0.100
	EKO20	23.2	2.7	0.045	31.1	2.5	0.121
	EKO100	23.2	2.6	0.044	27.6	2.9	0.092
	S35	28.6	2.7	0.043	32.2	2.7	0.108
	S50	29.3	2.7	0.043	32.6	2.7	0.109
	STV35	29.3	2.7	0.043	32.1	2.6	0.108
	ETA	13.8	2.6	0.039	30.6	2.5	0.104

Limit and guide concentrations (mg l^{-1}) set by EU directives and Slovenian regulations: **Sediment** (river) **25 mg l^{-1}**; **Nitrate (NO_3^-)** in drinking water **50 mg l^{-1}** and in surface water **14,08 - 30,8** (very good state) and **28,6 - 41,8 mg l^{-1}** (good state); **Total phosphorus (TP)** for salmonid waters **0,2 mg l^{-1}** and for cyprinid waters **0,4 mg l^{-1}**.

Table 10. Impacts of the alternative scenarios on the average annual concentration (mg l^{-1}) of the sediment, nitrate and total phosphorus

Following the trend of afforestation of agricultural land, the ETA scenario could become practicable, under which all grassland (18 %) would be overgrown by forest. However, such a scenario is not viable, since larger farmers round up their vineyards and olive groves and reduce overgrowth. However, this process is considerably slower than natural afforestation, which has affected the water cycle and erosion processes in the last decade (Globevnik, 2001). Sediment reductions in the catchment are expected with progressive land abandoned with afforestation and with parallel establishment of buffer zones on larger agriculturally rounded areas. The negative effect of erosion buffer zones is an exclusion of a certain percentage of agricultural land from agricultural production. At 3 m wide buffer zones on 1 ha of land (10,000 m^2) the loss of the land in production would be 12 % (1,200 m^2). An important element, which partially contributes to increased sediment loads in the river Dragonja are cliffs and steep eroded slopes without vegetation, which are eroded at the foothills by the river and torrential tributaries.

To achieve improvements in water quality in the two research catchments the use of a combination of several measures and a close cooperation with all key stakeholders (environmental, agricultural, spatial planning) would be necessary.

7. Conclusions

The application of the SWAT model in the Reka and Dragonja catchment has demonstrated that SWAT is able to represent the hydrological behaviour of this heterogeneous catchments and rivers. Within the constraints of the available data the model was able to represent the sediment and nutrients loads, concentrations and cumulative distributions. However, there are a number of issues that the model results can demonstrate as important in the diffuse water pollution control with agri-environmental measures.

1. Research process can demonstrate that because of the lack of monitoring and limited data on sediment, N and P concentrations, proper calibration or validation of the model would not be possible. Mixed sampling frequency on a monthly or fortnightly basis can provide the basis for imprecise estimates of nutrient loadings in rivers.
2. Although the simulated crops in the model can grow well and therefore taking up nutrients appropriately, the actual on-site spatial distribution of crops, crop rotations and actual management practices (sowing, harvest and fertiliser application dates and rates) are usually not known. These uncertainties further combine with those uncertainties in the spatial and attribute soil data, which can have an important influence on overall contribution to pollution and successful implementation of environmental measures.
3. As an important element of the catchment modelling is detailed analysis of point sources as in certain study areas can represent prevailing source of N and P in the watercourses.
4. Temporal aggregation of model outputs can improve the performance metrics for all the river outputs, including NO_3^-. Temporal aggregation is appropriate to simplify model outputs for those variables, which are adequately simulated at daily level and underpinned by appropriate process representation and model parameterization. This demonstrates the importance of ascertaining the reasons for the use of temporal aggregation in modelling studies.
5. There are important limitations to the treatment of edge of field filter strips within SWAT, which may over-estimate their efficiency of the EVP scenario. The SWAT algorithms relate the fraction of the nutrient load trapped by the buffer to the buffer width, so that additional factors such as slope, vegetation type, soil type and presence of under-drainage are not included. SWAT simulates reduction in pollutant transport across the entire length of a buffer strip, while in reality, as surface flow can concentrate at certain points alongwith buffer strips. SWAT assumes that buffer strips capture the range of particle sizes equally. However, buffer strips may trap coarser sediment with lower P concentrations, suggesting that the finer fraction, enriched in TP, may preferentially pass through the buffers towards river channels.
6. Base flow represents an important pathway for the transport of dissolved contaminants from the landscape to surface water receptors. The delivery of surface water targets will require the integrated management of land, groundwater base flow and surface water systems. However, SWAT has all the tools and options for setting the initial conditions

in the model, which can lead to appropriate modelling of nutrients pathways and to account for the nutrient lag times in the groundwater.

7. Physical landscape spatial variability within catchments (topography, soils, land use, land management etc.) have important influence on the model results. This means that pollutant sources and loads are not evenly distributed in space. Rather than impose blanket agri-environmental measures in the model, it is better to target key source areas or HRU combinations that deliver excessive loads.
8. The scenarios assume that all farmers in the catchment take up the structural measures or the changes in land use and management uniformly. However, field work shows that this is not the case. A close cooperation with all key stakeholders on local, regional, national and transnational level and financial support, like EU Common Agriculture Policy, which enable areas to develop in a sustainable way, is necessary.

At the end of this chapter we would like to increase awareness that model results and their interpretation by the modeller must lead to constructive discussion, which aims to achieve and maintain good water quality in research catchments, which is the objective of the Water Framework Directive and other legislation related to water.

8. Acknowledgments

Financial support for this study was provided by the Slovenian Research Agency founded by the Government of the Republic of Slovenia. Contract number: 1000-06-310163.

9. References

Abbaspour, K.C.; Yang, J.; Maximov, I.; Siber, R.; Bogner, K.; Mieleitner, J.; Zobrist, J. & R. Srinivasan (2007). Modelling hydrology and water quality in the pre-alpine/alpine Thur watershed using SWAT. *Journal of Hydrology,* Vol.333, pp. 413-430

Arnold, J.G.; Srinivasan, R.S.; Muttiah, R.S. & Williams, J.R. (1998). Large area hydrological modelling and assessment Part I: Model development. *Journal of the American Water Resources Association*, Vol.34, No.1, pp. 73–89

Bockstaller, C.; Guichard, L.; Makowski, D.; Aveline, A.; Girardin, P. & Plantureux, S. (2009). Agri-environmental indicators to assess cropping and farming systems – A review. *Agronomy for Sustainable Development*, Vol.28, pp. 139–149

Bowatte, S.; Tillman, R.; Carran, A. & Gillingham, A. (2006). Can phosphorus fertilisers alone increase levels of soil nitrogen in New Zeland hill country pastures? *Nutrient Cycling in Agroecosystems*, Vol.75, pp. 57–66

Bracmort, K.S.; Arabi, M.; Frankenberger, J.R.; Engel, B.A. & Arnold, J.G. (2006). Modelling long-term water quality impact of structural BMPs. *Agricultural Society of Agricultural and Biological Engineers*, Vol.49, No.2, pp. 367–374

Buda, A.R.; Kleinman, P.J.A; Srinivasan, M.S.; Bryant, R.B. & Feyereisen, G.W. (2009). Effects of Hydrology and Field Management on Phosphorus Transport in Surface Runoff. *Journal of Environmental Quality*, Vol.38, pp. 2273–2284

Čarman, M.; Mikoš, M. & Pintar, M. (2007). Različni vidiki erozije tal v Sloveniji = Different aspects of soil erosion in Slovenia. In: *Strategija varovanja tal v Sloveniji*: *Zbornik referatov,* M. Knapič, (Ed.), 39–50, Pedološko društvo Slovenije = Slovenian Soil science Society, Ljubljana

Di Luzio, M.; Arnold, J. G. & Srinivasan, R. (2005). Effect of GIS data quality on small watershed streamflow and sediment simulations. *Hydrolgical Processes*, Vol.19, No.3, pp. 629–650

Dillaha, T.A.; Reneau, R.B.; Mostaghimi, S.& Lee, D. (1989). Vegetative filter strips for agricultural nonpoint source pollution control. *Transactions of the ASAE*, Vol.32, No.2, pp. 513–519

Dymond, R.; Lohani, V.; Kibler, D.; Bosch, D.; Rubin, E.J.; Dietz, R.; Chanat, J.; Speir, C.; Shaffer, C.A.; Ramakrishnan, N. & Watson, L.T. (2003). From landscapes to waterscapes: A PSE for landuse change analysis. *Engineering with Computers*, Vol.19, pp 9–25

Everard, M. (2004). Investing in sustainable catchments. *Science of the Total Environment*, Vol.324, pp. 1–24

EUSOILS (2004). Nature and extent of soil erosion in Europe, European Commisson, 2. June 2011, Available from http://eusoils.jrc.ec.europa.eu/esdb_archive/pesera/pesera_cd/

Frame, J. (1992). *Improved grassland management*. Farming Press Books, Wharfedale

Garen, D.C. & Moore, D.S. (2005). Curve number hydrology in water quality modeling: uses, abuses, and future directions. *Journal of the American Water Resources Association*, Vol.41, No.6, pp. 1491–1492

Gassman, P.W.; Reyes, M.R.; Green, C.H. & Arnold, J.G. (2007). The soil and water assessment tool: Historical development, applications, and future research direction. *Transactions of the ASABE*, Vol.50, No.4, pp. 1211–1250

Glavan, M. & Pintar, M. (2010). Impact of point and diffuse pollution sources on nitrate and ammonium ion concentrations in the karst-influenced Temenica river. *Fresenius Environmental Bulletin*, Vol.19, No.5A, pp. 1005–1014

Glavan, M.; White, S. & Holman, I. (2011). Evaluation of river water quality simulations at a daily time step - Experience with SWAT in the Axe Catchment, UK. *CLEAN - Soil, Air, Water*, Vol.39, No.1, pp. 43–54

Globevnik, L. (2001). *Celosten pristop k urejanju voda v porečjih = An integrated approach to water management in river basins*. Doctoral thesis. University of Ljubljana, Ljubljana

Harmel, R.D.; Potter, S.; Ellis, P.; Reckhow, K.; Green, C.H. & Haney, R.L. (2006). Compilation of measured nutrient load data for agricultural land uses in the US. *Journal of American Water Resources Association*, Vol.42, pp. 1163–1178

Hatch, L.K.; Mallawatantri, A.; Wheeler, D.; Gleason, A.; Mulla, D.; Perry, J.; Easter, K.W.; Smith, R.; Gerlach, L. & Brezonik, P. (2001). Land management at the major watershed-agroecoregion intersection. *Journal of Soil and Water Conservation*, Vol.56, No.1, pp. 44–51

Hejzlar, J.; Anthony, S.; Arheimer, B.; Behrendt, H.; Bouraoui, F.; Grizzetti, B.; Groenendijk, P.; Jeuken, M.; Johnsson, H.; Lo Porto, A.; Kronvang, B.; Panagopoulos, Y.; Siderius, C.; Silgram, M.; Venohrd, M. & Žaloudíka, J. (2009). Nitrogen and phosphorus retention in surface waters: an inter-comparison of predictions by catchment models of different complexity, *Journal of Environmental Monitoring*, Vol.11, pp. 584–593

Henriksen, H. J.; Troldborg, L.; Nyegaard, P.; Sonnenborg, O. T.; Refsgaard, J. C. & Madsen, B. (2003). Methodology for construction, calibration and validation of a national hydrological model for Denmark. *Journal of Hydrology*, Vol.280, pp. 52–71

Johnes, P.J. (2007). Uncertainties in annual riverine phosphorus load estimation: Impact of load estimation methodology, sampling frequency, baseflow index and catchment population density. *Journal of Hydrology*, Vol.332, pp. 241-258

Khan, F.A & Ansari, A.A (2005). Eutrophication: An Ecological Vision. *The Botanical Review*, Vol.71, No.4, pp. 449-482

Komac, B. & Zorn, M. (2007). Probability modelling of landslide hazard. *Acta geographica Slovenica*, Vol.47, No.2, pp. 139-169

Kronvang, B.; Behrendt, H.; Andersen, H.; Arheimer, B.; Barr, A.; Borgvang, S.; Bouraoui, F.; Granlund, K.; Grizzetti, B.; Groenendijk, P.; Schwaiger, E.; Hejzlar, J.; Hoffman, L.; Johnsson, H.; Panagopoulos, Y.; Lo Porto, A.; Reisser, H; Schoumans, O.; Anthony, S.; Silgram, M.; Venohr, M. & Larsen, S. (2009a). Ensemble modelling of nutrient loads and nutrient load partitioning in 17 European catchments. *Journal of Environmental Monitoring*, Vol.11, pp. 572-583

Kronvang, B.; Borgvang, S. A. & Barkved, L. J. (2009b). Towards European harmonised procedures for quantification of nutrient losses from diffuse sources - the EUROHARP project. *Journal of Environmental Monitoring*, Vol.11, No.3, pp. 503-505

Krysanova, V. & Arnold, J.G. (2008). Advances in ecohydrological modelling with SWAT—a review. *Hydrological Sciences-Journal des Sciences Hydrologiques*, Vol.53, No.5, pp. 939-947

Kummu, M.; Sarkkula, J.; Koponen, J. & Nikula, J. (2006). Ecosystem Management of the Tonle Sap Lake: An Integrated Modelling Approach. *Water Resources Development*, Vol.22, No.3, pp. 497-519

Lohse, K.A.; Newburn, D.A.; Opperman, J.J. & Merenlender, A.M. (2008). Forecasting relative impacts of land use on anadromous fish habitat to guide conservation planning. *Ecological Applications*, Vol.18, No.2, pp. 467-482

Mihelič, R.; Čop, J.; Jakše, M.; Štampar, F.; Majer, D.; Tojnko S. & Vršič S. (2010). *Smernice za strokovno utemeljeno gnojenje = Guidelines for professionally justified fertilisation*. Ministry of Agriculture, Forestry and Food of Republic of Slovenia, Ljubljana

Montana Department of Environmental Quality (2005). *Flathead Basin Program, quality assurance project plane (QAPP)*. Land and Water Quality Consulting/PBS&J, Montana

Moriasi, D.N.; Arnold, J.G.; Van Liew, M.W.; Bingner, R.L.; Harmel, R.D. & Veith, T.L. (2007). Model evaluation guidelines for systematic quantification of accuracy in watershed simulations. *Transactions of the ASABE*, Vol.50, No.3, pp. 885-900

Nash, J. & Sutcliffe, J. (1970). River flow forecasting through conceptual models: I. A discussion of principles. *Journal of Hydrology*, Vol.10, pp. 374-387

Neal, C. & Heatwaite, A.L. (2005). Nutrient mobility within river basins: a European perspective. *Journal of Hydrology*, Vol.304, pp. 477-490

Neitsch, S.L.; Arnold, J.G.; Kiniry, J.R. & Williams, J.R. (2005). *Soil and water assessment tool theoretical documentation - Version 2005*. Texas Agricultural Experiment Station, Blackland Research Center, Agricultural Research Service, Grassland, Soil and Water Research Laboratory, Texas, Temple

Pedosphere (2009). Soil Texture Triangle Hydraulic Properties Calculator, 21. July 2010, Available from http://www.pedosphere.com/

Petek, F. (2007). Spreminjanje rabe tal v severnih Goriških brdi. *Geografski vestnik*, Vol.79, No.1, pp. 9-23

Ramos, M.C. & Martinez-Casasnovas, J.A. (2006). Nutrient losses by runoff in vineyards of the Mediterranean Alt Penede`s region (NE Spain). *Agriculture, Ecosystems and Environment*, Vol.113, pp. 356–363

Randhir, T.O. & Hawes, A.G. (2009). Watershed land use and aquatic ecosystem response: Ecohydrologic approach to conservation policy. *Journal of Hydrology*, Vol.364, pp. 182–199

Rusjan, S. (2008). *Hidrološke kontrole sproščanja hranil v porečjih = Hydrological controls of nutrient mobilization in watersheds*. Doctoral thesis. University of Ljubljana, Ljubljana

Saxton, K.E.; Rawls, W.J.; Romberger, J.S. & Papendick, R.I. (1986). Estimating generalized soil-water characteristics from texture. *Soil Science Society of America Journal*, Vol.50, No.4, pp. 1031–1036

Schoumans, O.F.; Silgram, M.; Walvoort, D.J.; Groenendijk, P.; Bouraoui, F.; Andersen, H. E.; Lo Porto, A.; Reisser, H.; Le Gall, G.; Anthony, S.; Arheimer, B.; Johnsson, H.; Panagopoulos, Y.; Mimikou, M.; Zweynert, U.; Behrendt, H. & Barr, A. (2009). Description of nine nutrient loss models: capabilities and suitability based on their characteristics. *Journal of Environmental Monitoring*, Vol.11, pp. 506–514

van Griensven, A.; Meixner, T.; Grunwald, S.; Bishop, T.; Di Luzio, M. & Srinivasan, R. (2006). A global sensitivity analysis tool for the parameters of multi-variable catchment models. *Journal of Hydrology*, Vol.324, pp. 10–23

Volk, M.; Liersch, S. & Schmidt, G. (2009). Towards the implementation of the European Water Framework Directive? Lessons learned from water quality simulations in an agricultural watershed. *Land Use Policy*, Vol.26, pp. 580–588

Wagner, W.; Gawel J.; Furumai, H.; Pereira De Souza, P.; Teixeira, D.; Rios, L.; Ohgaki, S.; Zehnder, A.J.B. & Hemond, H.F. (2002). Sustainable watershed management: An international multi-watershed case study. *Ambio*, Vol.31, No.1, pp. 2–13

Wolfe, M.L. (2000). Hydrology. In: *Agricultural nonpoint source pollution:: Watershed management and hydrology*, W.F. Ritter & A. Shirmohammadi, (Ed.), Lewis Publishers, Boca Raton

6

Evolution of Water Quality in Romania

Ioan Oroian and Antonia Odagiu
University of Agricultural Sciences and Veterinary Medicine Cluj – Napoca Romania

1. Introduction

Globally, water is a renewable natural resource, but vulnerable and limited, so it must be treated as a natural heritage to be protected and defended. In our century, one of the largest global problems concerning water management, taking into account that the population of the planet is in continuous growing, is the crisis of drinking water. The structure of water resources is mainly represented by freshwater, which is a rather small percentage of total water on the planet, namely 2.5%, while the percentage of 70% constitutes the water on the surface of the Earth. The fresh water is directly accessible by springs, rivers, lakes, and ground water, the rest being found in glacial ice. It means that only 0.7% of the planet's water is available, as a source of survival for the current population (Dodds, 2002; http://www.anpm.ro/ Mediu/rapoarte, accessed 2011).

For these reasons, conservation, water saving and reuse, and not at least water quality are serious problems that concerns all states. In order to preserve water resources an maintain water quality at best standards by protecting water quality and quantity, states policies are elaborated in order to encourage the above mentioned desiderates by the application of economic stimuli, and by imposing penalties for those wastes or pollute the water (Meybeck, 2004).

Concerning Romania's case, the authorities confront with the same concerns regarding water quality as all other states, water quality being affected by a wide range of natural and human influences. If human influences concern the result of economical and domestic activities, the natural influences are geological, hydrological, and climatic (Wake, 2005; Shirodkar et al., 2009; Bulut et al., 2010; Odagiu, 2010; Odagiu et al., 2010). The Romanian particularities in the field are conferred by national geographical and economical specific. In this respect, we have to mention that because of the climate changes, especially in recent years, leading to increased drought phenomena, must be taken in view the need to manage water resources in a special manner in order to preserve this resource for future generations (Dodds, 2002; Blenckner, 2005; http://www.anpm.ro/ Mediu/rapoarte, accessed 2011). Another aspect, which must be taken into consideration, is that both economical and social realities recorded in last decade, imposed a better understanding of water quality evolution at national level, in order to find useful solutions for prediction models and reducing pollutants inputs of a large variety of sources (industry, agriculture, etc.).

Water resources of Romania are made up of surface water - rivers, lakes, river Danube (~ 90%) -, and groundwater (~ 10%). The main water resource of Romania is the inland rivers. Theoretically speaking, total water resource in Romania was in 2009 of 136,600,000 thousand m3 while that existing, according to the degree of development of river basin, is 40,482,841 thousand m3 under the terms of a national requirement of 12,265,698 thousand m3 (http://www.anpm.ro/ Mediu/rapoarte, accessed 2011). Taking into account only the contribution of inland rivers, from this point of view our country may be included in the category of countries with relatively low water resources in relation to the resources of other countries.

Generally speaking, monitoring water quality represents the activity of observations and standardized measurements and continues long-term awareness and evaluation of the parameters characteristic of water for household chores and defining the status and trend of the evolution of their quality, as well as permanent highlighting condition of water resources (Hirsch et al., 2009; Goyal et al., 2010; Odagiu, 2010).

In Romania, the quality of water is monitored according to the structure and the methodological principles of Integrated Monitoring System of Wastewater in Romania (S.M.I.A.R.), restructured in accordance with the requirements of European Directives. The inventory of the water pollutants is performed at regional level, based on the information delivered by the inventoried economical operators and statistical data collected and processed by regional competent authorities. The national system for monitoring water comprises two types of monitoring, in accordance with legislative requirements in the area: monitoring of supervisors having the role of assessing the status of all bodies of water in the river basin and operational monitoring (integrated monitoring displays) for bodies of water which have a risk not to fulfil the objectives of water protection (Oroian&Petrescu-Mag, 2011; http://www.anpm.ro/ Mediu/rapoarte, accessed 2011).

For the evaluation of chemical water quality overall, in each section, were calculated for each indicator, mean values, and these were compared with the limit values of the quality classes set by norm with five quality categories, resulting employment in one of the five quality categories. The indicators included in the Order of the Ministry of the Environment and Forests no. 161/2006 approving the Norms on the classification of the quality of surface waters in order to establish the ecological status of water bodies, were divided into five main groups (Oroian&Petrescu-Mag, 2011; http://www.anpm.ro/ Mediu/rapoarte, accessed 2011):

- the group "oxygen", which includes: dissolved oxygen, BOD5, COD - Mn, COD - Cr;
- the group "nutrients", which includes: ammonium, nitrites, nitrates, total nitrogen, orthophosphates, total phosphorus, chlorophyll a;
- the group "general ions, salinity", which includes: filterable dry residue, sodium, calcium, magnesium, total iron, total manganese, chlorine, sulphates;
- the group "metals", that contains: zinc, copper, chromium, arsenic; metals such as lead, cadmium, mercury, nickel, were assigned to the group of priority substances;
- the group "organic and inorganic micro pollutants", which includes: phenols, detergents, AOX, petroleum hydrocarbons; other substances, such as PAH, PCB, DDT, lindane, atrazine, tetrachloromethane, trichloromethane, tetrachloroethane, etc. were assigned the group priority substances.

Emphasizing the evolution of surface water, wastewater and ground water quality in Romania for the period of 5 years between 2005 and 2009, and testing the multiregression analyze model in order to predict this evolution, are the main objectives of our study.

2. Data collection

Usually, monitoring the quality of water resources at national level cannot be performed by the measurement of only one parameter, because of many reasons. The water quality indices are variable in time and space, and this needs complex monitoring activity involving the measurement of a series of chemical physical and biological parameters according to special patterns, which are changing over time function of external conditions. Some of the above mentioned indicators provide general information concerning water pollution, whereas others enable the direct tracking of water pollution sources.

The data concerning the main water pollutants were collected from annual reports elaborated by regional authorities and from public data delivered by the reports of the Ministry of Environment and Forests, and annual environmental reports of the National Agency of Environmental Protection.

3. Statistics

Basic statistics, correlation and multivariate calculations were carried out in order to give initial information about the water quality data. Calculation of means, Pearson correlations, and multiple correlations were performed using STATISTICA 7.0 software for Windows. The calculation of Pearson correlation coefficients were performed in order to evaluate the correlations between the levels of variables (water virtual pollutants within surface water, groundwater and wastewater), and multiple regression analysis was conducted in order to evaluate the interrelation between chemical pollutants from surface water, groundwater and wastewater, and predict their future evolution. All tests of significance and correlations were considered statistically significant at P values of < 0.05, < 0.01, and < 0.001 (Blenckner, 2005; Sojka et al., 2008; Kazi et al., 2009; Papaioannou et al., 2010).

During the period of five years that concerns the data collection, processing and analyze, the management of water quality in Romania was conducted and practiced, by authorities, according to requirements of the following EU directives: 60/2000/CEE, 75/440/EEC, 76/464/EEC, 91/676/EEC, 78/659/EEC, and 91/271/EEC, adopted and/or transposed in Romanian legislation (Oroian&Petrescu-Mag, 2011). They concern usual physical, chemical and ecological water indicators.

The main water pollutants identified during the period of 5 years and all water categories (surface water, groundwater and wastewater) were: nitrogen compounds, chlorine, Fe, P, Fe, Mn, Cu, Cd, Zn, pesticides, oil products and detergents.

4. The quality of surface waters

The summary of the quality of surface waters by 2005 (Fig. 1a) was the result of processing the raw data deriving from the physico-chemical analyses of water sampled from 825 monitoring flowing water sections and 97 lakes (http://www.anpm.ro/Mediu/rapoarte, 2005,accessed 2011), and performed in accredited laboratories. It was distributed as follows: 31.40% Ist quality water, 46.10% IInd quality water, 15.80% IIIrd quality water, 3.70% IVth quality water, and 3.10% Vth quality water.

From the point of view of monitoring activities developed at large national scale, the main components of the surface water are the flowing water (rivers) and lakes (Mihăiescu et al.,

2010). The results of monitoring activities performed on these water categories may be summarized for flowing water distribution by five qualities (Fig. 1b) as follows: 29.80% Ist quality water, 46.40% IInd quality water, 16.50% IIIrd quality water, 5.10% IVth quality water, and 2.20% Vth quality water. Concerning the lake water, it was distributed only by four qualities (Fig. 1c), the Vth quality water missing, as follows: 47.10% Ist quality water, 33.30% IInd quality water, 17.60% IIIrd quality water, and 2.00% IVth quality water.

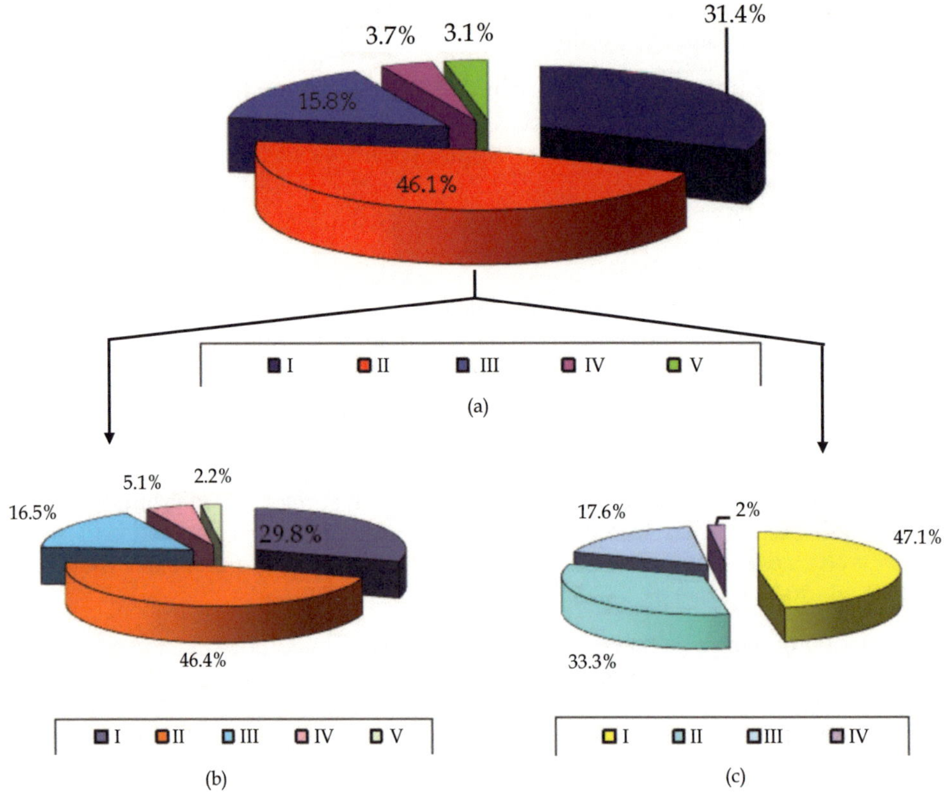

Fig. 1. (a) The surface water distribution by 5 qualities in 2005. (b) The flowing water distribution by 5 qualities in 2005 (c) The lake water distribution by 4 qualities in 2005

By 2006, the summary of the quality of surface waters (Fig. 2a) was, as the same as in 2005, the result of processing the raw data deriving from the physico-chemical analyses of water sampled from 825 monitoring flowing water sections and 102 lakes (http://www.anpm.ro/Mediu/rapoarte, 2006, accessed 2011). The results of the analysis demonstrates that the water quality was also divided by five categories, but the water of first quality was with 0.80% higher in this year, compared to previous: 25.80% Ist quality water, 41.50% IInd quality water, 17.20% IIIrd quality water, 12.10% IVth quality water, and 3.40% Vth quality water. The IVth and Vth quality water summarize 15.50%, and this represents a bigger share compared to 2005, when the share of the most polluted water at national scale was reported to be 6.80%.

The flowing water distribution, in 2006, by five qualities (Fig. 2b) was: 25.30% Ist quality water, 41.50% IInd quality water, 17.40% IIIrd quality water, 12.00% IVth quality water, and

3.80% Vth quality water. The share of the Ist quality water is with 4.50% smaller compared to previous year, while the share of low quality water (IVth and Vth) was 15.80%, also bigger (with 8.50%) compared to previous year (7.30%).

In this year, the lake water was distributed by five qualities (Fig. 2c) as follows: 22.30% Ist quality water, 47.10% IInd quality water, 15.20% IIIrd quality water, 9.00% IVth quality water, and 6.40% Vth quality water. We find that, in 2006, the lake water of low quality (IVth and Vth) occupies a bigger share (15.40%), compared to 2005, when Vth quality water was not reported, and IVth quality water represented only 2% from total analyzed lake water at national level, while the Ist quality water occupied a share with 23.80% smaller compared to 2005 (47.10%).

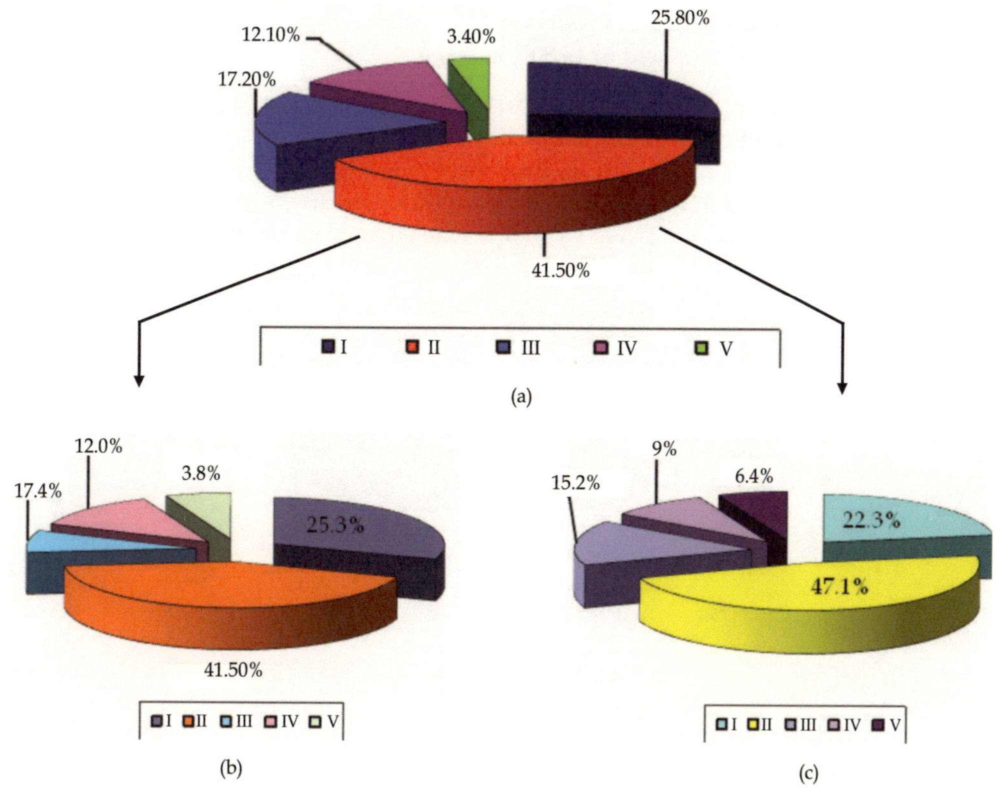

Fig. 2. (a) The surface water distribution by 5 qualities, in 2006. (b) The flowing water by 5 qualities, in 2006. (c) The lake water distribution by 5 qualities, in 2006

The same study, as for the years 2005 and 2006, was performed on 2007, and data reported for the summary of the quality of surface waters by 2007, (Fig. 3a) also resulted after processing the raw data deriving from the physico-chemical analyses of water sampled from almost the same numbers of flowing water sections and lakes, as previous years, 824 monitoring flowing water sections and 100 lakes, respectively (http://www.anpm.ro/Mediu/rapoarte, 2007, accessed 2011). The share of the Ist quality water was smaller in this year (23.80%) compared to previous years (which was bigger than 31%). The distribution of the surface water, lakes and rivers, in 2007, by quality categories may be summarized as follows: 23.80% Ist quality water, 45.00% IInd quality water, 18.80% IIIrd quality water, 7.90% IVth quality water, and 4.50% Vth quality water. For the low quality water (IVth and Vth) was reported a share of 12.40%, which means a bigger pollution of lake waters compared to 2005 (6.80%) but smaller compared with 2006 (15.50%).

Concerning the Romanian flowing water quality, in 2007, the National Agency for Environmental Protection reported the following distribution by five qualities (Fig. 3b): 23.40% Ist quality water, 44.50% IInd quality water, 21.30% IIIrd quality water, 7.20% IVth quality water, and 3.60% Vth quality water. The share of the Ist quality water, was in 2007 with 6.40% smaller compared to 2005, and with 1.90% compared to 2006. The same evolution was recorded concerning the IVth and Vth quality water, which was 10.80% from total monitored flowing waters, if the comparison is made with the values reported for 2005 (7.30%). If we report the low flowing water quality to 2007, we find that 5.00%variation was recorded (10.80% in 2007 compared to 15.80% in 2006).

The picture of lake water quality emphasize in 2007, the same distribution as in 2006, by five qualities (Fig. 3c), as follows: 45.00% Ist quality water, 27.00% IInd quality water, 13.00% IIIrd quality water, 11.00.00% IVth quality water, and 4.00% Vth quality water. The difference between the situation of Romanian lakes water quality in 2007 and previous years consists in smaller share (about 2%) of Ist quality water, compared to 2005 and 22.27%

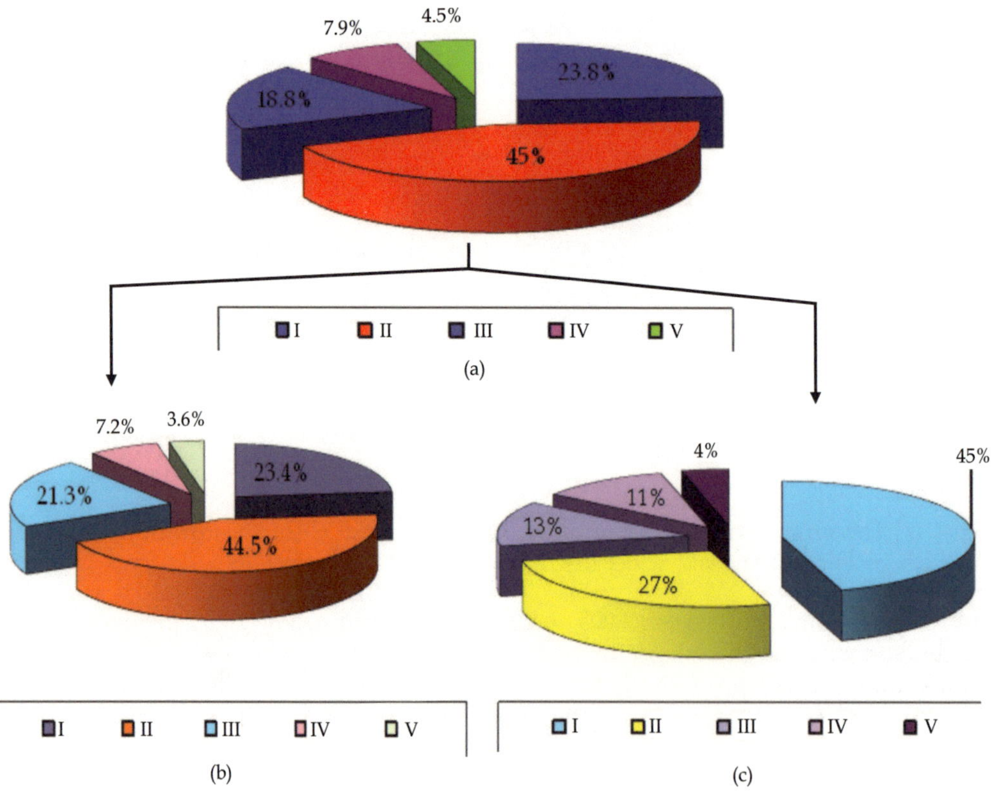

Fig. 3. (a) The surface water distribution by 5 qualities, in 2007. (b) The flowing water distribution by 5 qualities, in 2007. (c) The lake water distribution by 5 qualities, in 2007

compared to 2006, and much bigger concerning IVth and Vth quality water, which was 15.00% from total monitored lake waters, while in 2005 it was only 2% (Vth quality water was not reported), and 13.40% in 2006.

In 2008, the data concerning the summary of the quality of surface waters (Fig. 4a) was the result of processing the raw data deriving from the physico-chemical analyses of water sampled from 817 monitoring flowing water sections, less than previous years (825 in 2005 and 824 in 2006 and 2007) and 102 lakes. Globally, the quality of flowing water and lakes monitored at national level during the reference year 2008 was distributed as follows: 29.00% Ist quality water, 44.40% IInd quality water, 17.10% IIIrd quality water, 5.50% IVth quality water, and 3.90% Vth quality water (http://www.anpm.ro/Mediu/rapoarte, 2005, accessed 2011). Compared to 2007, the share of Ist quality water was bigger with more than 5%, and compared with 2006 with 3.20%, but if comparison is made with 2005, it was smaller (with 2.40%).

Concerning the flowing water, the distribution was by five qualities (Fig. 4b) as follows: 26.40% Ist quality water, 45.50% IInd quality water, 19.40% IIIrd quality water, 5.60% IVth quality water, and 3.10% Vth quality water. The most polluted water (IVth and Vth quality water) covered a share of 8.70%, less compared with 2007 (10.80%), and 2006 (15.80%), but more compared with 2005 (7.30%). For the Ist quality water was reported a bigger value

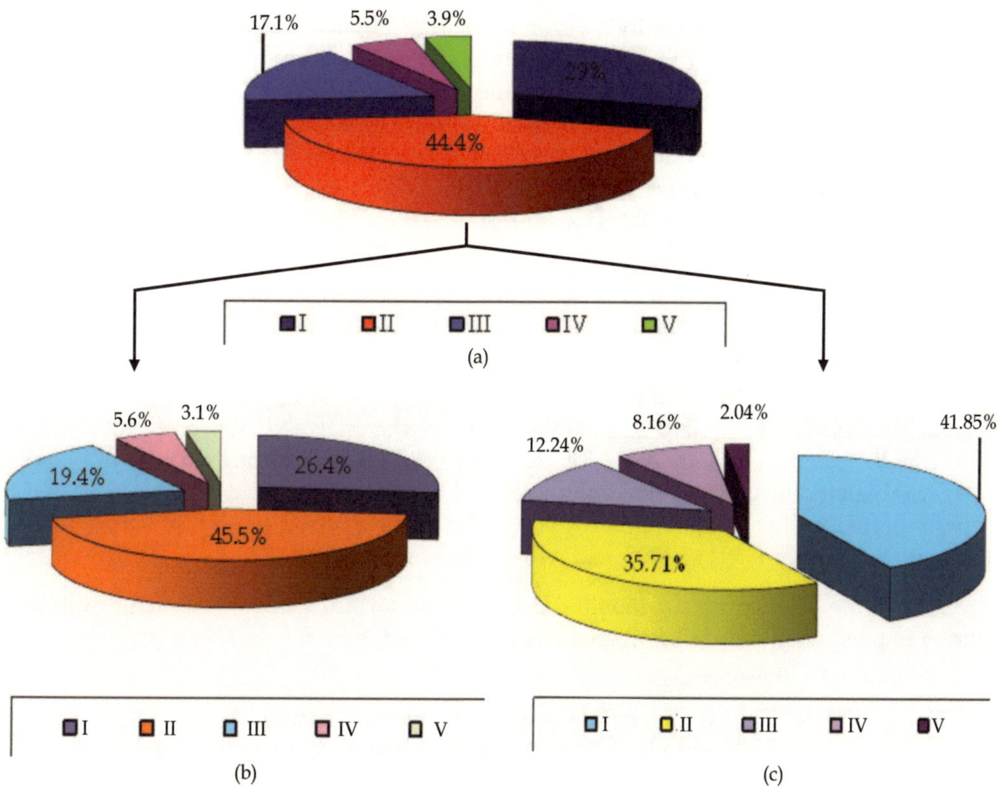

Fig. 4. (a) The surface water distribution by 5 qualities, in 2008 (b) The flowing water distribution by 5 qualities, in 2008. (c) The lake water distribution by 5 qualities, in 2008

compared with 2007 (with 3.00), and 2006 (with 1.10%) but smaller compared with 2005 (with 2.40%).

The water quality of the main lakes in Romania in relation to the water data analysis leads to the conclusion that out of the total of 102 (Fig. 4c), their water have different qualities, as follows: 41.85% Ist quality water, 35.71% IInd quality water, 12.24% IIIrd quality water, 8.16% IVth quality water, and 2.04% Vth quality water. The values concerning the Ist quality water, recorded in 2008, are bigger compared to 2006 (25.80%), but smaller compared with 2005 (47.10%), and 2007 (45.00%). Concerning the share of most polluted lakes (IVth and Vth quality water), their share was 10.20% in 2008, bigger compared to 2005 (2.00%), but smaller if comparison is made to 2006 and 2007, when the share of IVth and Vth quality water was reported as 15.80% and 15.00%, respectively.

In 2009, the data concerning the summary of the quality of surface waters (Fig. 5a) was the result of processing the raw data deriving from the physico-chemical analyses of water sampled from a similar number of monitoring flowing water sections as in 2008 (818 in 2009 compared to 817 in 2008), and 95 lakes. It was distributed as follows: 29.80% Ist quality water, 43.60% IInd quality water, 16.50% IIIrd quality water, 5.70% IVth quality water, and 4.30% Vth quality water (http://www.anpm.ro/Mediu/rapoarte, 2005, accessed 2011). The value of Ist quality water share was similar with 2008 (29.00%), and the same situation was also reported for IVth and Vth quality water.

The flowing water distribution by five qualities (Fig. 5b) was: 21.70% Ist quality water, 46.80% IInd quality water, 17.90% IIIrd quality water, 6.10% IVth quality water, and 3.50% Vth quality water. The share of the Ist quality flowing water in 2009 was the smallest reported for the period of 5 years we studied (29.80% in 2005, 25.80% in 2006, 23.40% in 2007, and 26.40% in 2008), while the share of 9.60% reported for IVth and Vth quality water was bigger compared to values reported in 2008 (8.70%) and 2005 (7.30%), but smaller compared to values reported in 2006 (15.80%), and 2007 (10.80%).

The water quality of the main lakes in Romania in relation to the water data analysis leads to the conclusion that out of the total of 95 (Fig. 5c), their water have different qualities, as follows: 34 had Ist quality water (35.80%), 42 had IInd quality water (44.21%), 14 had IIIrd quality water (14.70%), 4 had IVth quality water (4.20%), and 1 Vth quality water (1.10%). The share of the Ist quality lake water in 2009 (35.80%) was of bigger compared with 2006 (25.80%), but smaller compared to the other years of the analyzed period (47.10% in 2005, 45.00% in 2007, and 41.81% in 2008). The share of 5.30% reported for IVth and Vth quality water was bigger only compared to value reported in 2005 (2.00%), and smaller compared to values reported for the other analyzed years of the studied period (with 8.10% compared to 2006, with 9.70% compared to 2007, and with 4.90% compared to 2008).

If we study the distribution of monitored quality of surface waters in their assembly by entire 5 years studied period between years 2005 and 2009, by five qualities (Fig. 6) we find the evolution of its quality.

Thus, the biggest share of the Ist quality surface water was recorded in 2005 (31.40%), with similar values in 2008 (29.00%) and 2009 (29.80%), while the lowest value was recorded in 2007 (23.80%). For the IInd quality surface water there was reported the biggest share during entire studied period of five years, more than 40.00%. The biggest value was reported in 2005 (46.10%), with close values in 2007 (45.00%), and 2008 (44.40%), and smallest share in

2006 (41.50%). Concerning the IIIrd quality water, shares between 15.80% and 18.80% were reported during the studied monitoring period between 2005 and 2009. The biggest value of the IIIrd quality surface water was reported in 2007 (18.80%), while the smallest in 2005 (15.80%). Almost identical values were recorded in 2006 (17.20%), and 2008 (17.10%), while in 2009 a share of 16.50% was reported for the same surface water category.

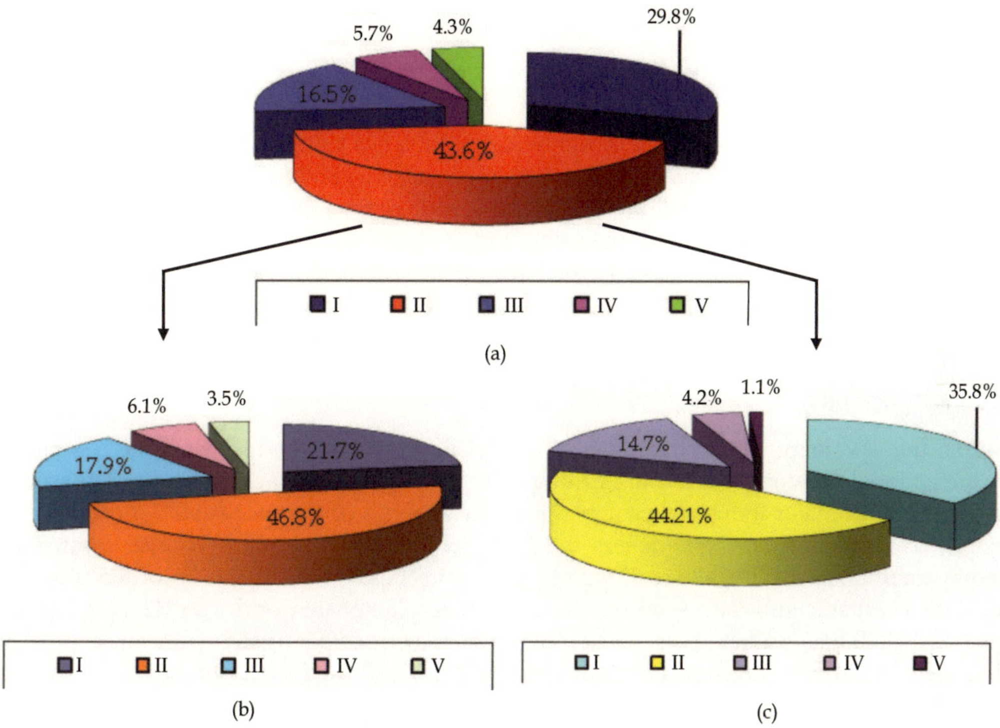

Fig. 5. (a) The surface water distribution by 5 qualities, in 2009 (b) The flowing water by 5 qualities, in 2009 (c) The lake water distribution by 5 qualities, in 2009

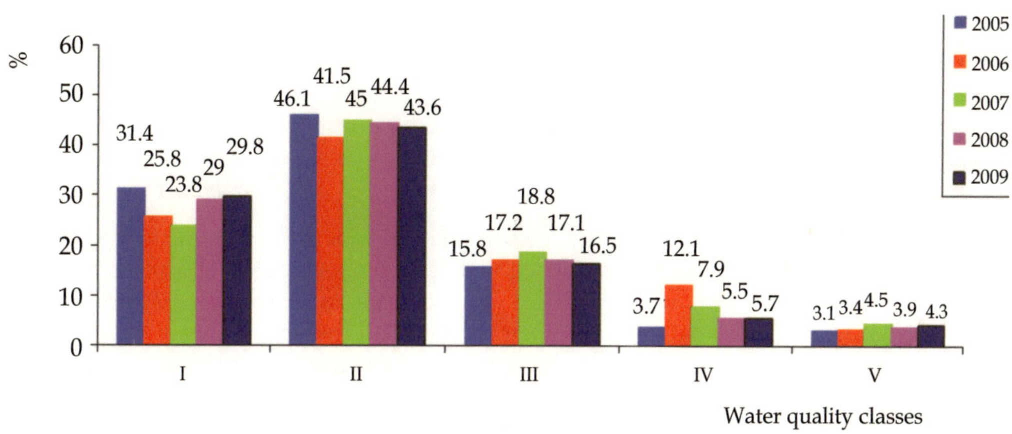

Fig. 6. The surface water distribution by 5 qualities, and by years, during the period of five years 2005 - 2009

The IVth quality surface water and Vth quality surface water represent the most polluted water category, with highest content in the main pollutants identified in the surface water (nitrogen compounds, chlorine, Fe, P, Fe, Mn, Cu, Cd, Zn, pesticides, oil products and detergents). The biggest values of the low quality surface water were recorded in 2006, 15.80 respectively, and the smallest in 2005, 6.80, respectively. The values of the IVth and Vth quality surface water reported in the other years of the studied period of 5 years (2005 - 2009) were: 12.40% in 2007, 9.40% in 2008, and 10.00% in 2009.

5. The quality of groundwater

It was emphasized that the natural groundwater regime has suffered over time, a number of quantitative and qualitative changes. These changes are due both to their use as a source of drinking water supply, execution of industrial and construction of water projects and hydrological improvements, and polluting factors (natural and antropic). Groundwater bodies are classified in two classes: good and poor, for both quantitative and chemical status. For the assessment of groundwater chemical status, the concentrations determined at the point of monitoring laid down in accordance with the water framework directive is compared with threshold values (TV) which are regarded as self-defence for good status of groundwater body.

In 2005 there were monitored a number of 1,947 drilling places, of which 1,664 are belonging to the national network and 283 are drillings performed with the aim of tracking pollution located around major industrial centres. From the analysis of the processed data consisting in physical and chemical parameters resulted from monitoring phreatic layer located in above mentioned drillings, most values over thresholds have been recorded for the indicators: organic substances, ammonium, nitrates, phosphates, and iron. Thus in 580 of analyzed drillings, values over thresholds were registered in organic substance; in 450 drillings, values over thresholds for nitrates, and in 85 drillings, for phosphates.

In 2006 there were monitored the same number of drilling places, 1,947 respectively, of which 1,664 are belonging to the national network and 283 are drillings performed with the aim of tracking pollution located around major industrial centres. From the analysis of the processed data consisting in physical and chemical parameters resulted from monitoring phreatic layer located in above mentioned drillings, most concerning have been recorded for the indicators: organic substances, ammonium, nitrates, phosphates, iron. Thus in 543 of analyzed drillings (30.00%) values over thresholds were registered in organic substance; in 437 drillings (22.40%) values over thresholds for nitrates, and in 78 drillings (4.00%) for phosphates. Compared with the previous year, in 2006 it was shown a trend of decrease of aquifers contamination with these substances, in overall.

In 2007, the drillings were monitored taking into consideration the new system for groundwater monitoring, implemented in 2006, which pursues closer and more concrete supervision in terms of water quality. In 2007 were monitored a number of 1,939 drillings, of which 1.687 are part of national hydro geological network (of which 28 are springs) and 252 are drillings performed in order to tracking pollution, located around major industrial centres. From the analysis of the processed data resulted as consequence of monitoring the physical and chemical parameters of the phreatic layer located in drillings, most values over the established thresholds concerned: organic substances, nitrogen, ammonium, phosphates, and iron.

In 2008, there have been monitored 1,899 drillings. Through the county public Health Offices, there were also monitored fountains, whose water is generally non-drinking, because of the overshoot recorded for ammonium, nitrates, and bacteriological indicators. These, fountains and are infested by infiltrations from non hydro - isolated sanitation groups, and from domestic waste and animal origin waste, originating in private households. From the analysis of the processed data resulted as consequence of monitoring the physical and chemical parameters of the phreatic layer located in drillings, most values over the established thresholds concerned: organic substances, nitrogen, ammonium, total hardness, chlorine, phosphates, and iron. Concerning the groundwater contamination with nitrates, overshoot of concentration has been recorded in 220 drillings, what represents 11.59% of total drillings monitored. Pollution is differentiated felt, existing areas in majority of river basin, in which, in the aquifer are found concentrations that lie far above the limit allowed, 50 mg/L. Another cause of unsatisfactory groundwater quality is the intense contamination of aquifers with ammonium and organic substances. Thus, in 466 of analysed drillings, values over threshold were recorded for the organic substance, and in 518 drillings, the ammonium indicator had values over admitted threshold.

In 2009, for the assessment of the quantitative status of bodies of groundwater it has been used the European Guide recommendations, prepared in the framework of the Common Implementation Strategy Framework, using the following criteria: hydric balance; connection with surface waters; influence on terrestrial ecosystems dependent on groundwater, intrusion of saline water or other intrusions. The good status of groundwater, involves a number of conditions set out in annex V of the Water Framework Directive (Directive 2000/60/EC). Additional conditions for chemical status and evaluation procedures are developed in the Groundwater Daughter Directive (Directive 2006/118/EC), transposed into national law by Governmental Decision no. 53/2009, for the approval of the national plan of groundwater protection against pollution and deterioration.

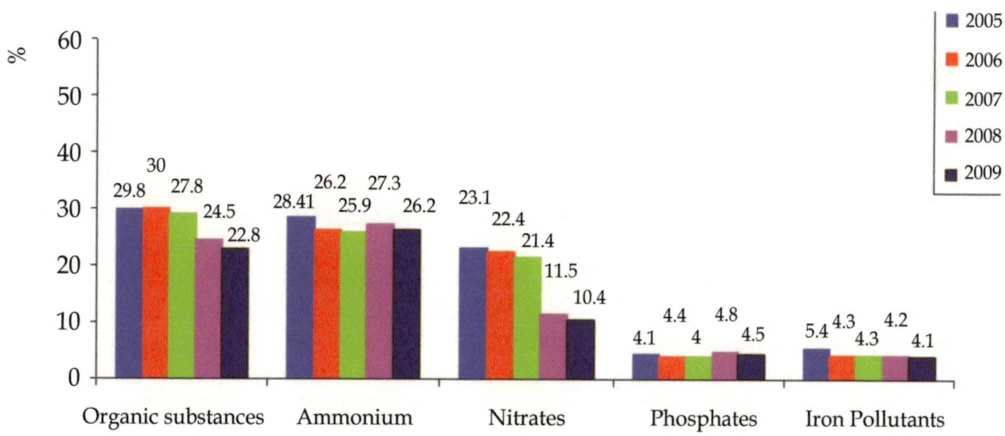

Fig. 7. The evolution of groundwater pollutants during studied five years period 2005 - 2009

Assessment of the status of bodies of groundwater was compiled on the basis of the comparison of chemical analysis carried out in 2009, with the threshold values (TV), values that have been determined for a number of 142 bodies of groundwater, from the 142 bodies established in Romania and which were published in the Order of the Ministry of Environment and Forests no. 137/2009. By applying the methodology and criteria for the assessment of groundwater bodies, the situation of those qualitative 142 bodies of groundwater shall be presented as follows: 109 bodies are in a good quality, 22 of groundwater bodies are in a poor quality, while 11 groundwater bodies are in a poor quality, locally.

6. The status of the wastewater

Insufficiently purified or even not at all purified wastewater discharge is one of the main causes of pollution and degradation of surface waters. Therefore, the main practical measures for the protection of the quality of surface waters is the purification of wastewater, which is collected and processed by a sewage system by sewage station, from where, as a rule, then are returned to the emissary.

The largest volumes of not purified and/or insufficiently purified wastewater resulted from units of the communal areas: households, heat and power plants, chemical processing, and then smaller volumes from units in the areas of trade and services for the population and from the extractive industry. Related to potential pollution, the highest share belongs to the fields of communal enterprises, and thermal electricity industry, extractive industry followed by businesses in the chemical industry, metallurgical, etc.

In 2005, the results of the monitoring activity performed on the main sources of wastewater, revealed the following realities: toward a total volume of the evacuated 3,886.126 million m3/year, 1,944.389 million m3/year, meaning 50.03%, is part of the waste water to be purified. Of the total volume of wastewater requiring treatment, 1,944.389 million m3/year, respectively, a volume of 351.400 million m3/year, (18.07%), was sufficiently purified (as scheduled). Otherwise 660.634 million m3/year, (about 34.9%), are not purified wastewater, and 848.482 million m3/year, about 44.9%, not completely purified wastewater. So, in 2006, about 79.8% of not or insufficiently purified wastewaters from the main sources of pollution, have reached the natural receptors, especially rivers. Compared with the total number of 1,035 investigated purification stations, installations, only 274 stations, representing 26.5%, have functioned properly, and the rest of 761 (73.5%) operated improperly.

The statistical analysis of the main sources of wastewater, according to the results of the monitoring activity carried out in 2006, revealed the following global issues: toward a total volume of the evacuated 3,586.126 million m3/year, 1,891.622 million m3/year, meaning 52,7%, is part of the waste water to be purified. Of the total volume of wastewater requiring treatment, 1,891.622 million m3/year, respectively, a volume of 382.506 million m3/year, (20.2%), was sufficiently purified (as scheduled). Otherwise 660.634 million m3/year, (about 34.9%), are not purified wastewater, and 848.482 million m3/year, about 44.9%, not completely purified wastewater. So, in 2006, about 79.8% of not or insufficiently purified wastewaters from the main sources of pollution, have reached the natural receptors, especially rivers. Compared with the total number of 1,035 investigated purification stations, only 274 stations, representing 26.5%, have functioned properly, and the rest of 761 (73.5%) operated improperly.

The statistical analysis of the main sources of wastewater, according to the results of the monitoring activity carried out in 2007, revealed the following global issues: toward a total volume of the evacuated 4,985..065 million m3/year, 2,210.285 million m3/year, meaning 44.30%, is part of the waste water to be 498.668.506 million m3/year, (22.60%), was sufficiently purified (as scheduled). Otherwise 791.320 million m3/year, (about 35.80%), are not purified wastewater, and 919.083 million m3/year, about 41.60%, not completely purified wastewater. So, in 2007, about 77.40% of not or insufficiently purified wastewaters from the main sources of pollution, have reached the natural receptors, especially rivers. Compared with the total number of 1,348 investigated purification stations, only 410 stations, representing 30.40%, have functioned properly. The remaining 938 stations (69.60%) operated improperly, because of not enough treatment capacity, or due to the operating and maintenance problems (advanced physical and moral wear inefficiency of the biological treatment phase concerning the insurance of needed oxygen, lack of investments for modernization, etc).

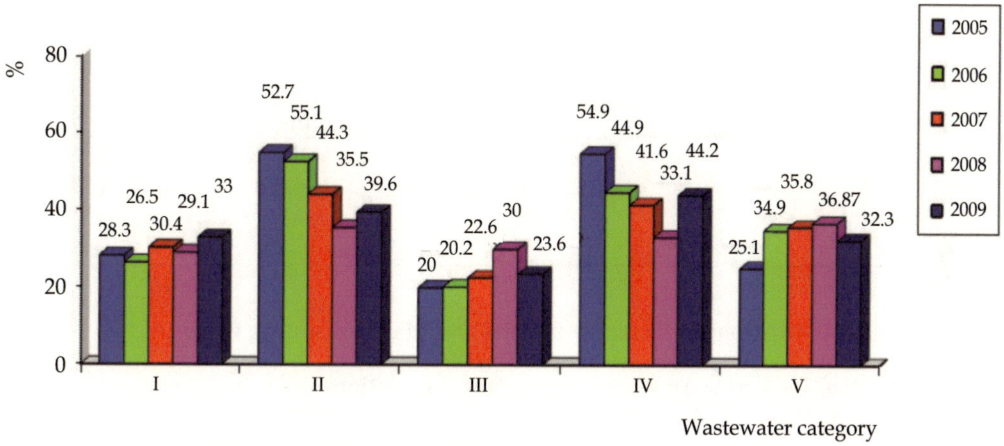

Fig. 8. The evolution of the wastewater categories during studied period of 5 years 2005 - 2009. I - water resulted from purification stations that functioned properly; II - wastewater to be purified; III - wastewater sufficiently purified (corresponding); IV - wastewater insufficiently purified; V - wastewater not purified

Compared with the total refused volume 5,254.565 million m3/year, 1,868.832 million m3/year (35.57% of the total) were wastewater to be purified. Of these, 560.623 million m3/year (30%) were sufficiently purified (corresponding), 689.145 million m3/year (36,87%) have been insufficiently purified wastewater and 619.064 million m3/year (33.13%) were not purified wastewater. Therefore, in 2008, a rate of 70% of wastewaters, not purified or insufficiently purified, from the main sources of pollution, has reached the natural receptors, especially rivers.

Statistical analysis of the situation of the main sources of waste water, according to the results of the monitoring carried out in 2009, revealed the global issues that are described below. Compared with the total volume of the refused water - 5,206.207 million m3/year, 2,058.899 millions m3/year (39.6% of the total) are wastewater to be purified. Of these, 485.438 million m3/year (23.60%) were sufficiently purified (corresponding), 909.019 millions m3/year (44.2%) were insufficiently purified wastewater and 664.442 million m3/year (32.2%) were not purified wastewater. Therefore, in 2009, 76.5% of not purified or insufficiently purified wastewaters from the main sources of pollution have reached the natural receptors, especially rivers. Compared with the total number of 1.363 investigated (urban and industrial) purification stations, only 445 stations, accounting for 33%, have functioned properly, and the remaining 9 stations, namely 67% operated improperly.

7. Correlations

In order to measure and establish the relationships between variables representing the main pollutants (nitrogen compounds, phosphates, Cl, S, Pb, Cd, Hg, As, oil products, pesticides) of the monitored groundwater, monitored surface water (Cu, Cd, Mg, and Zn), in Romania during analyzed five years period, the correlation coefficients were calculated.

7.1 Correlations between main pollutants of surface water

Very strong (0.858 - 0.921) and very significant correlations ($P < 0.001$) were identified, in 2005, between Cu, Cd, Mg, and Zn pollutants in surface water (Table 1), and according to the average value of the determination coefficient, 87.50% of our data conform to the linear relationship.

Issue	Cu (mg/L)	Cd (mg/L)	Mg (mg/L)	Zn (mg/L)
Cu (mg/L)	1.000	0.858***	0.892***	0.893***
Cd (mg/L)		1.000	0.867***	0.921***
Mg (mg/L)			1.000	0.889***
Zn (mg/L)				1.000

*$R^2 = 0.875$

Table 1. The correlation matrix *between metallic pollutants of surface water in 2005

Very strong (0.872 - 0.911) and very significant correlations ($P < 0.001$) were identified between Cu, Cd, Mg, and Zn pollutants in surface water (Table 2), and according to the average value of the determination coefficient, 89.10% of our data conform to the linear relationship.

Issue	Cu (mg/L)	Cd (mg/L)	Mg (mg/L)	Zn (mg/L)
Cu (mg/L)	1.000	0.872***	0.887***	0.897***
Cd (mg/L)		1.000	0.891***	0.911***
Mg (mg/L)			1.000	0.891***
Zn (mg/L)				1.000

*$R^2 = 0.891$

Table 2. The correlation matrix *between main pollutant components of surface water in 2006

Issue	Cu (mg/L)	Cd (mg/L)	Mg (mg/L)	Zn (mg/L)
Cu (mg/L)	1.000	0.858***	0.892***	0.863***
Cd (mg/L)		1.000	0.867***	0.921***
Mg (mg/L)			1.000	0.889***
Zn (mg/L)				1.000

*$R^2 = 0.882$

Table 3. The correlation matrix *between main pollutant components of surface water in 2007

In 2007, the relationship between Cu, Cd, Mg, and Zn pollutants in surface water (Table 3) was the same as those reported in previous years of the studied period (2005 - 2009). Between them was calculated very strong (0.858 - 0.921) and very significant correlations ($P < 0.001$) were identified between and according to the average value of the determination coefficient, 88.20% of our data conform to the linear relationship.

The correlation coefficients emphasized between Cu, Cd, Mg, and Zn pollutants in surface water by 2008 (Table 4), were positive very strong (0.861 - 0.918) and very significant ($P < 0.001$). The calculated average value of the determination coefficient justifies 88.60% the conformity of our data with the linear relationship.

Similarly with previous years of analyzed period of five years (2005 - 2009), the calculated correlations between the main metallic pollutants (Cu, Cd, Mg, and Zn) of the surface water in the last year of the studied interval of time, 2009, respectively (Table 5), revealed very strong (0.868 - 0.908) and very significant values ($P < 0.001$). The average value of the determination coefficient, 89.00% of our data conform to the linear relationship.

Issue	Cu (mg/L)	Cd (mg/L)	Mg (mg/L)	Zn (mg/L)
Cu (mg/L)	1.000	0.861***	0.899***	0.869***
Cd (mg/L)		1.000	0.875***	0.918***
Mg (mg/L)			1.000	0.895***
Zn (mg/L)				1.000

*$R^2 = 0.886$

Table 4. The correlation matrix* for main pollutant components of surface water data, in 2008

Issue	Cu (mg/L)	Cd (mg/L)	Mg (mg/L)	Zn (mg/L)
Cu (mg/L)	1.000	0.899***	0.898***	0.868***
Cd (mg/L)		1.000	0.893***	0.908***
Mg (mg/L)			1.000	0.875***
Zn (mg/L)				1.000

*$R^2 = 0.890$

Table 5. The correlation matrix*between main pollutant components of surface water in 2009

7.2 Correlations between main pollutants of groundwater

In groundwater, by 2005 (Table 6), very strong positive correlations, statistically very significant ($P < 0.001$) were identified between phosphates and nitrogen compounds (0.905; $R^2 = 0.819$), and also between Cl, S, Pb, Cd, Hg, As, oil products, pesticides (0.922 - 0.793), with an average $R^2 = 0.797$. Weak positive correlations, statistically not significant ($P > 0.05$) were identified between Cd, As, oil products, pesticides and phosphates, and also between Cl, S, Pb, Cd, Hg, As, oil products, pesticides and nitrogen compounds (Table 6). Average positive correlations were calculated between Cl and nitrogen compounds, statistically distinct significant ($P < 0.01$), and significant ($P < 0.05$) for S, Pb, Hg and nitrogen compounds, respectively (Table 6).

In 2006 (Table 7), too, very strong positive correlations, statistically very significant ($P < 0.001$) were identified between phosphates and nitrogen compounds identified in groundwater (0.912; $R^2 = 0.831$), and between Cl, S, Pb, Cd, Hg, As, oil products, pesticides (0.923 - 0.791), with an average $R^2 = 0.821$. Weak positive correlations, statistically not significant ($P > 0.05$) were identified between Cl, S, Pb, Cd, Hg, As, oil products, pesticides and phosphates (0.195 - 0.439), and also between S, Cd, Hg, As, oil products, pesticides and nitrogen compounds (0.354 - 0.498) while between Cl and nitrogen compounds was identified a positive (0.528) statistically significant moderate correlation ($P < 0.05$), and between Pb and nitrogen a statistically distinct significant ($P < 0.01$) moderate correlation coefficient of 0.583 (Table 7).

Issue	Nitrogen compounds	Phosphates	Cl	S	Pb	Cd	Hg	As	Oil products	Pesticides
Nitrogen compounds	1.000	0.905^{***}	0.654^{**}	0.543^{*}	0.553^{*}	0.412^{ns}	0.423^{*}	0.315^{ns}	0.413^{ns}	0.523^{ns}
Phosphates		1.000	0.250^{ns}	0.191^{ns}	0.278^{ns}	0.314^{ns}	0.322^{ns}	0.273^{ns}	0.389^{ns}	0.412^{ns}
Cl			1.000	0.793^{***}	0.812^{***}	0.845^{***}	0.889^{***}	0.878^{***}	0.909^{***}	0.913^{***}
S				1.000	0.832^{***}	0.891^{***}	0.799^{**}	0.803^{***}	0.903^{***}	0.886^{***}
Pb					1.000	0.872^{***}	0.923^{***}	0.925^{***}	0.904^{***}	0.796^{***}
Cd						1.000	0.893^{***}	0.835^{***}	0.902^{***}	0.798^{***}
Hg							1.000	0.887^{***}	0.908^{***}	0.825^{***}
As								1.000	0.895^{***}	0.805^{***}
Oil products									1.000	0.922^{***}
Pesticides										1.000

Table 6. The correlation matrix between main pollutant components of groundwater in 2005 (mg/L)

Issue	Nitrogen compounds	Phosphates	Cl	S	Pb	Cd	Hg	As	Oil products	Pesticides
Nitrogen compounds	1.000	0.912^{***}	0.528^{*}	0.432^{ns}	0.583^{**}	0.468^{ns}	0.381^{ns}	0.354^{ns}	0.462^{ns}	0.498^{ns}
Phosphates		1.000	0.195^{ns}	0.254^{ns}	0.259^{ns}	0.385^{ns}	0.364^{ns}	0.228^{ns}	0.315^{ns}	0.439^{ns}
Cl			1.000	0.791^{***}	0.812^{***}	0.845^{***}	0.889^{***}	0.878^{***}	0.909^{***}	0.913^{***}
S				1.000	0.832^{***}	0.891^{***}	0.799^{***}	0.803^{***}	0.913^{***}	0.886^{***}
Pb					1.000	0.872^{***}	0.923^{***}	0.915^{***}	0.914^{***}	0.796^{***}
Cd						1.000	0.893^{***}	0.835^{***}	0.912^{***}	0.798^{***}
Hg							1.000	0.887^{***}	0.918^{***}	0.825^{***}
As								1.000	0.895^{***}	0.805^{***}
Oil products									1.000	0.921^{***}
Pesticides										1.000

Table 7. The correlation matrix between main pollutant components of groundwater in 2006 (mg/L)

In the same year, 2007, in groundwater (Table 8), very strong positive correlations, statistically very significant ($P < 0.001$) were identified between phosphates and nitrogen compounds (0.915; R^2 = 0.837), and also between Cl, S, Pb, Cd, Hg, As, oil products, pesticides (0.920 - 0.798), with an average $R^2 = 0.803$. Weak positive correlations, statistically not significant ($P > 0.05$) were identified between Cl, S, Pb, Cd, Hg, As, oil products, pesticides and phosphates (0.182 - 0.422), and also between Cd, Hg, As, oil products and nitrogen compounds (0.322 - 0.457) while between S, Pb and nitrogen compounds were identified positive (0.525 and 0.541, respectively) statistically significant moderate correlation ($P < 0.05$), and between Cl and nitrogen a statistically distinct significant ($P < 0.01$) moderate correlation coefficient of 0.633 (Table 8).

Issue	Nitrogen compounds	Phosphates	Cl	S	Pb	Cd	Hg	As	Oil products	Pesticides
Nitrogen compounds	1.000	0.915***	0.633**	0.525*	0.541*	0.432 ns	0.457 ns	0.322 ns	0.405 ns	0.521*
Phosphates		1.000	0.214 ns	0.182 ns	0.233 ns	0.366 ns	0.322 ns	0.281 ns	0.383 ns	0.422 ns
Cl			1.000	0.799***	0.868***	0.858***	0.889***	0.894***	0.905***	0.915***
S				1.000	0.829***	0.883***	0.799***	0.812***	0.920***	0.899***
Pb					1.000	0.865***	0.913***	0.921***	0.912***	0.798***
Cd						1.000	0.893***	0.837***	0.921***	0.865**
Hg							1.000	0.897***	0.909***	0.828***
As								1.000	0.896***	0.831***
Oil products									1.000	0.914***
Pesticides										1.000

Table 8. The correlation matrix between main pollutant components of groundwater in 2007 (mg/L)

Issue	Nitrogen compounds	Phosphates	Cl	S	Pb	Cd	Hg	As	Oil products	Pesticides
Nitrogen compounds	1.000	0.925***	0.654**	0.543*	0.553*	0.412 ns	0.423 ns	0.315 ns	0.413 ns	0.523*
Phosphates		1.000	0.250 ns	0.191 ns	0.278 ns	0.314 ns	0.322 ns	0.273 ns	0.389 ns	0.412 ns
Cl			1.000	0.786***	0.812***	0.845***	0.889***	0.878***	0.909***	0.913***
S				1.000	0.832***	0.891***	0.799***	0.803***	0.920***	0.886***
Pb					1.000	0.872***	0.921***	0.975***	0.904***	0.796***
Cd						1.000	0.893***	0.835***	0.912***	0.798***
Hg							1.000	0.887***	0.918***	0.825***
As								1.000	0.895***	0.805***
Oil products									1.000	0.912***
Pesticides										1.000

Table 9. The correlation matrix between main pollutant components of groundwater in 2008 (mg/L)

In groundwater, by 2008 (Table 9), very strong positive correlations, and statistically very significant ($P < 0.001$) were identified between phosphates and nitrogen compounds (0.925; $R^2 = 0.856$), and also between Cl, S, Pb, Cd, Hg, As, oil products, pesticides (0.921 - 0.786), with an average $R^2 = 0.785$. Weak positive correlations, statistically not significant ($P > 0.05$) were identified between Cl, S, Pb, Cd, Hg, As, oil products, pesticides and phosphates (0.191 - 0.412), and also between Cd, Hg, As, oil products and nitrogen compounds, with values within the interval 0.315 - 0.423 (Table 9). Average positive correlations were calculated between Cl and nitrogen compounds (0.654), statistically distinct significant ($P < 0.01$). Between S, Pb, pesticides and nitrogen compounds were identified average correlation coefficients statistically significant ($P < 0.05$), with values within the interval 0.523 - 0.553 (Table 9).

Issue	Nitrogen compounds	Phosphates	Cl	S	Pb	Cd	Hg	As	Oil products	Pesticides
Nitrogen compounds	1.000	0.917***	0.629**	0.558*	0.541*	0.426ns	0.453ns	0.298ns	0.432ns	0.511*
Phosphates		1.000	0.217ns	0.185ns	0.238ns	0.325ns	0.388ns	0.235ns	0.325ns	0.461ns
Cl			1.000	0.795***	0.856***	0.865***	0.845***	0.835***	0.917***	0.920***
S				1.000	0.825***	0.887***	0.811***	0.822***	0.911***	0.921***
Pb					1.000	0.869***	0.901***	0.905***	0.908***	0.799***
Cd						1.000	0.900***	0.847***	0.920***	0.816***
Hg							1.000	0.891***	0.915***	0.836***
As								1.000	0.887***	0.827***
Oil products									1.000	0.921***
Pesticides										1.000

Table 10. The correlation matrix between main pollutant components of groundwater in 2009 (mg/L)

The correlation coefficients calculated in 2009 for the groundwater pollutants (Table 10), revealed very strong positive correlations, statistically very significant ($P < 0.001$) between phosphates and nitrogen compounds (0.917; $R^2 = 0.841$), and also between Cl, S, Pb, Cd, Hg, As, oil products, pesticides (0.921 - 0.795), with an average $R^2 = 0.783$. Weak positive correlations, statistically not significant ($P > 0.05$) were identified between Cl, S, Pb, Cd, Hg, As, oil products, pesticides and phosphates (0.185 - 0.461), and also between Cd, Hg, As, oil products and nitrogen compounds (0.298 - 0.453), while between Cl and nitrogen compounds, statistically distinct significant ($P < 0.01$) correlation was found (0.629), and average positive correlations (0.558, and 0.629 respectively), statistically significant ($P < 0.05$), between S, Pb and nitrogen compounds (Table 9).

8. The multiregression analyze

Correlations were also calculated between main pollutants of: surface water and ground water, surface water and wastewater, ground water and wastewater. Their values (0.695 ÷ 0,986) emphasize the strong interdependence between pollutants matrices. The correlations calculated between the main pollutants identified in national water supply during a five years period, 2005 - 2009, deliver a complex picture of the water quality in Romania. Surface water quality is most affected by the discharge of untreated or inadequately treated sewage. In this context, a key measure to protect surface water quality is to increase wastewater treatment, upgrading and improving the cleaning process (tables 11 - 15).

Issue	Groundwater												
	I	II	III	IV	V	VI	VII	VIII	IX	X	XI	XII	XIII
							Surface water						
I	0.983												
II		0.857											
III			0.899										
IV				0.856									
V					0.932								
VI						0.986							
VII							0.956						
VIII								0.973					
IX									0.956				
X										0.888			
XI											0.867		
XII												0.829	
XIII													0.863
							Waste water						
I	0.735												
II		0.699											
III			0.695										
IV				0.803									
V					0.788								
VI						0.695							
VII							0.743						
VIII								0.739					
IX									0.788				
X										0.803			
XI											0.768		
XII												0.731	
XIII													0.744

I – Nitrogen compounds; II – Phosphates; III – Cu; IV – Cd; V - Mn; VI – Zn; VII – Cl; VIII – S; IX – Pb; X – Hg; XI – As; XII - oil products; XIII - pesticides

Table 11. The correlation matrix between main pollutant components of surface water, groundwater and wastewater in 2005

Issue	Groundwater												
	I	II	III	IV	V	VI	VII	VIII	IX	X	XI	XII	XIII
Surface water													
I	0.977												
II		0.869											
III			0.903										
IV				0.835									
V					0.955								
VI						0.932							
VII							0.947						
VIII								0.961					
IX									0.924				
X										0.835			
XI											0.858		
XII												0.845	
XIII													0.821
Waste water													
I	0.733												
II		0.701											
III			0.733										
IV				0.783									
V					0.781								
VI						0.791							
VII							0.725						
VIII								0.767					
IX									0.731				
X										0.784			
XI											0.772		
XII												0.739	
XIII													0.751

I – Nitrogen compounds; II – Phosphates; III – Cu; IV – Cd; V - Mn; VI – Zn; VII – Cl; VIII – S; IX – Pb; X – Hg; XI – As; XII - oil products; XIII - pesticides

Table 12. The correlation matrix between main pollutant components of surface water, groundwater and wastewater in 2006

Issue	Groundwater												
	I	II	III	IV	V	VI	VII	VIII	IX	X	XI	XII	XIII
	Surface water												
I	0.958												
II		0.842											
III			0.917										
IV				0.851									
V					0.927								
VI						0.968							
VII							0.942						
VIII								0.959					
IX									0.963				
X										0.895			
XI											0.882		
XII												0.839	
XIII													0.884
	Waste water												
I	0.732												
II		0.711											
III			0.699										
IV				0.811									
V					0.758								
VI						0.698							
VII							0.699						
VIII								0.759					
IX									0.775				
X										0.783			
XI											0.772		
XII												0.748	
XIII													0.757

I – Nitrogen compounds; II – Phosphates; III – Cu; IV – Cd; V - Mn; VI – Zn; VII – Cl; VIII – S; IX – Pb; X – Hg; XI – As; XII - oil products; XIII - pesticides

Table 13. The correlation matrix between main pollutant components of surface water, groundwater and wastewater in 2007

Issue	Groundwater												
	I	II	III	IV	V	VI	VII	VIII	IX	X	XI	XII	XIII
Surface water													
I	0.972												
II		0.861											
III			0.902										
IV				0.931									
V					0.912								
VI						0.973							
VII							0.971						
VIII								0.986					
IX									0.942				
X										0.898			
XI											0.875		
XII												0.837	
XIII													0.898
Waste water													
I	0.729												
II		0.701											
III			0.712										
IV				0.721									
V					0.753								
VI						0.763							
VII							0.791						
VIII								0.73925					
IX									0.749				
X										0.769			
XI											0.734		
XII												0.758	
XIII													0.783

I – Nitrogen compounds; II – Phosphates; III – Cu; IV – Cd; V - Mn; VI – Zn; VII – Cl; VIII – S; IX – Pb; X – Hg; XI – As; XII - oil products; XIII - pesticides

Table 14. The correlation matrix between main pollutant components of surface water, groundwater and wastewater in 2008

Issue	Groundwater												
	I	II	III	IV	V	VI	VII	VIII	IX	X	XI	XII	XIII
							Surface water						
I	0.963												
II		0.861											
III			0.908										
IV				0.831									
V					0.947								
VI						0.985							
VII							0.955						
VIII								0.928					
IX									0.943				
X										0.896			
XI											0.875		
XII												0.901	
XIII													0.859
							Waste water						
I	0.729												
II		0.697											
III			0.705										
IV				0.792									
V					0.773								
VI						0.714							
VII							0.751						
VIII								0.749					
IX									0.793				
X										0.759			
XI											0.739		
XII												0.749	
XIII													0.763

I - Nitrogen compounds; II - Phosphates; III - Cu; IV - Cd; V - Mn; VI - Zn; VII - Cl; VIII - S; IX - Pb; X - Hg; XI - As; XII - oil products; XIII - pesticides

Table 15. The correlation matrix between main pollutant components of surface water, groundwater and wastewater in 2009

The results of the multiregression analyze applied to water quality prediction in the future, show that only 0.07% of the original variability cannot be explained when dependent variable was represented by the nitrogen compounds, 0.05% when metals represented the dependent variable, 0.09% when phosphorus and/or phosphates were the dependent variable, and 0.11% when pesticides are dependent variable. This emphasizes the accuracy of this prediction model for explaining the approached water pollutants evolution.

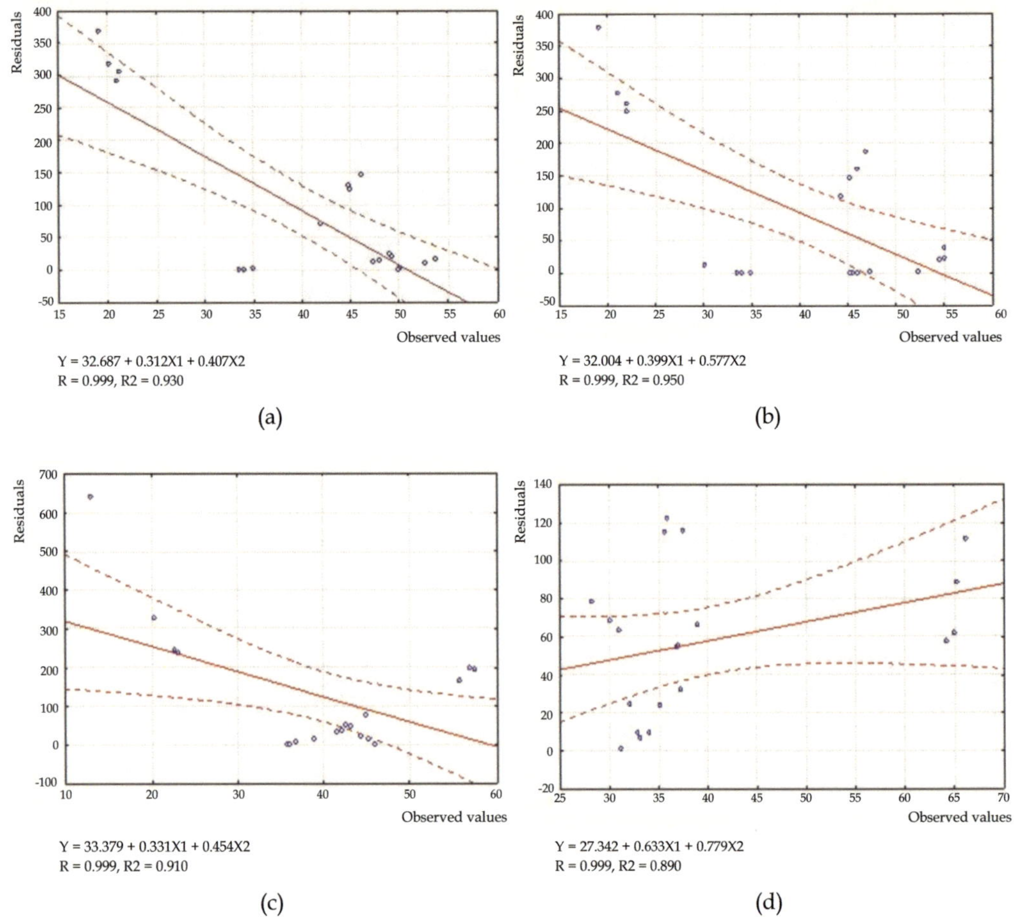

Fig. 9. Scatter diagram depicting the multiple correlation analysis between variables. (a) correlation between nitrogen compounds and Cl, S, Pb, Cd, Hg, As, oil products, pesticides, phosphates (or P); (b) correlation between Pb, Cd, Hg, As, and Cl, S, oil products, pesticides, nitrogen compounds, phosphates (or P); (c) correlation between phosphates (or P) and nitrogen compounds, Cl, S, Pb, Cd, Hg, As, oil products, pesticides; (d) correlation between pesticides nitrogen compounds, Cl, S, Pb, Cd, Hg, As, oil products, phosphates (or P)

9. Conclusion

The main water pollutants identified during the analyzed time interval and all water categories (surface water, groundwater and wastewater) were: nitrogen compounds, Fe, P,

Fe, Mn, Cu, Cd, Zn, pesticides and detergents. The surface waters are mainly contamined with metals (Cu, Cd, Mn, Zn), and contamination ranges between 15 - 22% of analyzed water bodies. The groundwater pollution frames within 20 - 25% of total ground water and the main pollutants are: NO_2, NH_4, P, PO_4^{2-}, Cl, S, Pb, Cd, Hg, As, oil products, pesticides. 70.2 - 76.5% of total analyzed wastewater resulted from the main sources of pollution, have reached the natural receptors, especially rivers, not cleaned or insufficiently purified. Strong correlation (0.925) was identified between phosphates and nitrogen compounds in groundwaters. The same trait (strong correlation) can be attributed to the correlations between Cu, Cd, Mn, Zn (0.858 ÷ 0.921) in surface waters and NO_2, NH_4, Cl, S, Pb, Cd, Hg, As, oil products, pesticides (0.793 ÷ 0.921) in groundwater. Correlations were also calculated between main pollutants of: surface water and ground water, surface water and wastewater, ground water and wastewater. Their values (0.695 ÷ 0,986) emphasize the strong interdependence between pollutants matrices.

Major issues that should be addressed in future research include the ability to simulate regional water quality and its sensitivity to social and economical realities. Research needs to be undertaken on the role of environmental management, particularly in view of increase the potential for diminishing the wastewater content in harmful pollutants, and extension of cleaning process of these waters.

10. References

Blenckner, T. (2005). A conceptual model of climate-related effects on lake ecosystems. *Hydrobiologia,* Vol. 533 No. 1-3 (January 2005) pp. 1 - 14. ISSN: 0018-8158 (print version), ISSN: 1573-5117 (electronic version)

Bulut, V. N.; Bayram, A.; Gundogdu, A.; Soylak, M.; Tufekci, M. (2010). Assessment of water quality parameters in the stream Galyan, Trabzon, Turkey. *Environmental Monitoring & Assessment,* Vol. 165 No. 1 - 4 (June 2010) pp. 1 - 13, ISSN: 0167-6369 (print version), ISSN: 1573-2959 (electronic version)

Dodds, W.K. 2002. *Freshwater Ecology. Concepts and Environmental Applications.* Academic Press, ISBN 0-12-219135-8, USA

Goyal, M., Dhar, D.N., Singh, S. K. (2010). Physicochemical methods for monitoring seasonal variations in Cl−, F−, and Fe++ in underground water in Unnao District in Uttar Pradesh (India). *Environmental Monitoring & Assessment,* Vol. 171 No. 1-4 (December 2010) pp. 425 - 427, ISSN: 0167-6369 (print version), ISSN: 1573-2959 (electronic version)

Kazi, T. G.; Arain, M. B.; Jamali, M. K.; Jalbani, N.; Afridi, H. I.; Sarfraz, R. A. (2009). Assessment of water quality of polluted lake using multivariate statistical techniques: A case study. *Ecotoxicology and Environmental Safety,* Vol. 72 No. 2 (February 2009) pp. 301 - 309, ISSN: 0147-6513

Hirsch, R.M.; Hamilton, P.A.; Miller, T.L. (2006). U.S.Geological Survey perspective on water-quality monitoring and assessment. *Journal of Environmental Monitoring* Vol. 8 No. 5 (May 2006) pp. 512 - 518, ISSN (printed): 1464-0325, ISSN (electronic): 1464-0333

Meybeck, M. (2004). The global change of continental aquatic systems: dominant impacts of human activities. *Water Science and Technology,* Vol. 49 No. 7 (August 2004) pp. 73 - 83, ISSN: 0273-1223

Mihaiescu, T.; Mihaiescu, R.; Muntean, E.; Muntean, N. (2010). Study Regarding Hydrochemical Classification of the main Lakes from Fizeş Watershed (Romania), *ProEnvironment* Vol. 3 No. 5 (June 2010) pp. 72 - 87, ISSN: 1844-6698 (print version), ISSN: 2066-1363 (electronic version)

Odagiu, A. (2010). Municipality of Cluj-Napoca - The Quality of Wastewaters. Note 1. Monitoring Disolved Oxygen, *ProEnvironment* Vol. 3 No. 5 (June 2010) pp. 78 - 83, ISSN: 1844-6698 (print version), ISSN: 2066-1363 (electronic version)

Odagiu, A.; Oroian, I.; Mihăiescu T.; Şotropa A.; Petrescu-Mag,I.V.; Burduhos P.; Balint, C. (2010). Municipality of Cluj-Napoca - The Quality of Wastewaters. Note 2. Monitoring Turbidity, *ProEnvironment* Vol. 3 No. 5 (June 2010) pp. 84 - 88, ISSN: 1844-6698 (print version), ISSN: 2066-1363 (electronic version)

Oroian, I.Gh., Petrescu-Mag, M.R. (2011). *Environmental Law and Legislation*, Bioflux, ISBN 978-606-8191-04-1, Cluj-Napoca, Romania [in Romanian]

Papaioannou, A., Athina Mavridou, C. Hadjichristodoulou, P. Papastergiou, Olga Papp, Eleni Dovriki, I. Rigas (2010). Application of multivariate statistical methods for groundwater physicochemical and biological quality assessment in the context of public health. *Environmental Monitoring & Assessment*, Vol. 170 No. 1 - 4 (November 2010) pp. 87 – 97, ISSN: 0167-6369 (print version), ISSN: 1573-2959 (electronic version)

Shirodkar, P. V., Mesquita, A., Pradhan, U. K., Verlekar, X. N., Buba, M. T., & Vethamony, P. (2009). Factors controlling physicochemical characteristics in the coastal waters off Mangalore - A multivariate approach. *Environmental Research*, Vol. 109 No. 2 (July 2009) pp. 245 – 257, ISSN: 0013-9351

Sojka, M., Siepak, M., Ziola, A., Frankowski, M., Murat-Blazejewska, S., Siepak, J. (2008). Application of multivariate statistical techniques to evaluation of water quality in the Mala Welna River (Western Poland). *Environmental Monitoring & Assessment*, Vol. 147 No. 1-3 (December 2008) pp. 159 – 170, ISSN: 0167-6369 (print version), ISSN: 1573-2959 (electronic version)

Wake, H. (2005). Oil refineries: a review of their ecological impacts on the aquatic environment. *Estuarine, Costal and Shelf Science* Vol. 62 No. 4 (March 2005) pp. 131 – 140, ISSN: 0272-7714

Environmental Report of the National Romanian Agency of Environmental Protection by the year 2005, http://www.anpm.ro/Mediu/rapoarte [in Romanian]

Environmental Report of the National Romanian Agency of Environmental Protection by the year 2006, http://www.anpm.ro/Mediu/rapoarte [in Romanian]

Environmental Report of the National Romanian Agency of Environmental Protection by the year 2007, http://www.anpm.ro/Mediu/rapoarte [in Romanian]

Environmental Report of the National Romanian Agency of Environmental Protection by the year 2008, http://www.anpm.ro/Mediu/rapoarte [in Romanian]

Environmental Report of the National Romanian Agency of Environmental Protection by the year 2009, http://www.anpm.ro/Mediu/rapoarte [in Romanian]

7

Ecological Tools for the Management of Cyanobacteria Blooms in the Guadiana River Watershed, Southwest Iberia

Helena M. Galvão et al.*
Center for Marine and Environmental Research (CIMA), Universidade do Algarve, Gambelas Campus, Faro, Portugal

1. Introduction

Strong water demand for irrigation, energy and drinking water production is responsible for an increasingly regulation of freshwater flow patterns and watersheds. In this context, the construction of dams allows water storage but seriously restricts freshwater flow downstream. Due to scarcity of freshwater resources, reservoir water management often promotes high hydraulic residence. This may cause strong impacts on biological components of aquatic ecosystems, influencing the development of cyanobacteria blooms and aggravating their harmful impacts.

Aquatic cyanobacteria, a group of relatively slow growing photosynthetic organisms, are stimulated by high water residence times as well as increased temperatures and low N : P ratios, conditions that usually limit the growth of other competing phytoplankton groups (Carmichael et al., 1996; Chorus & Bartram, 1999; Kawara et al., 1998; Kononen et al., 1998; Paerl, 2008). Cyanobacteria blooms have been repeatedly associated with eutrophication processes (Berg et al., 1987; Carmichael et al., 1988; Codd, 2000; Chorus, 2005; Druvietis, 1997; Pinckney et al., 1998), but they might also dominate under oligotrophic conditions (Galvão et al., 2008; Havens et al., 2003; Mez et al., 1997; Sivonen & Jones, 1999).

Cyanobacteria blooms management became an emergent priority as a result of worldwide surveys of aquatic ecosystems affected by massive cyanobacteria blooms and their serious health and ecosystem risks (Blaha et al., 2009). Indeed, cyanobacteria are able to produce a wide range of secondary metabolites which are toxic to humans and wildlife, generally referred as cyanotoxins. From a toxicological perspective, cyanotoxins are classified as

* Margarida P. Reis[1], Rita B. Domingues[1], Sandra M. Caetano[1], Sandra Mesquita[1], Ana B. Barbosa[1], Cristina Costa[1], Carlos Vilchez[3] and Margarida Ribau Teixeira[2]

[1]*Center for Marine and Environmental Research (CIMA), Universidade do Algarve, Gambelas Campus, Faro, Portugal*

[2]*Center for Environmental and Sustainability Research (CENSE), Universidade do Algarve, Gambelas Campus, Faro, Portugal*

[3]*International Center for Environmental Research (CIECEM), University of Huelva, Huelva, Spain*

hepatotoxins, neurotoxins, cytotoxins, and dermatotoxins (Wiegand & Pflugmacher, 2005). According to their chemical structure, these toxic compounds are peptides, heterocyclic (alkaloids) or lipidic compounds (Sivonen & Jones, 1999). Effects of toxins on humans can be triggered mainly by direct skin contact or consumption of contaminated water. Furthermore other potential routes of exposure have been documented including aerosol inhalation, contaminated food ingestion and dialysis (Chorus & Bartram, 1999; Dunn, 1996; Jochimsen et al., 1998; Pouria et al., 1998). Additional problems related to cyanobacteria bloom episodes in raw water sources used for drinking water production include noxious effects such as bad taste and odour, due to the presence of geosmin and 2-methylisoborneol (Jähnichen et al., 2011). Cyanobacteria may also produce a wide range of currently unknown toxins with great toxicological significance (Blaha et al., 2009). Thus, cyanobacteria blooms constitute a key concern for drinking water production, and are also relevant for establishing water quality management policies (e.g., Water Frame Directive, WFD; Directive 2000/60/CE of 23 October 2000).

Phytoplankton is recognized as an essential biological element in monitoring programs used to define the ecological quality and health of aquatic environments. In the scope of the WFD, phytoplankton is used to classify trophic state of aquatic ecosystems (Domaizon et al., 2003), as well as to determine the effectiveness of management, restoration programs and environmental legislation (Brierley, et al., 2007). Phytoplankton biomass and composition, along with trophic state indices (TSI) and physical-chemical variables, are essential to establish freshwater ecological status (Carlson, 1977; Reynolds et al., 2002).

The need for translating complex biological information into Multimetric Indicators of Ecological Condition, required by water managers, has led to the development and testing of multiple ecological indices. According to the Evaluation Guidelines adopted by the United States, (Jackson et al., 2000), selected ecological indicators used for ecological classification should: (1) be easily obtained through standardized well-documented methods; (2) provide relevant information in terms of specific management concerns; (3) allow for temporal and spatial variability, without losing discriminant capacity, and (4) maintain reliability. Despite the great effort put into sampling and analytical methods standardization, we consider that indices recently adopted to evaluate ecological status of surface waters are still far from complying with all these criteria.

The Guadiana River watershed (Fig. 1) is the fourth largest river basin in the Iberian Peninsula (67480 km^2), and is located in a semi-arid region with a Mediterranean climate. Annual precipitation averages ca. 500 mm, and the hydrographic regime is torrential, with concentrated rainy periods and a prolonged dry season, usually from May to September. The Mediterranean climate irregularity is also expressed in strong interannual variability, with intense rainy years alternating with years of extended droughts (Daveau, 1987). Managing water availability under such demanding conditions lead to the construction of hundreds of dams, from which almost 90 have a volume capacity over 1 hm^3. Reservoir water management strategies are strongly limited by increasing water demands for irrigation and drinking water production, causing severe restriction of freshwater flow. Recent construction of the large Alqueva dam further increased flow regulation.

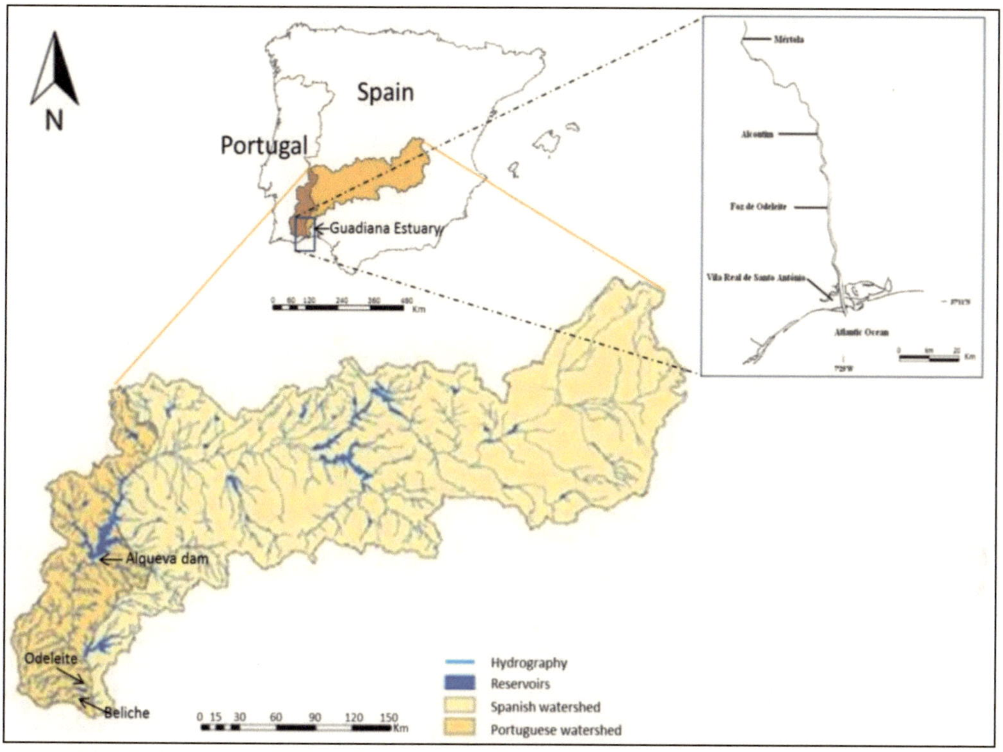

Fig. 1. Guadiana River water basin and location of study sites: Guadiana estuary and adjacent freshwater reservoirs (Alqueva, Odeleite and Beliche)

The main goal of this work is to evaluate recent water management strategies adopted for the Guadiana watershed, comparing different criteria used to classify ecological status and potential. The study is based on long-term ecological data series, and addresses two distinct case studies: (i) the Guadiana estuary (1997-2009); and (ii) adjacent Beliche and Odeleite freshwater reservoirs (2003-2010).

2. Guadiana estuary study

2.1 Study area, sampling strategy and methods

The Guadiana estuary is a mesotidal system (tidal amplitude: 1.3-3.5 m), with a length of 70 km, a maximum width of ca. 550 m, an average depth of 6.5 m, and an average water residence time of 12 days (Domingues & Galvão, 2007; Vasconcelos et al., 2007). The upstream saltwater intrusion is usually located close to Alcoutim (ca. 38 km from river mouth), whereas tidal influence extends to Mértola (ca. 70 km from river mouth; see Fig. 1). The lower estuary ranges from partially stratified to well-mixed, whereas the upper estuary is generally well-mixed (Cravo et al., 2006; Morais et al., 2009; Oliveira et al., 2006; Rocha et al. 2002). A series of dams has severely restricted its freshwater flow (ca. 75 %), and the recent construction of the large Alqueva dam (ca. 150 km upstream from river mouth) increased flow regulation up to 81% of the total catchment area (55 000 km^2) starting in 2003 (Galvão et al., 2008). Since human activity in the Guadiana watershed is mostly agriculture and the main anthropic pressure is associated to dams, the Guadiana is considered one of the best conserved but also most vulnerable estuaries of the Iberian Peninsula (Vasconcelos et al. 2007).

At different stations (see Fig. 1), vertical profiles of water temperature and salinity were determined in situ using a YSI 556 MPS probe. Vertical profiles of photosynthetically active radiation (PAR) intensity were determined using a LI-COR radiometer. Light extinction coefficient (k_e, m^{-1}) was calculated using an exponential function (eq. 1), where I_z represents the light intensity at depth Z (m) and I_0 is the light intensity at the surface:

$$I_z = I_0 e^{-KeZ} \quad (1)$$

Subsurface water samples (ca. 0.5 m) were collected at different sampling stations (Alcoutim and Mértola) for determination of dissolved inorganic nutrients and phytoplankton variables. For nutrient concentration, samples were immediately filtered through cellulose acetate filters (Whatman, nominal pore diameter 0.2 µm) to acid-cleaned vials. Ammonium (NH_4^+), phosphate (PO_4^{3-}) and silicate (DSi) were determined upon arrival to the laboratory, while samples for nitrate (NO_3^-) where frozen (-20°C) until analysis. All nutrient analyses were made in triplicate, according to the spectrophotometric methods described by Grasshoff et al. (1983), using a spectrophotometer Hitachi U-2000 for ammonium, phosphate and silicate, and an autoanalyzer Skalar for nitrate and nitrite.

Chlorophyll *a* concentration was determined spectrophotometrically using glass fiber filters (Whatman GF/F, nominal pore diameter = 0.7 µm). Chlorophyll *a* was extracted overnight at 4°C with 90% acetone; after centrifugation, absorbance of the supernatant was measured in the spectrophotometer Hitachi U-2000 at 750 and 665 nm, before and after addition of HCl 1 M (Parsons et al., 1984).

Phytoplankton composition (including cyanobacteria), abundance and biomass were determined using epifluorescence (Haas, 1982) and inverted microscopy (Utermöhl, 1958). Samples for enumeration of pico- (<2 µm) and nanophytoplankton (2 - 20 µm) were preserved with glutaraldehyde (final concentration 2%) immediately after collection, stained with proflavine and filtered (1 - 5 mL, depending on the amount of suspended matter) onto black polycarbonate membrane filters (Whatman, nominal pore diameter 0.45 µm). Preparations were made within 24 h of sampling using glass slides and non-fluorescent immersion oil (Cargille type A), and then frozen (-20°C) in dark conditions, to minimize loss of autofluorescence. Enumeration was made at 787.5x magnification using an epifluorescence microscope (Leica DM LB). Samples for enumeration of microphytoplankton (>20 µm) were preserved with acid Lugol's solution (final concentration ca. 0.003%) immediately after collection, settled in sedimentation chambers (2 - 10 mL, depending on the amount of suspended matter; sedimentation time = 24 hours) and observed at 400x magnification with an inverted microscope (Zeiss Axiovert S100). A minimum of 50 random visual fields, at least 400 cells in total and 50 cells of the most common genus were counted.

For microcystin - LR (MC-LR) determination, 1.5 to 2 L water samples were filtered through Whatman GF/F filters, which were frozen until extraction with 20 mL 75 % (v/v) methanol. High performance liquid chromatography (HPLC) was carried out in a Dionex Summit equipment with photodiode array detector (PDA) and Chromeleon 6.3 software, using a C18 column (Merck Purospher STAR RP18 endcapped, 3 µm particles, LiChro-CART, 55 mm x 4mmm) kept at 40°C. As a mobile phase, acetonitrile and Milli-Q water were used containing

0.05% (v/v) TFA (trifluoroacetic acid) in a 25 : 75 ratio. Extract was then evaporated in a rotary evaporator (50-54 °C). Chromatograms were analyzed between 180 and 900 nm, with main detection at 238 nm for absorption spectrum characteristic of MC-LR. Purified MC-LR (Sigma) was used as standard, and results are expressed in MC-LR equivalents per volume of sample (Meriluoto & Codd, 2005; Ribau-Teixeira & Rosa, 2005; Sobrino et al., 2004). Samples collected during 1999 were analyzed for MC-LR using both ELISA and HPLC techniques in Dr. Wayne Carmichael´s laboratory, Wright State University, Ohio, U.S.A and results confirmed in Prof. Vitor Vasconcelos´ laboratory, Universidade do Porto, Portugal.

It is to be noted that the HPLC technique applied for MC-LR determination was carried out in different years by different technical staff and/or students (Master´s, PhD & post-doctoral fellows) in different specialized laboratories. During 1999, MC-LR was analyzed in parallel in two laboratories (Dr. W. Carmichael, Wright State, U.S.A., and Prof. V. Vasconcelos, Universidade do Porto, Portugal). From 2002 onwards, MC-LR analysis was performed in the Environmental Technology Lab., Universidade do Algarve with Prof. M. J. Rosa (2002-2003) and Dr. M. R. Teixeira (2004-2009). Therefore, although not considered significant in this study, slight variations in the extraction and HPLC methodology existed, as well as some adjustments in the interpretation of chromatograms.

2.2 Results

Monthly mean river flow at Pulo do Lobo (ca. 85 km upstream from river mouth) and total monthly rainfall at Alcoutim measured from 1996 to 2009 (Fig. 2) revealed four distinct river

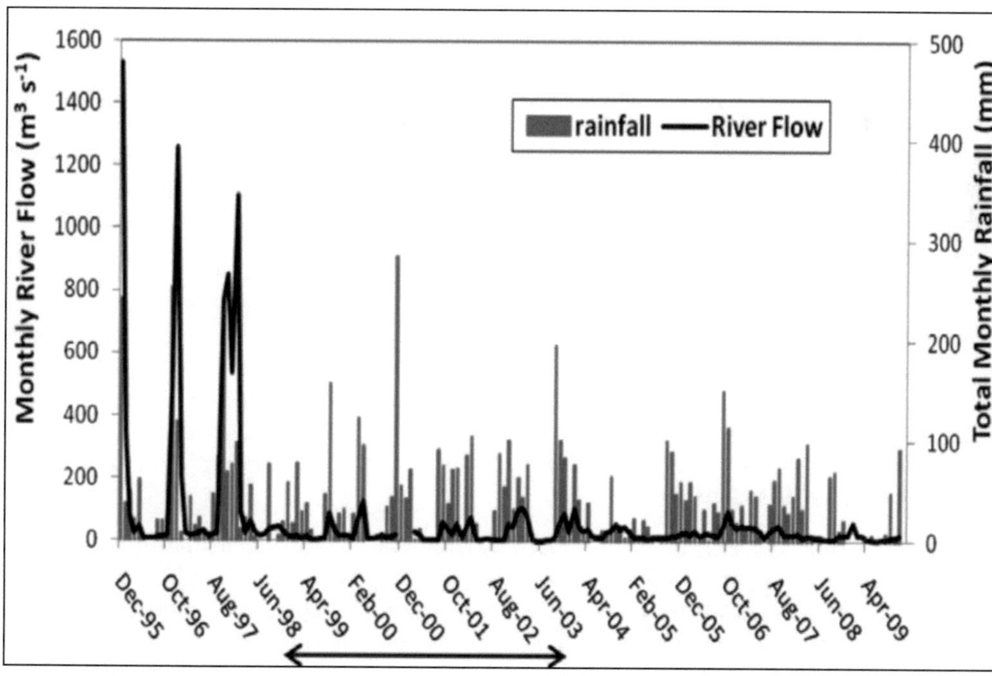

Fig. 2. Monthly mean river flow ($m^3 s^{-1}$) at Pulo do Lobo and total monthly rainfall (mm) at Alcoutim from 1996 to 2009 (data source: http://snirh.pt/). Arrow marks period of dam construction and filling

flow regimes: period before Alqueva dam construction (1996-1998), period during dam construction (1999 - 2000) and filling (2001 - 2003) and severely regulated river flow afterwards. Before Alqueva, river flow fluctuated widely from torrential winters to dry summers typical of Mediterranean flood - drought rainfall regime. However, starting with construction and filling of Alqueva dam, river flow became abruptly restricted particularly during winter months despite heavy rainfall. Mean river flow during summer reached 20 - 25 $m^3 s^{-1}$ during 1997 - 1998 previous to Alqueva, and decreased below 10 $m^3 s^{-1}$ from 1999 to 2003 during Alqueva construction and filling. Afterwards, summer river flow increased to 10 - 15 $m^3 s^{-1}$ during 2004 - 2005 reaching 20 - 25 $m^3 s^{-1}$ during 2007 - 2008, but decreased back below 10 $m^3 s^{-1}$ during 2008 - 2009.

Box plots in Fig. 3A also revealed these oscillations with respect to median values and overall distribution during these periods. Light extinction coefficient (Ke, see Fig. 3C), which is tightly correlated to sediment load, was generally low during 1997- 1998 previous to Alqueva; however, during dam construction in 1999 and part of 2000, light extinction reached maximum values in extremely turbid waters with very high sediment load (Suspended Particulate Material, SPM, peak values of 140 mg L^{-1}; data not shown). After this period of dam construction and extensive soil movement, waters tended to clear with Ke values decreasing from 2000 to 2010, but with wide intra-annual fluctuations due to winter summer oscillations in rainfall and river flow. It is to be noted that the composition of SPM varied markedly between river mouth (Vila Real Sto António) and upper estuary (Alcoutim and Mértola). In the lower estuary, SPM was mainly composed by quartz, which contributed minimally to light attenuation in the water column. On the contrary, in the middle and upper estuarine regions, SPM was mostly dominated by clays (Machado et al., 2007), which usually play an important role in light absorption.

Nitrate concentration, the predominant form of total dissolved inorganic nitrogen (ca. 65% - 89% of total inorganic N), showed a decreasing trend after construction of Alqueva dam (see Fig. 3B). Nitrate annual means during 1996-2001 in Alcoutim ranged between 65.0-73.6 µM, whereas mean NO_3^- was 56.2 µM in 2002, further decreasing to 30.4 ± 17.3 µM in 2005 (Barbosa et al., 2010), and remaining relatively low (32.23 ± 20.7 µM) during 2007-2009 (Domingues, 2010).

As referred in previous studies (Barbosa et al., 2010), silicate concentration was usually correlated with rainfall and river flow, and negatively correlated to chlorophyll *a*. Both DSi and nitrate exhibited seasonality with higher values during winter and lower values between midspring and summer. In contrast with nitrate, DSi exhibited an obvious increase during the period of the Alqueva dam filling (2002–2003) that led to a significant increase in the Si:N and Si:P molar ratios (data not shown), and a subsequent decline after its completion from 2004 to 2010 (see Fig. 3B).

Chlorophyll *a* (Fig. 4A) and total cyanobacteria abundance (Fig. 4B) in the upper estuary revealed a sharp collapse in 1999 during Alqueva construction, increasing during dam filling (2000 - 2001). Afterwards, chlorophyll *a* decreased again markedly from 2002 to 2010. Cyanobacterial abundance since Alqueva dam completion did not recuperate to high values observed previously (1997-1998). Furthermore, potentially toxic species, such as *Microcystis* spp., which were previously abundant exceeding WHO alert level 2 (≥100 000 cells mL^{-1}) during 1997 and 1998, have remained at very low densities if not practically absent from water samples collected in the upper estuary after Alqueva dam completion.

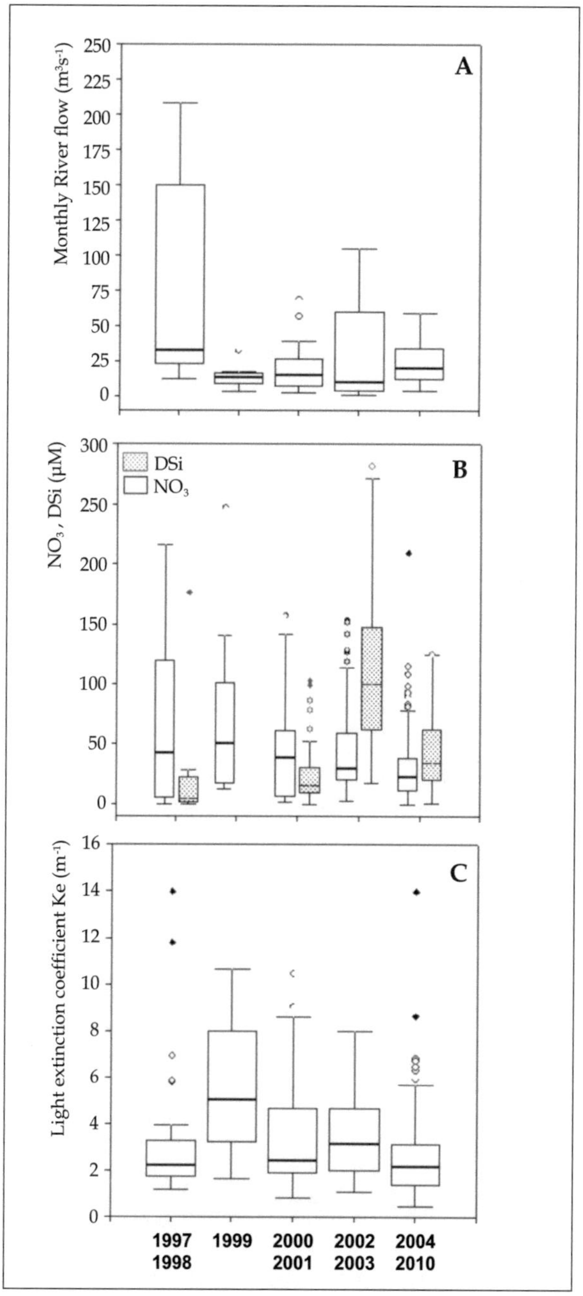

Fig. 3. Box and whisker plots showing the distribution of monthly Guadiana river flow (A), subsurface nitrate (NO_3^-) and silicate (DSi) concentration (B), and light extinction coefficient, Ke (C) in the Guadiana upper estuary, binned into different periods. Median value is represented by the line within the box, 25th to 75th percentiles are denoted by box edges, 5th to 90th percentiles are depicted by the error bars, outliers are indicated by circles, and extreme values by diamonds. Extreme values of monthly river flow (maximum 1258 m^3s^{-1}, year 1997) were omitted for clarity

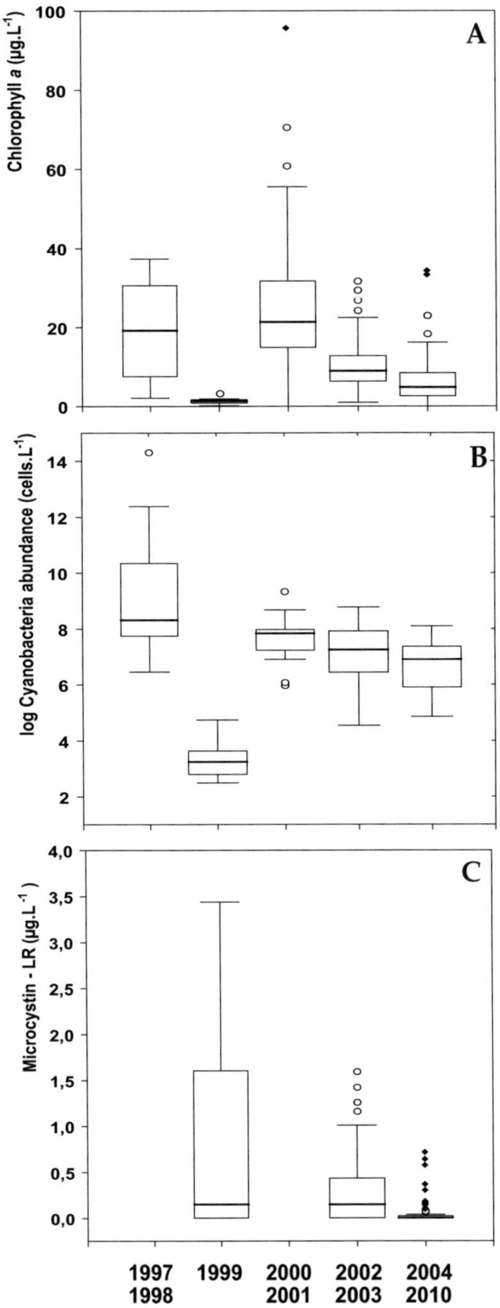

Fig. 4. Box and whisker plots showing the distribution of chlorophyll *a* concentration (A), total cyanobacteria abundance (B), and microcystin-LR concentration (C) in the Guadiana upper estuary, binned into different periods. Median value is represented by the line within the box, 25th to 75th percentiles are denoted by box edges, 5th to 90th percentiles are depicted by the error bars, outliers are indicated by circles, and extreme values by diamonds. An extreme chlorophyll *a* value (216.0 µgL^{-1}, year 2001) was omitted for clarity

Furthermore, the number of taxa observed not only in cyanobacteria, but also in phytoplankton populations has declined significantly. During summer of 1997 and 1998 the following genera were observed in numbers >1000 cells mL^{-1}: *Microcystis, Anabaena, Oscillatoria, Merismopedia, Lyngbya, Gomphosphaeria, Coelosphaerium, Syenchococcus,* and several unidentified species of *Chroococcales.* In contrast, during 2007 - 2009, besides unidentified Chroococcales, only *Planktothrix* could be occasionally identified.

Total cyanobacterial abundance included abundant small chroococcoid species and more rare large filamentous forms, and was not correlated with chlorophyll a (see Fig. 5).

As for microcystin - LR (MC-LR) concentrations in suspended particulate material (Fig. 4C), distribution in different periods showed highest values during 1999, often surpassing the 1 μg L^{-1} limit for drinking water (WHO 1998 guidelines). Yet, in 1999 both cyanobacteria abundance and chlorophyll *a* reached overall minimum values observed in the study period from 1997 to 2010. MC-LR decreased from 2002 onwards with concentrations frequently below detection limit. In fact, after 2004, MC-LR concentrations in the particulate fraction never surpassed 1 μg L^{-1}.

The variation of MC- LR concentration over time (see Fig. 6) revealed the same decreasing trend as box plots in Fig. 4C. The frequency of samples where microcystins were non-detectable increased over time particularly during 2004 and 2005. Gap years (2000, 2001, 2006 and 2007) are due to lack of funding for regular monitoring in estuarine waters.

Microcystin concentration during the study period was not correlated to total cyanobacteria abundance or chlorophyll *a* (see Fig. 7).

2.3 Discussion

In past published reports dealing with the microbial ecology of the Guadiana estuary (Domingues et al., 2005; Domingues & Galvão, 2007; Rocha et al., 2002), the impact of Alqueva dam construction was predicted to increase eutrophication conditions and possibly promote cyanobacterial blooms and associated cyanotoxins. In fact, this has not been observed during the seven-year period after dam completion. Not only cyanobacteria, but overall phytoplankton abundance, biomass and chlorophyll *a* concentrations have decreased markedly and have remained at low levels even in the upper estuary, where peak chlorophyll maxima usually occurred.

Typical estuarine phytoplankton succession observed in the Guadiana estuary from diatoms in early spring, to chlorophytes and finally cyanobacteria in late summer and fall was driven by nutrient regime with high winter loads of nitrogen and phosphorus discharged downriver, and silica depletion after the spring diatom bloom (Rocha et al., 2002). These authors also referred that cyanobacteria dominated the chlorophyll maximum zone in the upper estuary in late summer- early fall, due to warm waters, reduced sinking and grazing, as well as N limitation with low N:P ratio. Nitrogen limitation during summer increased in the period after Alqueva in the upper estuary (Barbosa et al., 2010). Additionally, nutrient enrichment experiments performed during 2008 clearly demonstrated that phytoplankton growth was nitrogen limited (Domingues et al., 2011).

Contrary to more stringent nitrogen limitation, the improved light regime with lower extinction coefficients should have promoted overall phytoplankton growth from 2003

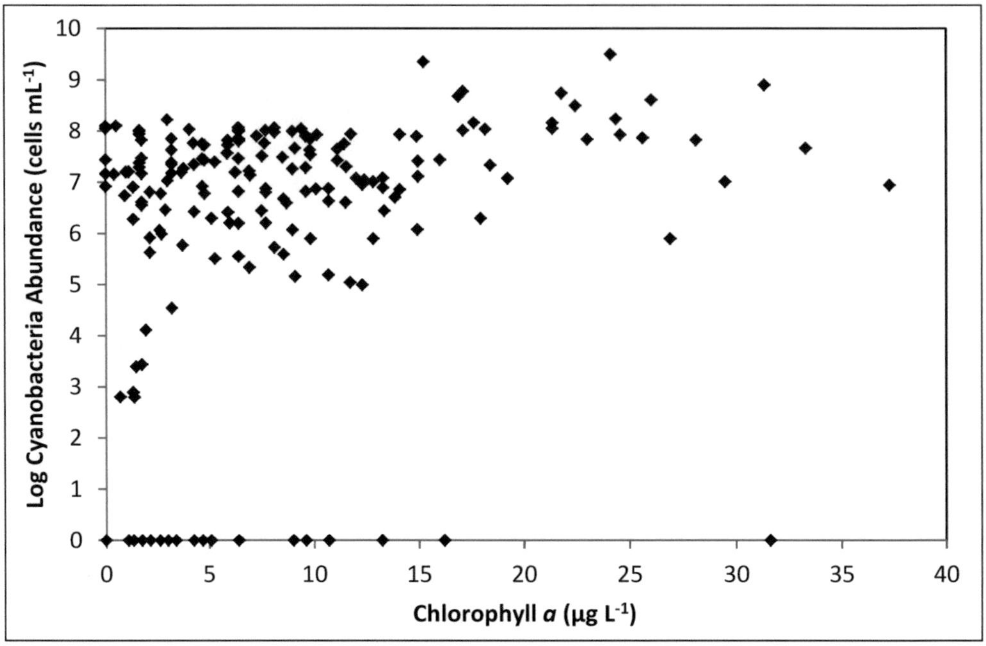

Fig. 5. Log of total cyanobacteria abundance (cells mL^{-1}) and chlorophyll *a* concentration ($\mu g\ L^{-1}$) in upper Guadiana estuary (pooled data from Alcoutim and Mértola) from 1997 to 2009. Zero abundance values not log transformed

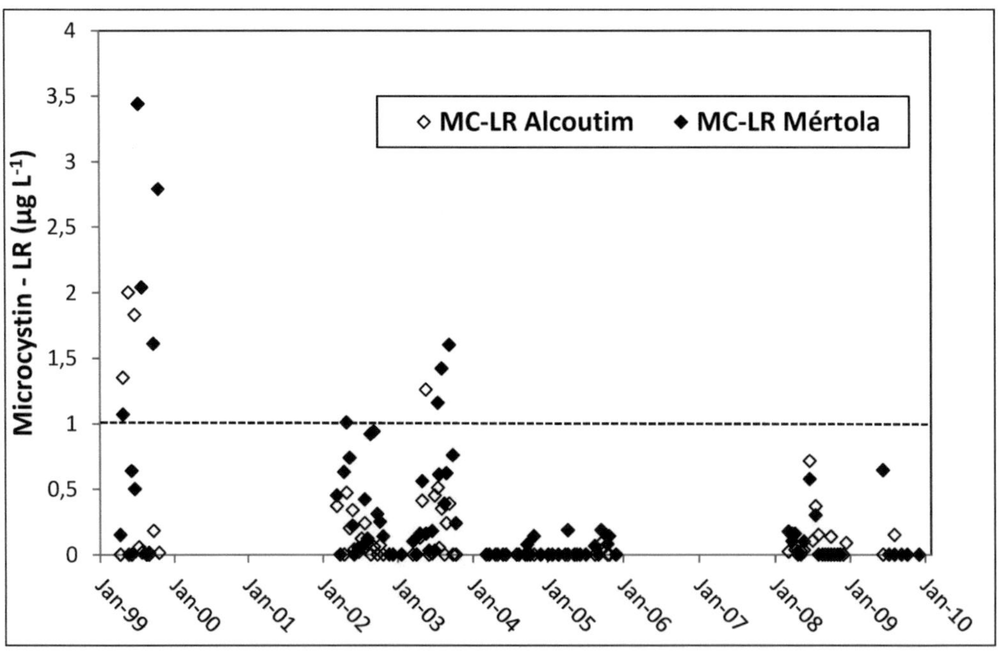

Fig. 6. Microcystin-LR concentration in particulate fraction ($\mu g\ L^{-1}$) over time in upper Guadiana estuary (Alcoutim and Mértola) from 1996 to 2009. Dashed line indicates the 1 $\mu g\ L^{-1}$ limit for drinking water (WHO 1998 guidelines)

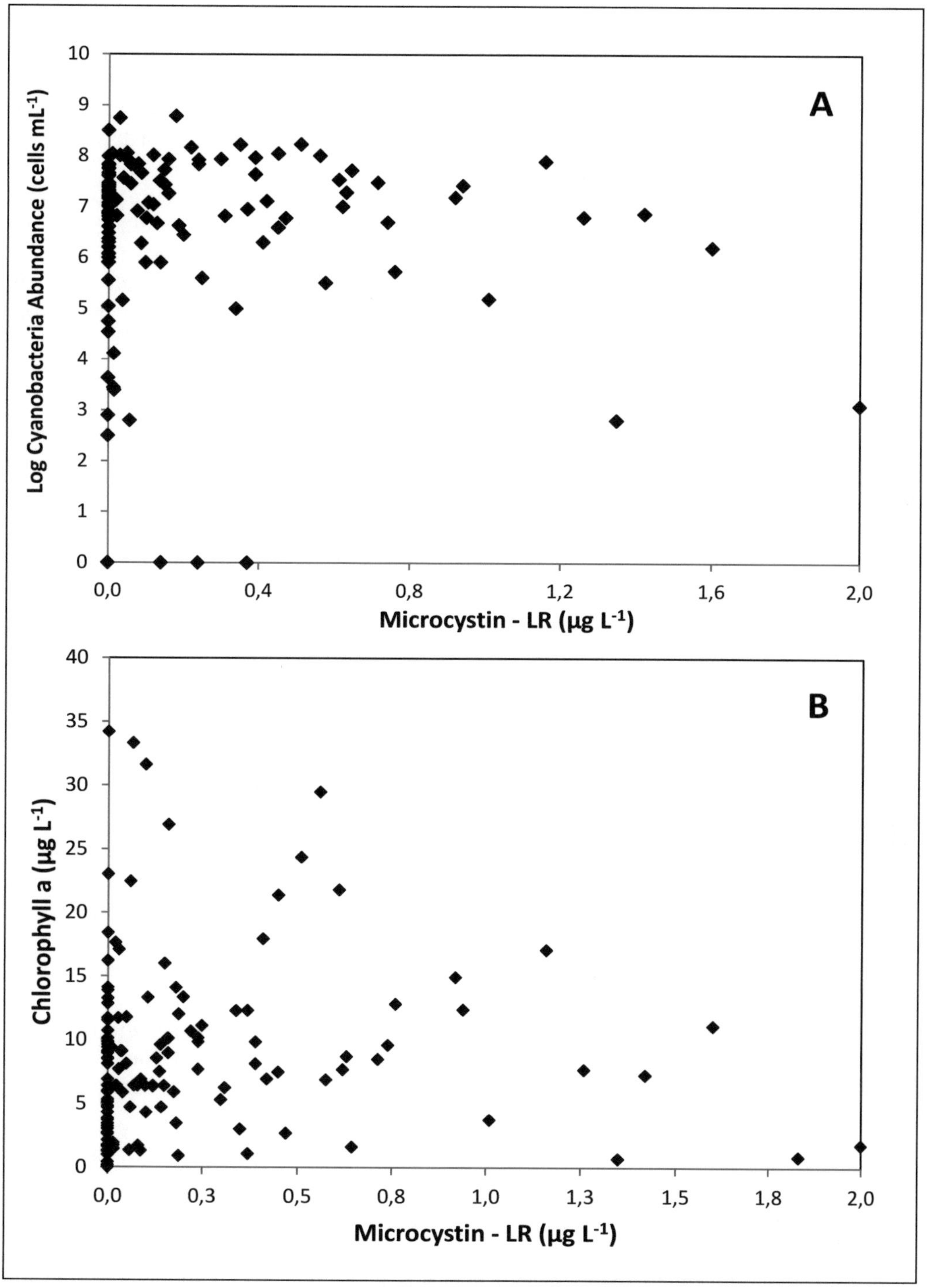

Fig. 7. Log of total cyanobacteria abundance (**A**) and chlorophyll *a* (**B**) versus microcystin concentration in particulate fraction ($\mu g\ L^{-1}$) in upper Guadiana estuary (Alcoutim and Mértola) from 1997 to 2009. Zero values not log transformed

onwards, after Alqueva dam completion. In view of decreasing trend in chlorophyll and overall phytoplankton abundance from 2003 to 2010, nitrogen availability appeared to play a preponderant role rather than light in this turbid estuary. This shift from light to nutrient limitation was probably the most determinant trend for phytoplankton observed after Alqueva (Barbosa et al., 2010). However, since chroococcoid cyanobacteria have higher affinity for nutrients due to small size, and most filamentous forms as well as some chroococcoid species have nitrogen fixing potential, these photosynthetic prokaryotes should have been less affected by lower nitrogen availability than larger non-nitrogen fixing eukaryotic phytoplankton. Freshwater reservoirs created by dams do not retain only water but also suspended particulate material, including planktonic microorganisms. In consequence, not only are nutritional regimes affected downstream, but also freshwater microbial populations with complex life cycles, such as cyanobacteria. Filamentous cyanobacteria in response to environmental forcing can produce different cell types which are adapted to nitrogen fixation, nutrient storage and reproductive strategies such as winter dormancy and dispersal. Thus, freshwater reservoirs by retaining these morphotypes could seriously affect not only bloom formation but also species composition downstream.

The lack of correlation between chlorophyll and cyanobacteria abundance could be simply explained by predominance of small chroococcoid cells with reduced chlorophyll content. Consequently, poor or absent correlation between chlorophyll and microcystin concentrations should also be expected.

As previously described by Galvão et al. (2008), microcystin concentration were generally not correlated with cyanobacteria abundance or biomass in natural waters (freshwater reservoirs and Guadiana river, South Portugal), since different strains and/or species could produce microcystins at different rates depending on cell cycles and environmental conditions, which has also been documented in laboratory analyses (eg. Kameyama et al., 2004; Rapala et al., 1997; Saker et al., 2005).

Furthermore, in all temperate estuaries, cyanobacteria accumulate and thrive in the chlorophyll *a* peak (Cloern, 1987; Pearl et al., 2006; Pinckney et al., 1998), directly upstream from the turbidity maximum. Restricting river flow can cause perturbations of estuarine circulation, particularly in terms of location and intensity of the turbidity maximum, which in turn will affect the chlorophyll *a* maximum in the upper estuary (Cloern, 1987, 1999). Thus, cyanobacteria decline cannot be simply explained by any one environmental driver, but rather in terms of estuarine circulation. Nutrients tend to be regenerated in the turbidity maximum and phytoplankton bloom directly upstream from this zone, benefiting in this interface between nutrient enriched and clear waters. Unfortunately, how seriously the huge Alqueva reservoir has affected estuarine circulation and the turbidity maximum in the Guadiana estuary has not yet been assessed.

The Alqueva dam not only is the largest dam in the Guadiana watershed but due to its location affects most strongly the estuarine section of the river. In spite of efforts by the Alqueva water management authorities to maintain "ecological" river flow this is not compulsory according to existing Portuguese water resources legislation. Ecological river flow can be broadly defined as the flow necessary to conserve and maintain natural aquatic (freshwater) ecosystems. In Portugal, this is very simply calculated as a value > 2.5 to 5% of the modular water flow to be maintained throughout the year, if conditions permit. Different studies have recently challenged this approach proposing more careful analyses of natural river flow variations and applying them to flow regulation by dams (Alves & Bernardo, 1998; Alves & Gonçalves, 1994; Chicharo et al., 2006; Chicharo et al., 2009; Wolanski et al., 2008). Flow Incremental Methodology, and other hydrological or ecohydrological approaches, would ensure that natural variations in freshwater flow would be mimicked by dam discharge, albeit dampened. Finally, monitoring of environmental impact usually considers either endangered or economically important vertebrate species existing in the freshwater zone. Yet, marine and freshwater micro- and macroorganisms need also to be considered in terms of whole ecosystem impact. In fact, microorganisms, such as cyanobacteria, appeared to be sensitive indicators of estuarine ecosystem perturbation in this study. Thus, it is proposed that photosynthetic prokaryotes should be used as indicators of "good" estuarine water quality rather than just "bad".

2.4 Conclusion

This 13-year study of the Guadiana estuary in Southern Portugal, directed towards assessing the impact of dam construction on cyanobacteria populations in the freshwater zone, revealed that phytoplankton abundance, chlorophyll *a* and diversity decreased markedly from 2003 onwards after Alqueva dam completion. This declining trend in phytoplankton could be explained by both light limitation during dam building coupled with more stringent nitrogen limitation after dam completion. Interestingly, cyanobacteria abundance, diversity and microcystin concentration exhibited an even more pronounced decrease, which could not be attributed to any monitored environmental factors, but instead to perturbations in overall estuarine circulation. The collapse in cyanobacteria populations in the upper estuary warrants a more careful approach towards maintaining ecological river flow in dam discharge. Future research in the Guadiana estuary should address not only the impact of restricted river flow on estuarine circulation, turbidity maximum and associated chlorophyll peaks, as well as provide more adequate approaches towards maintaining an ecological river flow, possibly using cyanobacteria as an indicator of good water quality.

3. Guadiana reservoirs management

3.1 Cyanobacteria management in reservoirs

Water management in the Guadiana River watershed is a complex transnational problem and has been object of negotiations between Portugal and Spain for decades now. The last bilateral Agreement assured the integrated management of water and territory, covering quantitative and qualitative features, stipulating minimum flows (under normal rainfall conditions), and foreseeing the permanent exchange of hydrologic and environmental data and information (Mendes, 2010). Environmental laws, in both countries, ensure public access to the monitoring data, allowing for international comparison of water quality in different parts of the watershed.

Normal and drought conditions in the Guadiana catchment have been modelled in multiple hydrologic studies (e.g. Brandão & Rodrigues, 2000), allowing for better management in terms of water availability. Nevertheless, water quality concerns have been mostly ignored in water reservoir management decisions. Impaired sewage treatment and agro-industrial mal-practices have been repeatedly blamed for water quality deterioration in the Guadiana river basin, both in Portugal and Spain. Official reports for the Portuguese part of the watershed specifically accuse illegal sewage discharges from pig production farms and olive mills of being responsible for high nitrate concentrations (PBH, 2001) As generally accepted, the mere existence of a new dam contributes to water quality degradation, since new populations and activities are attracted to the watershed. The main water user in the Guadiana watershed is agriculture, using 90 to 95% of the consumed water (PBH, 2001). It is thus expected that newly introduced crops after the start of the Alqueva dam irrigation system, in particular the new intensive olive tree orchards, will have strong impacts in future water quality.

A diagnosis of the actual ecological status of the catchment based on reliable methods and classification indices is therefore crucial. In Spain, the Confederácion Hidrográfica del Guadiana (CHG) and, in Portugal, the Administração da Região Hidrográfica do Alentejo, conducted a diagnostic snapshot classification, based on monitoring surveys from 2005 and 2006 for the Spanish part of the catchment, and on 2009-2010 data for the Portuguese watershed. Classification results, based on multiple indices, are available online (CHG, 2006, 2007-2008, 2009; ARH Alentejo, 2011).

According to these official reports, Guadiana reservoirs fall into diverse typologies, but the majority of them behave as warm monomyctic lakes, that remain stratified during the dry season and mix the water column in winter. As expected in result of high hydraulic residence time and elevated temperatures, these freshwater reservoirs are dominated, at least in the summer, by potentially toxic cyanobacteria from the genera *Pseudanabaena, Anabaena, Planktothrix, Oscillatoria., Geitlerinema., Aphanizomenon, Merismopedia, Microcystis, Woronichinia, Synechocystis,* and *Aphanocapsa*. (CHG, 2009) Toxic species of these cyanobacteria may reach high densities forming harmful algal blooms (HABs). In fact, the term water bloom originally referred to surface scums of cyanobacteria, but has since been applied to almost any planktonic population (not even necessarily algal) with densities significantly above the normal (Reynolds, 2006).

Managing these cyanobacteria harmful algal blooms (CHABs) has become a major concern in view of the potential health impacts both through drinking water or farming products consumption (Edwards et al., 1992; Hoegar et al., 2005).

CHABs management might involve prevention actions and/or mitigation solutions. Numerous techniques have been developed for these purposes, but as stated by Perovich et al. (2008) most of them have not been explicitly evaluated and optimized for use in the control of CHABs, particularly when toxins are present.

Prevention techniques rely on CHAB association with eutrophication processes and aim to control CHAB through nutrient limitation or decreasing hydraulic residence time. These techniques include watershed protection or restoration, through adequate sewage treatment implementation, promotion of farming good practices, particularly in the use of fertilizers and pesticides; erosion control; stimulation of margin riparian vegetation as well as controlled surface water discharges. Nutrient input reduction by controlling point sources has had success in several CHABs managing cases (Piehler, 2008), but it is now acknowledge that restoration efforts seldom bring aquatic communities back to the diversity and composition they used to bear before suffering human impacts (Jacquet et al., 2004)

Mitigation enforcement, by control and removal of an installed bloom, might rely on techniques such as the addition of algicides, the introduction of fish schools, surface scums elimination or water column mixing (Piehler, 2008). Such techniques often bring about unexpected results (Jacquet et al., 2004).

The only technique used both for prevention as well as mitigation of CHABs is the reduction of water residence time, through surface water discharges. It was well known in the 1960s (Odum, 1971) that the type of water discharges, and specially the height of water column, at which they were performed, strongly influenced plankton assemblages both up- and downstream from a reservoir. While surface release mainly exports warmer water and their plankton communities, bottom discharge introduces downstream cold, nutrient enriched water, keeping the warmer plankton rich waters inside the reservoir (Wright, 1967). This means that in reservoirs with bottom water flow, slow growing picoplankton, including cyanobacteria, is given the opportunity to develop blooms, instead of being rapidly flushed downstream. Water extraction for drinking water production tends to use water at a medium height of the water column, avoiding both the surface plankton, and the bottom metal enrichment. As observed in the Algarve reservoirs (Reis, unpublished data) withdrawing cold water from the hypolimnion and maintaining a floating inoculum of warm temperature selected cyanobacteria at the surface is transforming reservoirs into bioreactors like structures, favoring the occurrence of prolonged summer blooms. This water management technique tends to enhance stratification, delaying water column mixing.

As acknowledged by increasing awareness for the need of establishing an ecological flow, reservoir water management plays a key role in downstream river ecology, but also in upstream ecology.

Rapid changes in the water level in response to summer increased water demand seriously hinders the installation of riparian vegetation, challenging some prevention techniques. Thus, caution should be taken when applying the same ecological criteria to reservoirs as for lakes, as advocated by the European WFD. In fact, in most natural lakes excess water overflows into effluent streams, exporting phytoplankton and accumulating nutrients in bottom colder water. On the contrary, in a semi-arid region most reservoirs managers seldom let water level rise enough to cause superficial overflow, and regulate water flow by smaller continuous discharges at mid-height of the dam wall.

While classifying reservoirs as Heavily Modified Water Bodies (HMWB) the WFD allows for hydro-morphological pressures upon their ecological status, pressures to which natural lakes are not subjected. Reference conditions for establishing the ecological potential of the HMWBs should be given by reference conditions for the ecological status of natural lakes of the same eco-region, but reference conditions bearing natural lakes in a semi-arid region are scarce. In fact, there are no natural lakes in Southern Portugal.

3.2 Ecological tools foreseen in the European Water Frame Directive

In the scope of the WFD implementation, the Guadiana watershed is included in the Mediterranean Region. The Geographical Intercalibration Group for this region (Med GIG) was responsible for establishing boundary values for the Med GiG Member State classification systems. Submitted values were adopted through the European Commission Decision of 30 October 2008 (2008/915/EC).

For their intercalibration exercise, the Med GIG agreed on using chlorophyll *a* and total biovolume as phytoplankton biomass indicative parameters, and elected three phytoplankton composition metrics, namely the contribution of cyanobacteria to total phytoplankton biovolume, the General Algal Index (GAI - Catalàn et al. 2003) and the Mediterranean Phytoplankton Trophic Index (MedPTI - Marchetto et al. 2007). Although recognizing strong limitations of the dataset used and the fact that not all tipologies of Mediterranean lakes and HMWBs were covered, actual law enforcement stipulates that ecological potential of the reservoirs in the Guadiana watershed should be classified, according to the proposed metrics. However, application of such phytoplankton composition metrics to CHABs management has yet to be assessed.

The following case study constitutes an effort to evaluate the Med GIG selected ecological indicators when applied to water management strategies for the Guadiana watershed. Different phytoplankton metrics determined for two reservoirs and compared in order to assess their usefulness in CHAB management, taking in consideration the EPA Guidelines for Evaluation of Ecological Indicators (Jackson et al., 2000)

3.3 Study area: Beliche and Odeleite reservoirs

Beliche and Odeleite reservoirs are located on two small affluent streams to the Guadiana estuary (Fig. 8), and were built for purposes of drinking water production. They are interconnected by an underground water channel, with sluices that are operated by the managing authorities, whenever they need to transfer water from Odeleite to Beliche reservoir. Together these reservoirs constitute the raw water source for 230,000 inhabitants of eastern Algarve, a province on the south coast of Portugal. Since Algarve constitutes an important national and international tourism destination, summer population more than

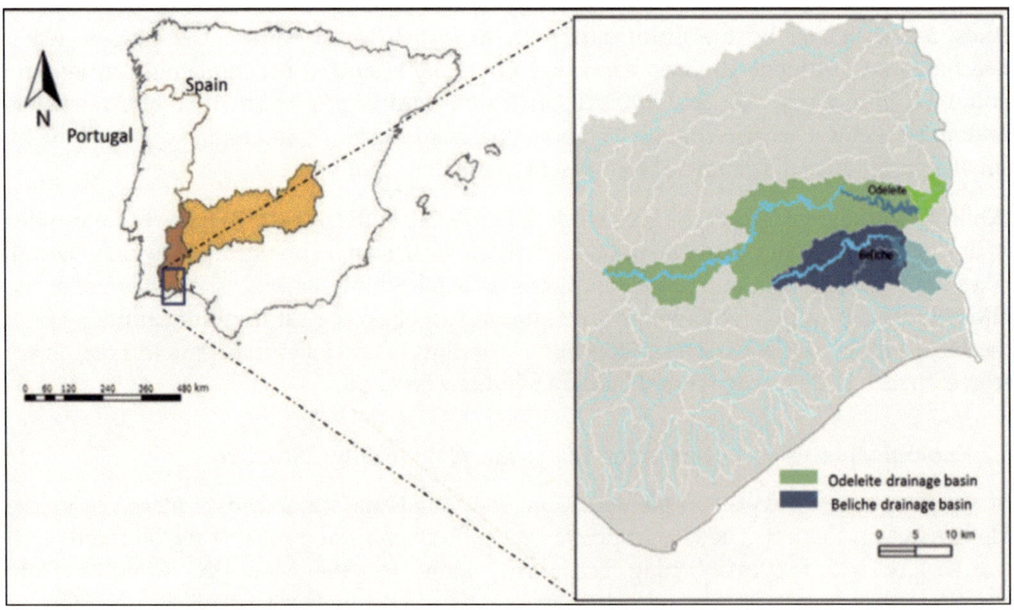

Fig. 8. Location of Beliche and Odeleite drainage basin

doubles and puts water demand at its peak during the high season (June through September), coinciding with low precipitation and high atmospheric temperature. Table 1 presents important features of these two reservoirs, where the most notorious aspects are the ones relative to their catchment area. While other reservoirs from the Guadiana watershed contributed to extensive development of irrigation (eg. Alqueva), promoted by the use of EU subsidies that encouraged high value intensive crop production (Varela Ortega et al. 1998, Varela-Ortega et al. 2003), the catchment areas of these two streams do not support any significant economic activity (Fig. 8).

Historical reasons linked to deforestation and farming mal-practices, back in the 30s of the XX century, led to extensive soil erosion, agriculture relinquishment and human desertification. Indeed, these catchment areas are the poorest counties in Portugal, with population densities around 20 habitants /Km^2. Human settlements are small villages concentrated downstream of the catchment area. In order to rehabilitate the landscape there has been a large investment in replanting pine woods with the objective of developing new soil and future stimulation of natural vegetation. Apart from some small goat herds in the Odeleite watershed and extensive cropping of sparse almond trees, there are no human impacts, no sewages, no pig style farms, no intensive or extensive farming. From a CHAB prevention point of view, it is difficult to point out what could be improved for the protection of their drainage basins.

Reservoir	**Beliche**	**Odeleite**
Stream	Ribeira de Beliche	Ribeire de Odeleite
Watershed	Guadiana	Guadiana
Catchment area (km^2)	98	347
Latitude (mean)	37º 16' 35''	37º 19' 52''
Longitude (mean)	-7º 30' 33''	-7º 29' 11''
Year of closure	1986	1996
Max. water column height (m)	30	30
Total volume (x 10^6 m^3)	48	130
Flooded surface (ha)	292	720
Mean annual precipitation (mm)	644	722
Min. stored water volume (x 10^6 m^3) / Date	7 / Sept 2006	40 / Nov 2005
Max. stored water volume (x 10^6 m^3) / Date	46 / May 2010	130 / May 2010

Table 1. Beliche and Odeleite reservoirs location and some main features. (Data source: http://snirh.pt/)

3.4 Methods

These reservoirs have been monitored for standard physical, chemical and microbiological water quality, including, since 2003, determination of phytoplankton biomass and composition. Monthly surface and bottom samples were taken, from 2003 to 2009, at Choça Queimada tower for Odeleite reservoir, and at the extraction tower of the Beliche reservoir. During 2009 and 2010, a new sampling site in the middle of the lake, 500m upstream from the dam wall, was added according to new guidelines from the European Med GIG (INAG, 2009). At each of these sampling sites, vertical profiles were determined *in situ* using a YSI

650 MDS probe, for water temperature, dissolved oxygen, pH and conductivity. Nutrient concentrations analysis were performed at the accredited (EN 17025) water analysis laboratory from Administração da Região Hidrográfica do Algarve (ARH Algarve), who was also responsible for all the sampling campaigns. Phytopigments including chlorophyll *a* were analyzed at Huelva University (Forján et al., 2008) through HPLC, according to Young et al. (1997). Microcystin detection and quantification was performed according to Carmichael & An (1999) using the micro-ELISA kit Microcystin Plate Kit from Adgen - Agrifood Diagnostic. Phytoplankton composition was determined by the same methods referred in 2.1. Phytoplankton biovolumes following European guidelines were calculated on the basis of predefined 3-dimensional shapes and their respective stereometric formulas as recommended by Edler (1979a, 1979b) and Hillebrand et al. (1999), according to the CEN/TC230/WG2/TG3 N108 Water Quality and Olenina et al. (2006).

Berger Parker dominance index was determined by calculating the proportion of the most abundant species over the total phytoplankton cell density (Magurran, 1988). Carlson Trophic State Index (TSI) was calculated for chlorophyll *a* values and both for total phosphorus (TP) and soluble reactive phosphorus (SRP) concentrations, according to Carlson (1977). Contribution of cyanobacteria to total phytoplankton is given by the percentage of total biovolume attributed to cyanobacteria. Catalán Index for Algal Groups (InGA) was determined by using biovolume proportions of colonial and non-colonial algal groups (Catalán et al., 2003). The MedPTI index was calculated according to Marchetto (2009).

3.5 Results and discussion

3.5.1 Hydrometric features

Monitoring data for the last 7 years included a severe 18 months drought from 2004 to 2006, and an exceptional rainy year in 2010. As seen in Fig. 9, water level at Odeleite reservoir had to be lowered in November 2006 for maintenance works, originating intense bottom and surface discharges. Surface overflow was released through the stream bed into the Guadiana estuary, but bottom discharged water flowed through the underground channel into Beliche reservoir, causing more surface and bottom discharges at this reservoir. Water volumes discharges at surface and at bottom of both reservoirs are enlisted in Table 2, revealing how water level regulation in both reservoirs is interconnected. Apart from water extraction for municipal consumption, no water outflow occurred from February 2004 to March 2006, during the drought (see Table 2). In fact, there is no ecological flow stipulated for these streams, such that, downstream from the dam walls, only estuarine water flows in during high tide.

Bottom discharges from Odeleite to Beliche, through the underground channel, induced mixing of water column, with ressuspension of sediment and nutrients. Indeed, both reservoirs did not behave as warm monomyctic lakes, but rather as artificially polimyctic, since water column mixed in the winter and also partially, whenever channel sluices were opened. Water level regulation and Beliche water withdrawal for drinking water were seemingly the main impacts on the water quality of these reservoirs. Nevertheless, sharp shifts in water level also affected margin vegetation, contributing to increased nutrient leaching from soils.

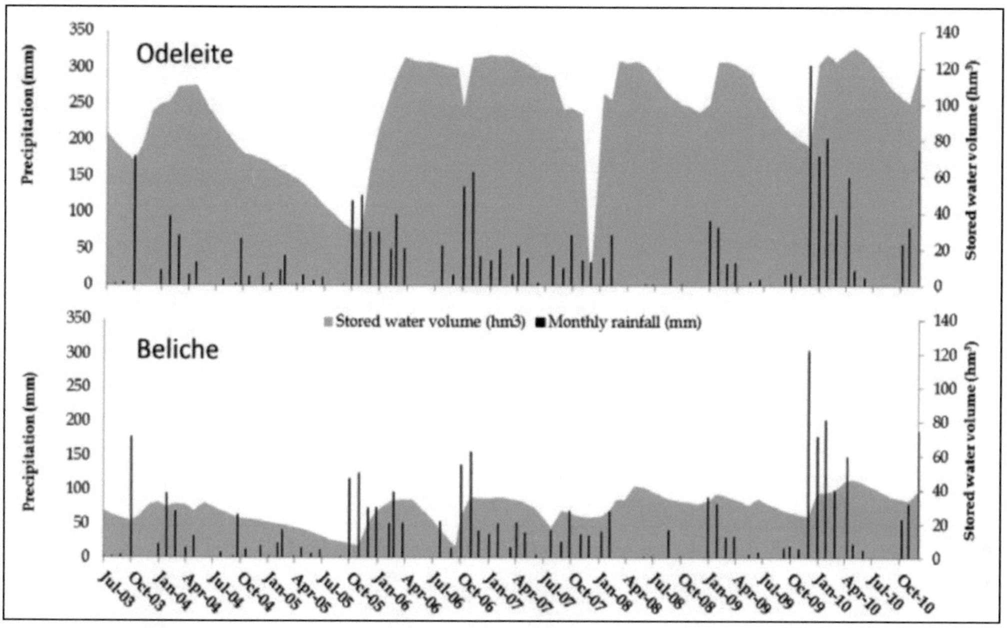

Fig. 9. Evolution of stored water volume (hm^3) and mean monthly rainfall from July 2003 to November 2010 in Odeleite and Beliche reservoirs. Data source: http://snirh.pt/ and http://www.drapalg.min-agricultura.pt/

Reservoir	Odeleite		Beliche	
Month/year	Bottom discharge (dam^3)	Surface discharge (dam^3)	Bottom discharge (dam^3)	Surface discharge (dam^3)
Jul/2003	0	0	13	0
Aug/2003	0	0	26	0
Sep/2003	0	0	13	0
Oct/2003	0	0	1	0
Nov/2003	1234	8722	1818	0
Dec/2003	3894	9821	382	0
Feb/2004	6648	0	1562	0
Mar/2006	5970	0	1409	0
Nov/2006	15196	19868	3536	1776
Dec/2006	8685	0	543	0
Feb/2007	2364	0	0	0
Aug/2007	74	0	0	0
Mar/2008	0	261	0	0
Apr/2008	6095	18388	0	1012
Feb/2009	8495	0	0	3502
Dec/2009	0	22450	0	5105
Dec/2010	0	29821	40	6714
Jan/2011	22528	0	0	2186
Mar/2011	50139	0	0	12577

Table 2. Surface and bottom water discharged from Odeleite and Beliche since 2003 (Data source: http://snirh.pt/)

3.5.2 Nutrient dynamics

Despite these hydrographical fluctuations, no nutrient accumulation or eutrophication trend was detected. Yearly turn-over of Dissolved Inorganic Nitrogen (DIN) and Soluble Reactive Phosphorus (SRP) was clear in Fig. 10, where water temperature at the surface can be used as reference for seasonal changes in nutrient dynamics in Beliche reservoir.

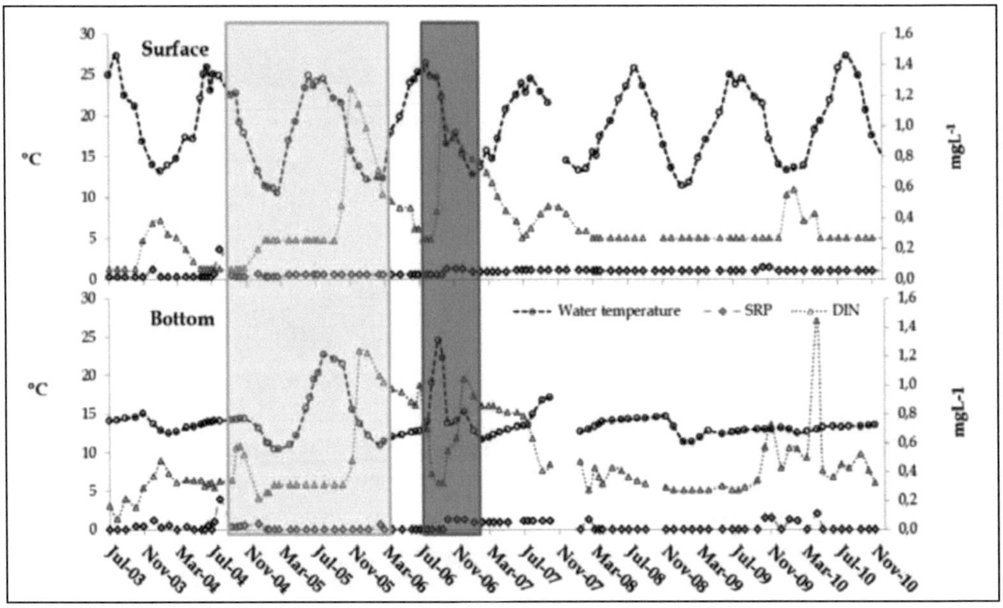

Fig. 10. Dissolved Inorganic Nitrogen (DIN; mg*L-1), Soluble Reactive Phosphorus (SRP; mg*L-1) and water temperature (°C) during 2003 -2010 in Beliche reservoir. Drought months are highlighted in light grey. Months were unusual Odeleite to Beliche discharges occurred are highlighted in darker grey

The increase in DIN levels during the drought years was fictitious (see light grey box Fig. 10), since water level was so low that surface and bottom samples were almost undistinguishable. Shallower depth allowed for oxygen diffusion to the bottom inhibiting deep summer denitrification. Unusual surface and bottom DIN levels occurred in November 2006 due to exceptional water transfer from Odeleite to Beliche. As stated in Galvão et al, 2008, management of the underground channel between the two reservoirs has been associated with conditions favoring blooms through bottom sediment and nutrient resuspension. Consequent water column mixing was revealed by similar bottom and surface temperatures. Comparing Fig. 9 with Fig. 10 also indicated that the increase in DIN concentration in March 2010 was linked to high precipitation levels. Overall low nutrient concentrations in both reservoirs were associated with oligotrophic conditions. In spite of phosphorus limitations and low median and mode values for DIN:SRP ratios, high average N:P ratios were observed, due to outlier values observed in 2005, 2006 and March 2010.

3.5.3 Phytoplankton dynamics

In terms of cell abundance, more than 80% of monthly water samples during last eight years from Odeleite and Beliche reservoirs, were dominated by cyanobacteria (Fig. 11), but data gathered during 2009 and 2010 showed diatom (Bacillariophycea) dominance in terms of biovolume proportion (Fig. 12) in at least 50% of the samples.

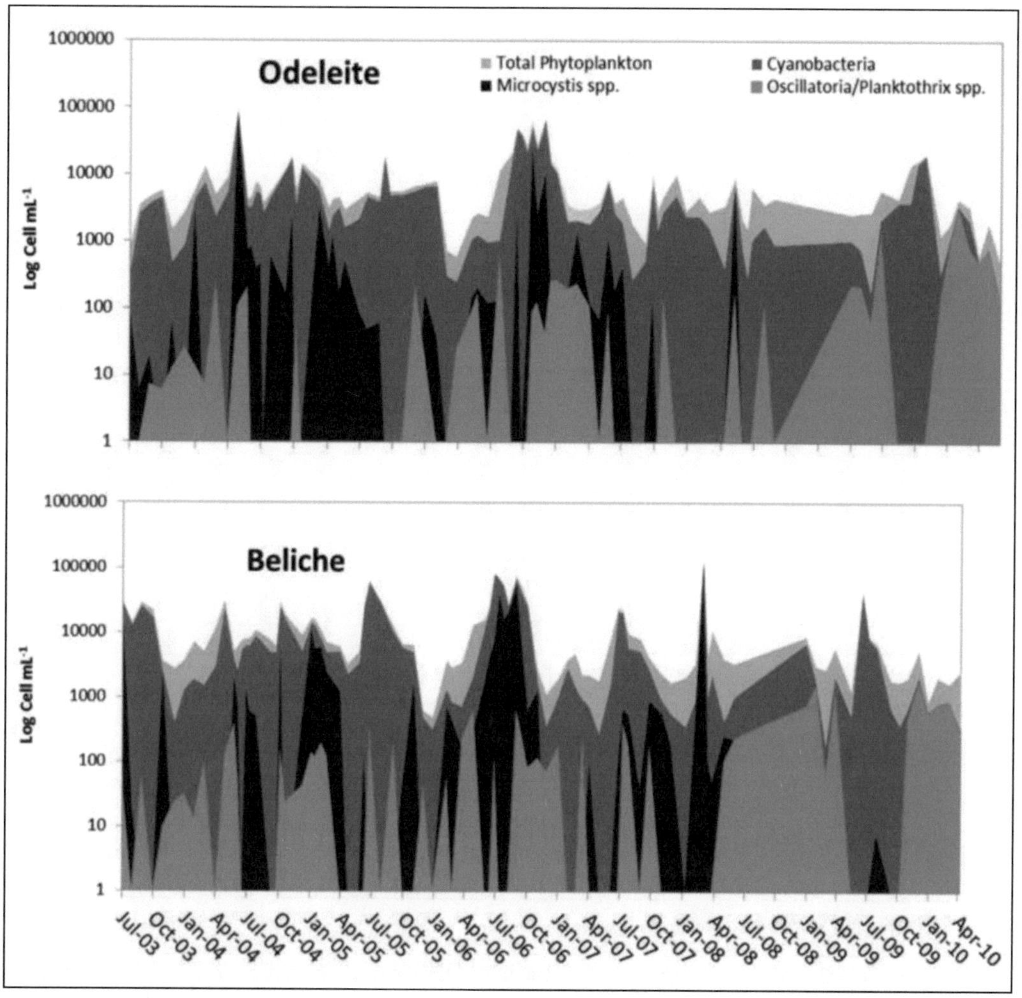

Fig. 11. Cyanobacteria dynamics in Odeleite and Beliche reservoirs. Total phytoplankton abundance (Log cells mL^{-1}) compared with total Cyanobacteria, *Microcystis* and *Oscillatoria/Planktothryx* spp. abundances

In terms of cell abundance, *Microcystis* spp. dominated both reservoirs until spring 2008, but Oscillatoriales dominated in terms of biovolume. Cyanobacteria cell densities above WHO alert level 1 of 2000 cells mL^{-1}, occurred in 62 to 63% of all samples, with episodes of *Microcystis* blooms in June 2004 for Odeleite and July 2004 for Beliche. In summer 2006, a *Microcystis* spp. bloom was toxic with microystin concentrations at the bottom of Beliche reservoir reaching 3.5 µg L^{-1}. Despite high cyanobacteria abundances, no significant levels of microcystins were detected under other bloom situations.

Biovolume proportions for main algal groups (Fig. 12) confirmed cyanobacterial dominance from August to October 2009 in Beliche and from October to December 2009 in Odeleite.

Summer bloom absence in 2010 could be linked to high water discharges in consequence of an exceptional rainy winter and spring. (see Fig. 9 and Table 2).Thus, during the study period both Beliche and Odeleite reservoirs were susceptible to CHABs.

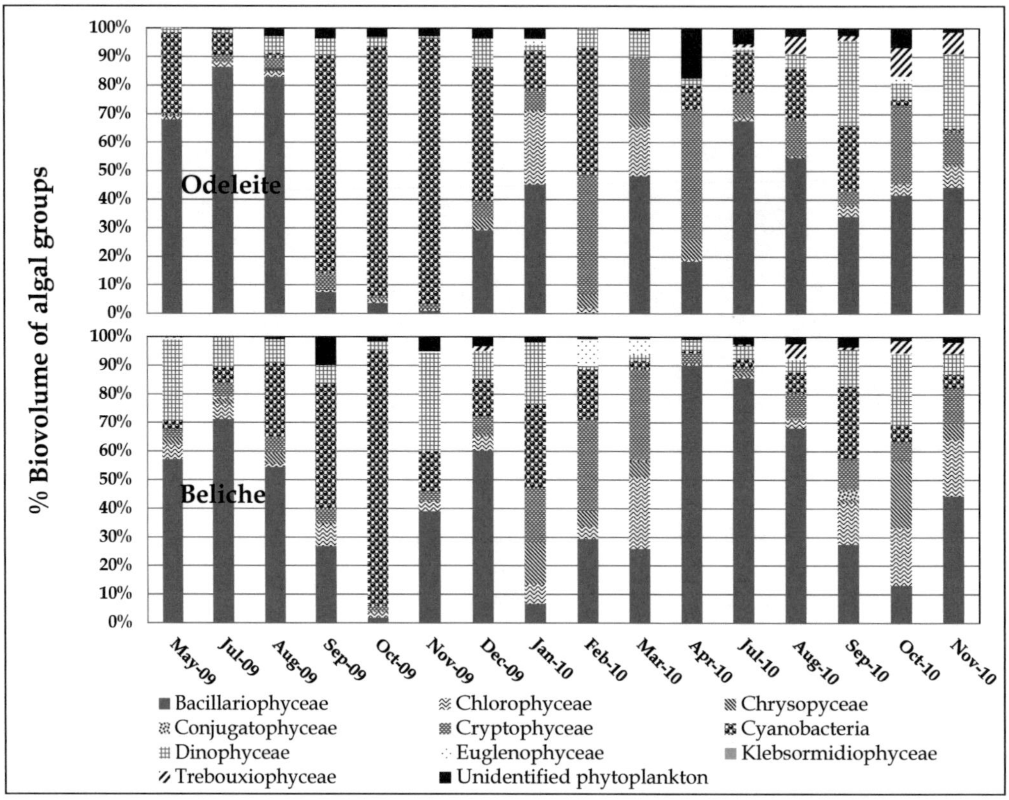

Fig. 12. Relative contribution (%) of main algal groups to total phytoplankton biovolume

3.5.4 Ecological indices

Table 3, compiles Carlson Trophic State Indices (TSI) calculated for Beliche and Odeleite based on Chlorophyll *a* (Chl-a) and on total phosphorus (TP) as well as for phytoplankton ecological index (MedPTI) proposed by Marchetto et al. (2009) for Italian deep lakes in the Mediterranean region. This index is based on the proportion of biovolumes of species, listed in Italian lakes, and should not be applied in situations where the biovolume of listed existing species does not exceed 70% of total phytoplankton biovolume. Since the contribution of Marchetto species to total phytoplankton biovolume reached 77% in Beliche during 2010 and 71% in Odeleite, the MedPTI index was also calculated for comparison. TSI based on transparency measured by Secchi depth was not calculated, since it has long been established that torrential hydrographic regimes promote high values for this index without any correlation with eutrophication. Total Phosphorus concentrations were also misleading, since bottom sediment resuspension promoted by artificially induced polimyctic behavior, released organic phosphorus unavailable for phytoplankton into the water column. Low chlorophyll *a* values in spite of high cellular cyanobacteria abundance, as referred previously, was due to low chlorophyll content of cyanobacteria.

Monthly values for the MedPTI index are illustrated in Fig. 13, with open circles and boxes representing non-valid values based on lower than 70% species contribution to total phytoplankton biovolume. This figure constituted a test to the robustness of the MedPTI index applied to Beliche and Odeleite.

A)

Reservoir/Index	TSI (Tota P)	TSI (Chl-a)	MedPTI
Beliche	30	47	3,05
Odeleite	19	47	2,90

B)

TSI classes	<30	30-40	40-50	50-60	60-70
MedPTI classes and upper limits	excelent	high - good (<2.77)	good - moderate (<2.45)	moderate - poor (<2.13)	poor - bad (1.81)

Table 3. A) Determined values for Carlson Trophic State Index (TSI) based on total P concentrations and on chlorophyll a content, and for phytoplankton composition MedPTI index. B) Classification boundary values for TSI and MedPTI with a color code to facilitate interpretation

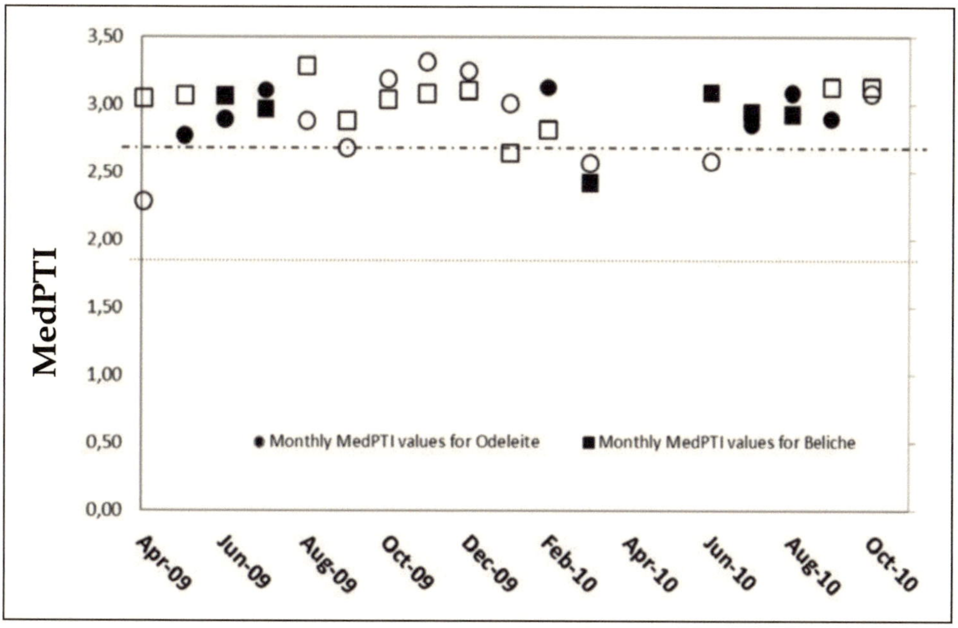

Fig. 13. Monthly values for MedPTI. Dashed lines indicate lower boundaries of "excellent" and "good" classifications. Open circles and boxes correspond to non-valid values for Odeleite and Beliche, respectively. See text for explanation

Since MedPTI was specifically developed for deep natural lakes in Italy, this index should probably not be applied to reservoirs without adjustments to existing species lists. Nevertheless, MedPT1 classifications obtained in this study seemed more consistent than those obtained using General Algal Group Index (InGA) proposed by Catalàn et al (2003).

Table 4 compiles values for InGA Catalàn et al. 2003). These authors recommended that the use of this multi-metric index of phytoplankton composition for ecological status classification should be calculated for late summer and fall samples. Trophic state metrics, as recommended by several water authorities in Portugal and in Europe, apply a color code for easy comparison of different classifications, which was used in Tables 3 and 4. In the case of Beliche and Odeleite reservoirs, fall values represented a worst case scenario and artificially attributed a good or moderate classification to waters that would otherwise be recognized as very good. Fig. 14 illustrates monthly variability of Catalàn InGA values.

Period	Beliche	Odeleite	InGA limits	InGA classes
2009-2010	0.167	0.237	>0.1	very good
2009	0.098	0.067	0.01-0.1	good
2010	0.196	0.403	0.005-0.01	moderate
October 2009	0.022	0.008	0.003-0.005	poor
October 2010	0.536	1.864	<0.003	bad

Table 4. Catalàn InGA values determined for Beliche and Odeleite reservoirs for several groups of samples

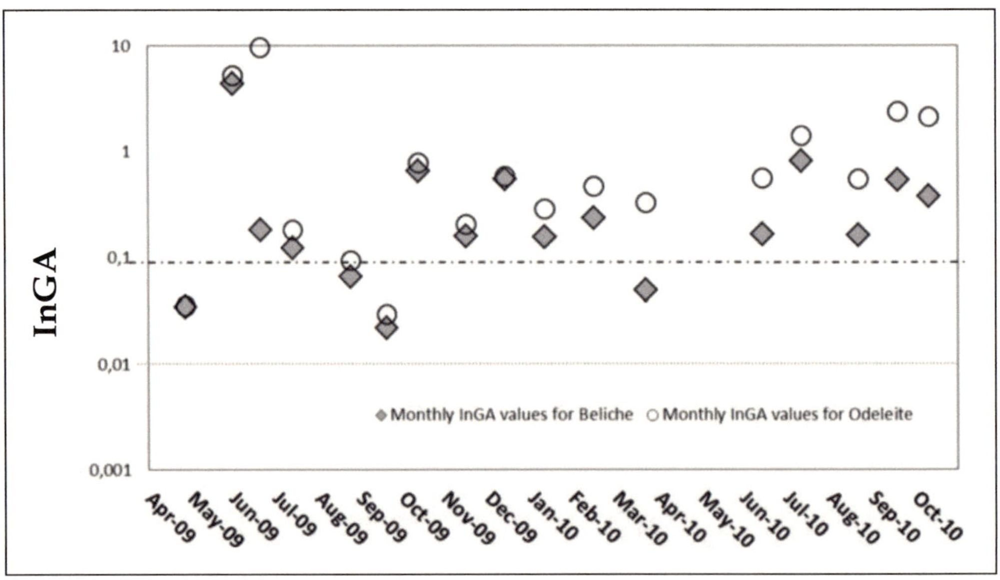

Fig. 14. Monthly values of the General Algal Groups Index (InGA) determined for Beliche and Odeleite Reservoir 2009-2010 data. Dashed lines indicate the lower limits for Very Good (0,1) and Good (0,01) classification

As discussed by Jackson et al. (2000), an index that does not allow for temporal and spatial variability, might lose reliability. The decreased InGA value calculated for summer 2009 (Fig. 14), reflected the weight of 4 given to cyanobacterial biovolumes in the index formula (Eq. 2)

$$InGa = \frac{1 + 2 * (D + Cnc) + Chnc + Dnc}{1 + 0.1 * Cr + Cc + 2 * (Dc + Chc) + 3 * Vc + 4 * Cia} \quad (2)$$

D	Dinophyceae	Cc	Chrysophyceae colonial
Cnc	Chrysophyceae non colonial	Dc	Bacillariophyceae colonial
Chnc	Chlorococcales non colonial	Chc	Chlorococcales colonial
DnC	Bacillariophyceae non colonial	Vc	Volvocales colonial
Cr	Cryptophyceae	Cia	Cyanobacteria

In fact, whenever a lake or reservoir has a cyanobacterial bloom, InGA value will indicate it, but so will simply calculating the relative contribution of cyanobacteria to total phytoplankton biovolume.

Another problem emerging from the application of multi-metric indices is the loss of interpretability. With InGA, high contributions to total biovolumes of Dinophyceae and non- colonial phytoplankton groups are assumed to improve the ecological status, whereas colonial forms and cyanobacteria worsen it. Thus, high biovolume proportions of non-toxic Chroococcales, are given the same negative weight as toxic filamentous cyanobacteria.

It is well known and accepted that different metrics applied to the same ecological condition can attribute different classifications. This is obviously linked to the information provided by the variables selected in each metric analysis. Criteria for the selection of these metrics should consider the prerequisites previously mentioned (see section 1.), such as: (i) obtainability, (ii) relevance in term of specific objectives, (iii) discriminant capacity, (iv) allowance of natural variability and (v) reliability.

Results obtained for dominance (Fig. 11) and cyanobacterial contribution (> 9%) to total phytoplankton biovolume, (Fig. 12), indicate eutrophication under WFD guidelines (JRC EC, 2009). Considering the oligotrophic state of both reservoirs, the reliability of such indicators should be reevaluated, at least for warm semi-arid regions.

Solimini et al. (2006) considered the contribution of cyanobacteria to total phytoplankton biomass, as a reliable and simple indicator of trophic state, based on the following assumptions: (i) most cyanobacteria species show a strong preference for eutrophic conditions, (ii) due to toxicity of some taxa, blooms can pose serious water quality, animal and human health risks, as well as environmental problems, and finally, (iii) the large contribution of cyanobacteria blooms to phytoplankton biomass.

This study contradicts the first assumption, since the genera and species typically linked to eutrophication were found associated to oligotrophy. The second assumption was partially verified, but toxin production seemed to be limited in oligotrophic conditions, despite high cell abundances. Potentially toxic cyanobacteria do not always produce cyanotoxins, so toxicity needs to be confirmed. Finally, the last assumption which links eutrophication to large contribution of cyanobacteria to phytoplankton biomass, should be re-assessed in Mediterranean regions, where, even under oligotrophic conditions, cyanobacteria are favoured in naturally warm waters.

3.6 Conclusion

This case study applied ecological multi-metric indices recommended by the European Commission (2008/915/EC) to two reservoirs of the Guadiana watershed, which have repeatedly developed cyanobacteria blooms in the summer, in association with high hydraulic retention. Values for the ecological potential measured by these indices ranged from Bad (> 9% contribution of cyanobacteria biovolumes) to Good (InGA) and Very Good (InGA and MedPTI). However, these indices do not provide insight on appropriate CHAB prevention and mitigation measures. Instead, long term monitoring of ecological data, was necessary to propose appropriate countermeasures. In fact, in these two reservoirs the only effective measure to prevent or mitigate CHABs was to reduce water residence time by discharging surface water.

4. Final considerations

The long-term study of the Guadiana estuary revealed unforeseen impacts in the aquatic microbial ecology after completion of the large Alqueva dam, causing in particular the collapse of natural cyanobacteria populations in the upper estuary. The sharp decline in photosynthetic prokaryotes, as well as in the phytoplankton community, could be attributed to overall perturbations in estuarine circulation, rather than any single or combined environmental drivers. Thus, regulation of dam discharges to maintain ecological river flow is essential to maintain estuarine primary productivity, using such ecological tools as abundance and diversity of cyanobacteria as sensitive indicators of "good" estuarine water quality. However, national environmental agencies and water resource authorities have yet to apply adequate ecohydrological approaches to river flow and dam discharge management, while policymakers seemingly lack the political will to enforce ecological river flow in Portuguese legislation.

On the other hand, monitoring of Beliche and Odeleite freshwater reservoirs assessed the usefulness of different ecological indicators. Aquatic ecologists have long presented a plea (e.g. Margaleff, 1974; Reynolds, 2002) for a better understanding of phytoplankton composition and dynamics in ecological studies. Multi-metric phytoplankton indices, such as recommended by the European Commission (2008/915/EC), attempt to translate complex biological information into user-friendly ecological classifications. These EC metrics might be useful for water policy purposes, but do not seem to have any utility in CHABs management. Ecological tools should clearly indicate the need for prevention or mitigation measures for CHAB management, which multi-metric indices fail to do. Instead, adequate ecological tools should rely on long-term multi-variate studies, which address the complexity of aquatic ecosystem function and dynamics.

5. Acknowledgments

The Guadiana estuary study was funded by a series of projects 16/REGII/6/96 and 15/REGII/6/96 (INTERREGII program), EMERGE (ODIANA regional program), project nr.

45 ("Risk evaluation of toxic blooms in lower Guadiana") from LEADER+ program, project DYNCYANO (PTDC/AMB/64747/2006) funded by the Portuguese Science and Technology Foundation (FCT). We also thank past post-doctoral fellows Dr. Alexandre Matthiensen, Dr. Carlos Rocha, and Dr. Cristina Sobrino for contributing to different parts of this study at different times, and the following Master´s and PhD students Rute Miguel, Pedro Mendes, Vânia Sousa and Teresa Cecílio, as well as research assistants Cátia Luis, Tânia Anselmo, Erika Almeida and Cátia Guerra. Collaboration from Dr. Wayne Carmichael and Prof. Vitor Vasconcelos was greatly appreciated.

R. Domingues acknowledges PhD (SFRH/BD/27536/2006) and Post-Doctoral (SFRH/BPD/68688/2010) fellowships from FCT, and S. Mesquita acknowledges PhD grant (SFHR/BD/18921/2004) and the research grant financed in the scope of the Southwest European Research Net (RISE) through the I2TEP UE Program- subprogram CIANOTOOLS.

Algarve reservoirs were studied in the scope of 4 research projects, namely CIANOALERTA I, II and III (2003-2008) funded through INTERREG IIIA, contracts nº SP5/P35/01, SP5/P19/02 and SP5/P138/03 and CIANOTOOLS financed in the scope of the Southwest European Research Net (RISE) through the POCTEP UE Program in collaboration with Algarve Regional Hydrographycal Administration and University of Huelva.

6. References

Admnistração da Região Hidrográfica do Alentejo (ARH Alentejo), (2011). Planos de gestão das bacias hidrográficas integradas nas regiões hidrográficas 5, 6 e 7-região hidrográfica 7. In: Relatório técnico para efeitos de participação pública, t09122/01, (Junho 2011), Acessed in August 1st 2011, Available from http://www.arhalentejo.pt/downloads/part_publi_pgrh/fase_final/RH7/t09122_01_PGBH_RH7_Relatorio_Tecnico.pdf

Alves, M. H. & Bernardo, J. M. (1998). Novas perspectivas para a determination do caudal ecológico em regiões semi-áridas. Proceedings of *Seminário sobre barragens e ambiente*. Comissão nacional Portuguesa da Grandes barragens, ISBN 92-894-5122-X, Porto, Maio de 1998

Alves, M. H. & Gonçalves, H. (1994). O caudal ecológico como medida de minimização dos impactes nos ecossitemas lóticos. Métodos para a sua determinação e aplicações. Proceedings of: *"Actas do 6ºSILUSB/1ºSILUSBA, Simpósio de Hidráulica e Recursos dos Países de Língua oficial Portuguesa"*, (Abril de 1994), Lisboa

Barbosa, A.B., Domingues, R.B. & Galvão, H.M. (2010). Environmental forcing of phytoplankton in a semi-arid estuary (Guadiana estuary, south-western Iberia): a decadal study of climatic and anthropogenic influences. *Estuaries and Coasts*, Vol. 33, No. 2, (March 2010), pp. 324-341, ISSN 1559-2723

Berg, K., Carmichael, W.W., Skulber, O.M., Benestad, C. & Underdal, B. (1987). Investigation of a toxic-water bloom of *Microcystis aeruginosa* (Cyanophyceae) in Lake Akersvatn, Norway. *Hydrobiologia*, Vol. 144, pp. 97–103

Blaha, L., Babica, P. & Maršálek, B. (2009). Toxins produced in cyanobacterial water blooms - toxicity and risks. *Intersisciplinary Toxicology*, Vol. 2, No. 2, (May 2009), pp.36-41

Brandão, C. & Rodrigues, R. (2000). Hydrological simulation of the international catchment of Guadiana River. *Physics and Chemistry of the Earth (B)*, Vol. 25, No. 3, pp. 329-339, ISSN 1464-1909

Brierley, B., Carvalho, L., Davies, S. & Krokowski, J. (2007). Guidance on the quantitative analysis of phytoplankton in Freshwater Samples. Report to SNIFFER (Project WFD80), Edinburgh, (December 2007), pp. 1-24

Carlson, R.E. (1977). A trophic state index for lakes. *Limnology and Oceanography*, Vol. 22, pp. 361-369.

Carmichael, W.W. & An, J. (1988). Using and enzyme linked immunosorbent assay (ELISA) and a protein phodphatade inhibition assay (PPIA) for the detection of microcystins and nodularins. *Natural Toxins*, Vol. 7, pp. 377–385

Carmichael, W.W. (1996). Toxic Microcystis and the environment. In: *Toxic Microcystis*. Watanabe, M.F., Harada, K.I., Carmichael, W.W. & Fujiki, H. (Eds). CRC Press, Boca Raton, Florida, ISBN 0-8493-7693-9

Carmichael, W.W., Min-Juan, Y., Zheng-Rong, H., Jia-Wan, H. & Jia-Lu, Y.(1988). Occurrence of the toxic cyanobacterium (blue-green alga) *Microcystis aeruginosa* in central China. *Archives Hydrobiology*, Vol. 114, pp. 21–30

Catalàn J., Ventura M., Munné A. & Godé L. (2003). Desenvolupament d'un índex integral de qualitat ecológica i regionalització ambiental dels sistemes lacustres de Catalunya. Agència Catalana del Agua, 177 pp.

CEN/TC230/WG2/TG3 N108 Water Quality - Phytoplankton biovolume determination by microscopic measurement of cell dimensions. European Center for Standartization, Accessed in August 1st 2011, Available from http://www.cen.eu/cen/Products/Pages/default.aspx

Chícharo, L., Ben Hamadou, R., Amaral, A., Range, P., Mateus, C., Piló, D., Marques, R., Chícharo, (2009). Application and demonstration of the Ecohydrology approach for the sustainable functioning of the Guadiana estuary (South Portugal). *Ecohydrology and Hidrobiology*, Vol. 9, No.1, pp. 55-71

Chícharo, L., Chícharo, M.A. & Ben-Hamadou, R. (2006). Use of a hydrotechnical infrastructure (Alqueva Dam) to regulate planktonic assemblages in the Guadiana estuary: basis for sustainable water and ecosystem services management. *Estuarine Coastal and Shelf Science*, Vol. 70, No. 1-2, (October 2006), pp. 3-18, ISSN 0272-7714

Chorus, I. & Bartram, J. (Eds.) (1999). Toxic cyanobacteria in water: a guide to their public health consequences, monitoring and management. E&FN Spon, ISBN 0-419-23930-8, London

Chorus, I. (2005). Current approaches to cyanotoxin risk assessment, risk management and regulations in different countries. Federal Environmental Agency, Dessau-Roßlau

Cloern, J. E. (1987). Turbidity as a control on phytoplankton biomass and productivity in estuaries. Continental Shelf Research, Vol. 7, No. 11-12, (November-December 1987), pp. 1367-1381, ISSN 0278-4343

Cloern, J.E. (1999). The relative importance of light and nutrient limitation of phytoplankton growth: a simple index of coastal ecosystem sensitivity to nutrient enrichment. *Aquatic Ecology*, Vol. 33, No. 1, (March 1999), pp. 3–16, ISSN 1386-2588

Codd, G.A. (2000). Cyanobacterial toxins, the perception of water quality, and the prioritization of eutrophication control. *Ecological Engeneering*, Vol. 16, pp.51–60

Comissão Europeia (2000) - Directiva 2000/60/CE do Parlamento Europeu e do Conselho de 23 de Outubro de 2000, que estabelece um Quadro de Acção Comunitária no Domínio da Politica da Água. Jornal Oficial das Comunidades Europeias, (Dezembro 2000). L 327, pp.1-72

Confederación Hidrográfica del Guadiana (CHG) (2006). Memoria de la cuenta del Guadiana 2006. Ministerio de Medio Ambiente, Medio Rural y Marino-Confederación Hidrográfica del Guadiana, Imprenta Moreno, S.L. - Montijo

Confederación Hidrográfica del Guadiana (CHG) (2007-2008). Memoria de la cuenta del Guadiana 2007-2008. Ministerio de Medio Ambiente, Medio Rural y Marino-Confederación Hidrográfica del Guadiana, Imprenta Moreno, S.L. - Montijo

Confederación Hidrográfica del Guadiana (CHG), (2009). Estado ecológico de las masas de agua de la conferaración hidrográfica del Guadiana, 2005-2006. In: Gráficas Paton. Dep. Legal, V-3712-2009, access in August 5th 2011, Available from: http://www.chguadiana.es/corps/chguadiana/data/resources/file/redes_control/informes_publicaciones/libro_guadiana_estado_ecologico.pdf

Cravo, A., M. Madureira, H. Felícia, F. Rita, & M.J. Bebianno. (2006). Impact of outflow from the Guadiana River on the distribution of suspended particulate matter and nutrients in the adjacent coastal zone. *Estuarine, Coastal and Shelf Science,* Vol. 70, pp. 63–75, ISSN 0272-7714

Daveau, S. (1987). Comentários e actuaçizações. In : *Geografia de Portugal II.* O ritmo climático e a paisagem. Edições São José da Costa, ISBN 972-9230-16-1, Lisboa

Domaizon, I., Wahl, B., Fehr, G. & Kroll, A. (2003). Water Resource Management for Important Deep European Lakes and their Catchment Areas EUROLAKES, D22: Microbial Biodiversity, Page 2 of 43 Accessed May 1st 2011, Available from http://www.hydromod.de/Eurolakes/results/D22.pdf

Domingues, R.B. & Galvão, H. (2007) Phytoplankton and environmental variability in a dam regulated temperate estuary. *Hydrobiologia,* Vol. 586, No. 1, (July 2007), pp. 117-134, ISSN 0018-8158

Domingues, R.B. (2010) Bottom-up regulation of phytoplankton in the Guadiana estuary. Ph.D. Thesis, Universidade do Algarve, Faro, 214 pp.

Domingues, R.B., Barbosa, A. & Galvão, H. (2005). Nutrients, light and phytoplankton succession in a temperate estuary (the Guadiana, south-western Iberia). *Estuarine, Coastal and Shelf Science,* Vol. 64, No. 2-3, (August 2005), pp. 249-260, ISSN 0272-7714

Domingues, R.B., Anselmo, T.P.,Barbosa, A.B., Sommer, U. & Galvão, H.M. (2011). Nutrient limitation of phytoplankton growth in the freshwater tidal zone of a turbid, Mediterranean estuary. *Estuarine, Coastal and Shelf Science,* Vol. 91, No. 2, (January 2011), pp. 282-297, ISSN 0272-7714

Druvietis, I. (1997). Observations on cyanobacterial blooms in Latvia's Inland. In: Harmful algae. Reguer,a B., Blanco, J., Fernández, M.L., Wyatt, T. (Eds). Xunta de Galicia, Intergovernamental Oceanographic Commision of UNESCO, Vigo

Dunn, J. (1996). Algae kills dialysis patients in Brazil. Br Med J, Vol. 312, pp. 1183-1184

Edler, L., (1979b). Phytoplankton counts. Results and analysis of the intercalibration experiments. Interim. Baltic Marine Environment Protection Commission, pp.20, Finland

Edler, L., (Ed.). (1979a). Recommendations for marine biological studies in the Baltic Sea. Phytoplankton and chlorophyll. *The Baltic Marine Biologists Publ.* No 5, pp. 1-38

Edwards, C., Beattie, K.A., Scrimgeour, C.M. & Codd, G.A. (1992) Identification of Anatoxin-a in Benthic Cyanobacteria (Blue-Green-Algae) and in Associated Dog Poisonings at Loch Insh, Scotland. *Toxicon*, Vol. 30, pp. 1165–1175.

Forján, E. D., Dominguez, M.J.V., Vilchez, C. L., Miguel, ., Costa, C. & Reis, M.P. (2008). Cianoalerta : estrategia para predecir el desarollo de cianobacterias tóxicas en embalses. *Ecossistemas XII*, Vol. 1, (January 2008), pp. 37-45

Galvão, H., Reis, M.P., Valério, E., Domingues, R.B., Costa, C., Lourenço, D.,Condinho, S., Miguel, R., Barbosa, A., Gago, C., Faria, N., Paulino, S. & Pereira, P. (2008) Cyanobacterial blooms in natural waters in Southern Portugal: a water management perspective. *Aquatic Microbial Ecology*, Vol. 53, No. 1, (September 2008), pp. 129-140, ISSN 0948-3055

Grasshoff, K., Ehrhardt, M. & Kremling, K. (Eds).(1983). Methods of Seawater Analysis. Verlag Chemie, ISBN 3527259988, Weinheim

Haas, L.W. (1982). Improved epifluorescence microscopy for observing planktonic micro-organisms. *Annalles de l'Institut Oceanographique de Paris*, Vol. 58, pp. 261-266, ISSN 0078-9682

Havens, K.E., James, R.T., East, T.L. & Smith, V.H. (2003). N:P ratios, light limitation, and cyanobacterial dominance in a subtropical lake impacted by non-point source nutrient pollution. *Environmental Pollution*, Vol. 122, 379-390

Hillebrand, H., Dürselen, C.-D., Kirschtel, D., Pollingher; U. & T. Zohary (1999). Biovolume calculation for pelagic and benthic microalgae. *Journal of Phycology*, Vol. 35, pp. 403-424.

Hoegar, S.J., Hitzfeld, B.C. & Dietrich, D.R. (2005). Occurrence and elimination of cyanobacterial toxins in drinking water treatment plants. Toxicology and Applied Pharmacology, Vol. 203, (April 2004), pp. 231-242, doi :10.1016/j.taap.2004.04.15

Instituto Nacional da Água, INAG. (2009). Manual para a avaliação da qualidade biológica da água. Protocolo de amostragem e análise para o Fitoplâncton. Ministério do Ambiente, do Ordenamento do Território e do Desenvolvimento Regional. Instituto da Água, I.P.

Jackson, L.E., Kurtz, J.C. & Fisher, W.S. (Eds). (2000). Evaluation Guidelines for Ecological Indicators. EPA/620/R-99/005. U.S. Environmental Protection Agency, Office of Research and Development, Research Triangle Park, NC

Jacquet, S., Briand, J.-F., Leboulanger, C., Avois-Jacquet, C., Druart, J.-C., Anneville, O. & Humbert, J.-F. (2004). The proliferation of the toxic cyanobacterium Planktothrix rubescens following restoration of the largest natural French lake (Lac du Bourget). *Harmfull Algae*, Vol. 4, pp.651-672

Jähnichen, S.; Jäschke, K.; Wieland, F.; Packroff, G. & Benndorf, J. (2011). Spatial-temporal distributoin of cell-bound and dissolves geosmin in Wahnbach Reservoir : causes and potential odour nuisances in raw water. Water Research, doi :10.1016/Jwatres.2011.06.043

Jochimsen, E.M., Carmichael, W.W., An, J., Cardo, D.M., Cookson, S.T., Holmes, C. E. M., Antunes, M. B. d C., Filho, D. A de M., Lyra, T.M., Barreto, V.S.T., Azevedo, S.M.F.O. & Jarvis, W.R. (1998). Liver failure and death after exposure to microcystins at a hemodialysis Centre in Brazil. *New England Journal of Medicine*, Vol. 338, pp.873-878

Joint Research Center European Commission (JRCEC) (2009). Part 2: Lakes. Section : Phytoplankton composition, In : Water Framework Directive Intercalibration technical report. In : EUR 28838 EN/2, Poikane, S. (Ed), Office for Official Publications of the European Communities, Luxembourg. Accessed May 2011, Accessed 28 June 2010, Available from:
hppt://circaeuropa.eu/Public/jrc/irc/jrc_eewai/library ?l=/intercalibration/ intercalibration_2/technical_versions/tr_feb08/lakes/phytoplankton_composition /_EN_1.0_&a=d

Kameyama, K., Sugiura, N., Inamori, Y. & Maekawa, T. (2004). Characteristics of microcystin production cell cycle of Microcystis viridis. Environmental Toxicology, Vol. 19, No. 1, (February 2004), pp. 20–25, ISSN 1522-7278

Kawara, O., Yura, E., Fujii, S. & Matsumoto, T. (1998). A study on the role of hydraulic retention time in eutrophication of the Asahi river dam reservoir. *Water Science and Technology*, Vol. 37, pp. 245–252

Kononen, K., Kuparinen, J., Makela, K., Laanemets, J., Pavelson, J. & Nömmann, S. (1996). Initiation of cyanobacterial blooms in a frontal region at the entrance to the Gulf of Finland, Baltic Sea. *Limnol Oceanogr*, Vol. 41, pp. 98–112

Machado, A., Rocha, F., Gomes, C. & Dias, J. (2007). Distribution and composition of suspended particulate matter in Guadiana Estuary (Southwestern Iberian Peninsula). *Journal of Coastal Research*, Vol. SI50, pp. 1040-1045, ISSN 0749-0208

Marchetto, A., Padedda, B.M., Mariani, M.A., Lugliè,A. & Secchi, N. (2009). A numerical index for evaluating phytoplankton response to changes in nutrient levels in deep mediterranean reservoirs. *Journal of Limnology*, Vol. 68, No. 1, pp. 106-121

Margalef, R. (1974). Ecología. Omega, Barcelona

Margurran, A. E. (1988). Ecological diversity and its measurement. Princeton University Press, ISBN 0-691-08485-8, Great Britain

Mendes, A. (2010). Water Scarcicity and drought management in transboundary river basin: Convenio da Albufeira » In: Procceding from International Conference on Water Scarcity and Drought « Path to Climate Change Adaptation .Madrid 18th-19th February. Acessed in August 1st 2011, Available from:
http://www.conferenciasequia.es/web_en/index.php?id=193&p=4&im=6

Meriluoto J., & Codd G.A. (Eds). (2005). Toxic Cyanobacterial Monitoring and Cyanotoxin Analysis. Abo Akademi University Press, ISBN 951-765-259-3, Finland

Mez, K., Beatie, K.A., Codd, G.A., Hanselmann, K., Hauser, B., Naegeli, H. & Preisig, H.R. (1997). Identification of a microcystin in benthic cyanobacteria linked to cattle deaths on alpine pastures in Switzerlan, fur. *Journal of Phycology*, Vol.32, pp.111-117

Morais, P., Chícharo, M.A. & Chícharo, L. (2009). Changes in a temperate estuary during the filling of the biggest European dam. *Science of the Total Environment*, Vol. 407, No. 7, (March 2009), pp. 2245-2259, ISSN 0048-9697

Odum, E. (1971). Fundamentals of Ecology, W.B. Saunders Company, ISBN: 0721669417

Olenina, I., Hajdu, S., Edler, L., Andersson, A., Wasmund, N., Busch, S., Göbel, J., Gromisz, S., Huseby, S., Huttunen, M. ,Jaanus, A. , Kokkonen, P., Ledaine, I. & Niemkiewicz,

E. (2006). Biovolumes and size-classes of phytoplankton in the Baltic Sea. HELCOM Balt.Sea Environ. Proc. No. 106, 144pp.

Oliveira, A., Fortunato, A.B. & Pinto, L. (2006). Modelling the hydrodynamics and the fate of passive and active organisms in the Guadiana estuary. *Estuarine, Coastal and Shelf Science*, Vol. 70, No. 1-2, (October 2006), pp. 76-84, ISSN 0272-7714

Paerl, H.W., Valdes, L.M., Adolf, J.E. & L.W. Harding Jr. (2006). Anthropogenic and climatic influences on the eutrophication of large estuarine ecosystems. *Limnology and Oceanography*, Vol. 51, pp.448- 462

Parsons, T.R., Maita, Y. & Lalli, C.M. (1984). A Manual of Chemical and Biological Methods for Seawater Analysis. Pergamon Press, Oxford, UK, ISBN 0-08-030287-4

Pearl, H. (2008). Nutrient and other environmental controls of harmful cyanobacterial blooms along thefreshwater–marine continuum. In: *Cyanobacterial Harmful Algal Blooms: State of the Science and Research Needs*. Hudnell HK. (Ed.), pp. 217 - 238, Springer, ISBN 978-0-387-75864-0, New York

Perovich, G., Dortch, Q. & Goodrich, J. (2008). Causes, prevention, and mitigation work group report. In: *Cyanobacterial Harmful Algal Blooms: State of the Science and Research Needs*. Hudnell H.K. (Ed.) pp. 185 - 216, Springer, ISBN 978-0-387-75864-0, New York

Piehler, M. (2008). Watershed management strategies to prevent and control cyanobacterial harmful algal blooms. In: Cyanobacterial Harmful Algal Blooms: State of the Science and Research Needs. Hudnell HK. (Ed.), pp. 259 - 273, Springer, ISBN 978-0-387-75864-0, New York

Pinckney, J.L., Paerl, H.W., Harrington, M.B. & Howe, K.E. (1998). Annual cycles of phytoplankton community-structure and bloom dynamics in the Neuse River Estuary, North Carolina. *Marine Biology*, Vol. 131, pp. 371-381

Plano da Bacia Hidrográfica do Guadiana (PBH), (2001). Plano da bacia hidrográfica do Guadiana, normas regulamentares. Ministério do ordenamento do Ambiente e do Ordenamento do Território, (Abril 2001), pp.114

Pouria, S., Andrade, A. De, barbosa, J., Cavalcanti, R.L., Barreto, V.T.S., ward, C.J., Preiser, W., poon, J.K., neild, G.H. & Cood, G.A. (1998). Fatal microcystin intoxication in haemodialysis unit in Caruaru, Brazil. *The Lancet*, Vol. 352, pp.21-26

Rapala, J., Sivonen, K., Lyra, C. & Niemelä, S.I. (1997). Variation of microcystins, cyanobacterial hepatotoxins, in Anabaena spp. as a function of growth stimuli. *Applied and Environmental Microbiology*, Vol. 63, No. 6, (June 1997), pp. 2206–2212, ISSN 1098-5336

Reynolds, C.S., Huszar, V.L., Naselli-Flores, L. & Melo, S. (2002). Towards a functional classification of freshwater phytoplankton. *Journal of Plankton Research*, Vol. 24, No. 5, pp. 417-425

Reynolds, C. S. (2006). The Ecology of Phytoplankton, Cambridge University Press, ISBN 10 0-511-19094-8, New York

Ribau Teixeira M. & Rosa, M.J. (2005). Microcystins removal by nanofiltration membrane. *Separation and Purification Technology*, Vol. 46, No. 3, (November 2005), pp. 192-201, ISSN 1383-5866

Rocha, C., Galvão, H. & Barbosa, A. (2002). Role of transient silicon limitation in the development of cyanobacteria blooms in the Guadiana estuary, south-western

Iberia. *Marine Ecology Progress Series*, Vol. 228, (March 2002), pp. 35-45, ISSN 0171-8630

Saker, M.L., Fastner, J., Dittmann, E., Christiansen, G. & Vasconcelos, V.M. (2005). Variation between strains of the cyanobacterium Microcystis aeruginosa isolated from a Portuguese river. *Journal of Applied Microbiology*, Vol. 99, No. 4, (October 2005), pp. 749–757, ISSN 13645072

Sivonen, K. & Jones, G. (1999). Cyanobacterial toxins. In : *Toxic cyanobacteria in water : a guide to their public health consequences, monitoring and management*. Chorus, I. & Bartram, J. (Eds), pp.41-111, E&FN Spon, London

Sobrino, C., Matthiensen, A., Vidal, S. & Galvão, H. (2004). Occurrence of microcystins along the Guadiana estuary. *Limnetica*, Vol. 23, No. 1-2, pp. 133-143, ISSN 0213-8409

Solimini, A.G., Cardoso, A.C. & Heiskanen, A.-S. (Eds). (2006). Indicators and methods for the ecological status assessment under the Water Framework Directive. Linkages between chemical and biological quality of surface waters. EUR22314. Joint research Center, European Commission. Accessed in May 1st 2011, Available from https://www.tcd.ie/Zoology/research/research/freshwater/documents/EUR22314.pdf

Utermöhl, H. (1958). Zur Vervollkommung der quantitativen Phytoplankton-Methodik. Internationale Vereinigung für Theoretische und Angewandte *Limnologie*, Vol. 9, pp. 1-38, ISSN 0368-0770

Varela-Ortega, C., J.M. Sumpsi, A. Garrido, M. Blanco & E. Iglesias. (1998). Water pricing policies, public decision making and farmers' response: implications for water policy. *Agricultural Economics*, Vol. 19, No.1-2, pp.193-202

Varela-Ortega, C., Sumpsi J.M. & M. Blanco. (2003). Análisis económico de los conflictos entre el regadío y la conservación de humedales en el Alto Guadiana. In: *Conflictos entre el desarrollo de las aguas subterráneas y la conservación de los humedales*. Coleto, C., Martínez-Cortina L. & Llamas, M.R. (Eds), Fundación Marcelino Botín. Mundiprensa, Madrid

Vasconcelos, P.R., Reis-Santos, P., Fonseca, V., Maia, A., Ruano, M., França, S., Vinagre, C., Costa, M.J., Cabral, H. (2007). Assessing anthropogenic pressures on estuarine fish nurseries along the Portuguese coast: a multi-index and conceptual approach. *Science of the Total Environment*, Vol. 374, No. 2-3, (March 2007), pp. 199-215, ISSN 0048-9697

Wiegand, C. & Pflugmacher, S. (2005). Ecotoxicological effects of selected cyanobacterial metabolites a short review. Toxicology and Applied Pharmacology, Vol. 203, (March 2005), pp. 201-218, ISSN 0041-008X

Wolanski, E., Chicharo, L. & Chicharo, M.A. (2008). Estuarine Ecohydrology. In: *Ecological Engineering*. Jørgensen, S.E., Fath, B.D. (Eds). Vol. 2 of Encyclopedia of Ecology, Elsevier, Oxford, ISBN 13: 978-0-44-452033-3

World Health Organization (WHO). (1998). Cyanobacterial toxins: Microcystin-LR. In: *Guidelines for drinking-water quality*. World Health Organization, Geneva, ISBN 0-419-23930-8

Wright, J.C. (1967). Effect of impoundments on productivity, water chemistry and heat budgets of rivers. Proceedings of *The reservoir fishery resources symposium, Southern Division American Fisheries Society*, Athens, University of Georgia

Young, A.J., Orset, S. & Tsavolos, A. (1996). Methods for carotenoids analysis. In: *Hanbook for Photosynthesis 1st edition*. Pessarakli, M. (Ed). Marcel Dekker, New York, USA

8

Ecosystem Technologies and Ecoremediation for Water Protection, Treatment and Reuse

Tjaša Griessler Bulc[1], Darja Istenič[2] and Alenka Šajn-Slak[3]
[1]*University of Ljubljana, Faculty of Health Sciences, Department of Sanitary Engineering, Ljubljana,*
[2]*LIMNOS Company for Applied Ecology Ltd.*
[3]*CGS plus Ltd.*
Slovenia

1. Introduction

Despite the access to safe drinking and sanitary water, which is a precondition for human health and well-being, water quality is still seriously threatened by point and non-point sources of pollution, originating mostly from urban and rural areas. Due to rapid development, the problems associated with urbanization do not delay to appear: water scarcity, food insecurity and pollution (Esrey, 2000). Most people in Europe do have access to drinking water of good quality, but on the other hand there are one billion people worldwide with limited or no access to uncontaminated water (Jenssen et al., 2004). Conventional sewer systems use considerable amounts of valuable drinking water for flushing and transporting toilet waste. In the processes, huge amounts of fresh water, up to 50,000 liters per year per person, are contaminated and deemed unfit for other purposes. A massive flow of nutrients, drained from rural and urban areas, mixes with fresh waters. These nutrients take the form of excreta and are usually disposed into deep lakes or pits, rivers, and coastal waters. The excreta are toxic for many forms of aquatic life (*e.g.* fish and coral reefs), they cause eutrophication, reduce biodiversity, affect human health and soil quality (Esrey, 2000). Accretion of excreta also causes accumulation and release of toxic substances like heavy metals and micro-pollutants. On the other hand, research findings show that the world's reserves of commercial phosphate will exhaust in fifty to hundred years and, as predicted, the production of phosphorus will reach its peak around 2030 (Cordell et al., 2009). It is obvious that wrong flow of nutrients causes the loss of soil fertility and unnecessary water pollution at the same time.

However, agriculture is beside sewage still recognized as one of the major sources of nutrient loading and a significant factor in terms of ecological quality (Iital et al., 2008). According to OECD (2006) pollution from agriculture have been declining in recent years , diffuse pollution of ground and surface waters with excess nitrogen and phosphorus remains the most severe environmental problem of intensive agriculture (Herzog et al., 2008). Soluble reactive phosphorus originates from point sources (*e.g.* overflow from slurry tanks, farmyard cleaning) (Neal et al., 2008), meanwhile non-point sources of phosphorus are caused by soil erosion, agricultural runoff, and drainage where phosphorus is mainly attached to soil particles (Simon & Makarewicz, 2009). In addition to nutrients, pesticides and heavy metals are also frequent pollutants originating from agriculture. It is reported that only 0.1% of pesticides applied to fields actually reaches the target out of 500 different used pesticides, while the rest enters the environment and contaminate soil, water and air (Arias-Estevez et al., 2008). The reason for inefficient application of pesticides and the resulting high emissions to the environment is inappropriate use of pesticides, including the use of unsuitable equipment for pesticide application, preventive use of the pesticide instead of obeying application programmes according to the crop growth, and application before the rainfall (Appleyard & Schmoll, 2006). Pollution of water bodies with pesticides usually coincides with nitrate and bacteria pollution. Further on, there is little information available about the fate, the behaviour, and the potential effects of xenobiotics in the environment (Žegura et al., 2009). Nevertheless, both water and soil pollution has to be considered holistically, including the synergistic effects of pollutants; namely, in most cases of pollution with non-degradable or slowly degradable pollutants, such as heavy metals and xenobiotics, sediments are the final recipient of these substances that consequently accumulate there. A problem arises when toxic substances re-enter the biological mass flows and integrate into food chains, which may represent hazard to numerous organisms.

The human perception of non-limited water and soil resources and the assumption that the environment can assimilate the wastes that we produce from using these resources, leads to a linear flow of resources and waste that are not reconnected. The linear attitude that regards resources and wastes must be therefore changed towards a circular one, advancing towards to a recycling society. The concept of ecological sanitation therefore provides a "recycle" philosophy of dealing with what in the past has been regarded as waste and wastewater (Werner et al., 2000). The incentives for wastewater reuse/recycling are becoming ever stronger with increasing pressures on drinking water supplies. As a reaction, water reclamation, recycling and reuse are now recognized worldwide as the key constituent of the efficient management of water resources. An increasing number of novel systems integrating decentralized treatment approaches, source separation and nutrient recycling have evolved in recent years (Jenssen et al., 2009). With such an approach we can minimize water pollution while ensuring rational water consumption and its reuse for irrigation, groundwater recharge or even direct reuse to the benefit of agriculture (Werner et al., 2000,). Recycling by the recovery of phosphorous from waste products and the efficient use of phosphatic mineral fertiliser and manure in agriculture are the major opportunities of increasing its life expectancy. As Vinnerås (2002) said, 80-90% of plant nutrients (nitrogen, phosphorus and potassium) in wastewater are present in the toilet waste and if these nutrients are reclaimed by safe methods, they can be applied locally as fertilizer in sustainable agriculture.

An important aspect in water consumption and reuse as well as in pollution of natural water bodies is also management of stormwater. Stormwater runoff generates as a result of precipitation on impervious surfaces, from where it flushes different pollutants. It also presents a hydraulic load for the receiving water body causing erosion and floods. Stormwater is characterized by containing relatively low, but not insignificant pollutant concentrations. This characteristic of stormwater creates difficulties in treatment of runoff water because rather low pollutant levels in large volumes of water need to be reduced to yet lower concentrations. However, in the short period of first flush event high concentrations of pollutants can occur. In order to protect natural water bodies against pollution and physical damage caused by stormwater runoff, the water has to be retained and treated. Due to the dispersed origin and the big quantities of runoff water that have to be controlled, the strategy of nowadays stormwater treatment systems is towards a large number of low-cost decentralized facilities. A proper retention and treatment of stormwater enables reuse of treated water for different purposes including toilet flushing, watering gardens and parks, carwash etc. which can significantly reduce the consumption of drinking water. Stormwater systems are frequently located in parks and recreational zones, and thus need to be planned in consideration of urban and landscape architecture. They often represent a pleasant wetland or pond element in urban parks and residential areas and as such give an added value to the area. Many systems for stormwater retention and treatment enable percolation of stormwater to the underground and thus recharge of the aquifers, which are otherwise disconnected from the recharge by precipitation due to impervious surfaces.

2. Approaches to water and pollution management

Sources of data for this chapter are EC official web page (www.ec.europa.eu) and European Environment Agency web page (www.eea.europa.eu).

The EU legislation and international agreements have extensively addressed the pollution of aquatic ecosystems in the last three decades, to mention in particular the Urban Wastewater Treatment Directive (91/271/EEC), the Nitrates Directive (Directive 91/676/EEC) and IPPC Directive (96/61/EC), the Bathing Waters Directive (76/160/EEC & 2006/7/EC) and the Water Framework Directive (WFD; 2000/60/EC). Data for this chapter are taken from the EC official web page (ec.uropa.eu).

The main requirements of the Water Framework Directive are to reach good ecological and chemical status of all inland, transitional and coastal waters by 2015. All pollutants and their associated anthropogenic activities must be addressed on river basin scale to ensure that good status is attained and maintained. Moreover, the WFD requires the removal or substantial reductions in the discharge of hazardous substances to water bodies. The adoption of the WFD has renewed debate on how the European Union's Common Agricultural Policy can contribute to achieving the goal of "good status" for all water bodies. Since 2000, a shift of the policy from a strictly production-oriented system towards a tool to support sustainable development has occurred (Agenda 2000).

Agri-environmental measures (AEM) have been a significant move towards achieving a good status for all water bodies, in particular with regard to nutrient losses. In the context of ecosystem technologies (ET) and ecoremediations (ERM), the AEM are the most effective legislative tool, which was first introduced into EU agricultural policy during the late 1980s. Since 1992, the application of agri-environment programmes has been compulsory for Member States in the framework of their rural development plans, whereas they remain optional for farmers.

The commitments included in national/regional agri-environmental schemes are:

- environmentally favourable extensification of farming;
- management of low-intensity pasture systems;
- integrated farm management and organic agriculture;
- preservation of landscape and historical features such as hedgerows, ditches and woods;
- conservation of high-value habitats and their associated biodiversity.

The AEM are implemented at the national, regional or local levels so that they can be adapted to particular farming systems and specific environmental conditions. Therefore, the agri-environment measures are a targeted tool for achieving environmental goals. The AEM are co-financed by Member States. For the period 2007-2013, the EU expenditure on AEM amounts to nearly 20 billion EUR or 22 % of the expenditure for rural development.

In Slovenia agricultural policy started to apply the first measures to support the environment-friendly ways of production in 1999 and after the adoption of the Slovenian agri-environmental programme in 2001. After joining the EU in 2004, the support under agri-environmental programme became part of the Rural Development Programme of the Republic of Slovenia. The area included in the implementation of AEM has been strongly increased after 1999 and in 2008 covered 323,043 ha (gross), or 247,420 ha (net). The share of area with one or several AEM (net) in the period 1999-2008 has increased from 0.6 % to 50.2 % of all utilised agricultural area.

However, only with adequate water monitoring can the potential impact of mitigation measures under the Water Framework Directive, Nitrate Directive, and AEM is assessed (Iital et al., 2008). Although some national studies regarding the impact of policy on pollutants concentration in waters already exist (Erisman et al., 2001) and some of them has been performed recently (Herzog et al., 2008), there is no international data base that compares the dynamic of implementation of national legislations concerning water quality and the level of water pollution deriving from agriculture in different countries.

In 2009, the European Commission introduced the White Paper on adapting to climate change, presenting the framework for measures and policies to reduce the European Union's vulnerability to the impacts of climate change. The White Paper underlines the need "to promote strategies which increase the resilience to climate change of health, property and the productive functions of land, inter alia by improving the management of water resources and ecosystems." Within this framework, Water Directors of EU Member States adopted in December 2009 a guidance document on adaptation to climate change in water management to ensure that the River Basin Management Plans (RBMP) are climate-proofed.

In spite of all these endeavours, the European Environment Agency indicated in its Environment State and Outlook Report 2010 that the attainment of EU water policy objectives is far from certain due to a number of old and emerging challenges. Therefore, the EU policy response to these challenges will be the Blueprint to Safeguard Europe's Water, aiming to ensure good quality water in sufficient quantities for all legitimate uses. The time horizon of the Blueprint is 2020 since it is closely related to the EU 2020 Strategy and, in particular, to planned Resource Efficiency Roadmap. The Blueprint will be the water milestone on the Roadmap. However, the groundwork supporting the Blueprint will take longer and will drive the EU policy until at least 2050.

3. Development of ecosystem technologies (ET) in sustainable water management

In its long history, nature has developed intense self-cleaning and buffering capacities. In the context of ecological sanitation and sustainable water management, the application of technologies that mimic healthy natural ecosystems became vital. These technologies aim to close the loop and return resources back to the source. Their basic characteristics, which can be utilised and further improved, are their high buffer and self-protective capacity as well as the provision of habitat diversity. Moreover, these systems have the remediation ability, provide for a high level of biodiversity and higher stability of ecosystems.

Ecosystem technologies or ecoremediations by definition comprise methods of protection and restoration of the environment through natural ecosystem processes. The establishment of ET provides sustainable solutions that contribute to the preservation of biodiversity, pollution reduction, enable nutrient recycling and reuse of material and can be applied in protected and sensitive areas. The functions of ET are based on aquatic, waterside, wetland as well as terrestrial ecosystems' characteristics, such as high water retention capacity, flood prevention, biodiversity as well as specific physical, chemical and biological processes for the reduction of diverse pollutants.

Most ET designs have its origin in the treatment wetlands (TW), ponds, and river restoration where, along with hydraulic, physical, chemical and microbiological processes, phytoremedation also plays an important role. The possibility of applying phytoremediation has become well recognized and integrated in ET development. Yet, the decision on a particular phytoremediation method depends not only on the type of pollutant and the polluted medium, but also on the objectives to be achieved, i.e. reduction, stabilization, sequestration, assimilation, detoxification, mineralization and decomposition.

By applying the ET, a local community or even a small society can play a significant role. Namely, the application of ecological sanitation has shown for the importance of a design of sustainable wastewater treatment systems as a "household-centred approach" that seeks to resolve environmental sanitation problems at the minimum practicable size (Schertenleib, 2001). In the context of our contemporary environment, it is important to bring sustainability to local communities or to a household level. The life style of local communities is an important factor in pollution quantity and quality and is aggravated by trans-boundary air, water and soil pollution. Pollution originating from local communities should be treated with the use of ET. ET should be seen as prophylactic and therapeutic measures to overcome local environmental problems. They include alleviation and adaptation of local communities at a time when climate change system and global changes affect common life in local communities. They could represent an innovative approach towards nature, space and environment protection based upon system thinking or, in other words, a holistic approach involving technologies aimed at regional and local community levels. Macro remediation includes region-specific complex problem solving as exemplified by integrated river basin management, coastal region management, management of water resources but encompassing also techniques for specific remediation of damaged local habitats, remediation of pollution hot spots, such as landfill sites, polluted waters, soils, etc., which strictly speaking should be regarded as micro remediation measures or local problem solving techniques. As said, from a spatial perspective one can distinguish between macro and micro remediation, while in the context of complex problem solving, one can also distinguish between social, natural and technical measures differentiating from integrated management to single techniques aimed, for instance, at wastewater treatment by means of TW, river restoration or co-natural reclamation of landfills, etc.

Although some concern still exists regarding the "technical" completeness of ET, its application is consistently enforced in practice as well as among the environmentally aware society. The ecosystem technologies are most useful in the remediation of persistent environmental contamination, impacts of point pollution as well as seasonal pollution due to tourism and non-point pollution caused by agriculture. They are also appropriate for protecting sensitive areas and for rational water management in dry areas (Griessler Bulc & Šajn-Slak, 2009).

3.1 Why ET?

Multi-functionality is an intrinsic feature of ET, which can be considered a flagship application of good ecological engineering/biotechnology.

- Water treatment: ET effectively treat a large variety of wastewater (sewage, gray water, agricultural, highway runoff, landfill leachate, wastewater from food processing and textile industry, composting facility runoff, etc.) and increase the self-cleaning capacity of natural or revitalized ecosystems.
- Water retention in the landscape: ET reduce hydraulic peaks by retaining water in the system and therefore prevent and mitigate floods and droughts. They can contribute to an improved water management, mitigate water abstractions and recharge groundwater.
- Saving energy: ET can provide their services with very little or no energy input if designed accordingly.
- Enhanced biodiversity: ET create a new habitat for wildlife and can contribute to an increased biodiversity in a barren landscape (e.g. spawning ground for frogs and toads, breeding sites for birds etc.).
- Biomass production and nutrient recycling; if designed for this purpose, ET can recycle nutrients from runoff to a large degree and convert them to biomass which can be used as energy or raw material source (*e.g.* thermal insulation).
- Recreation: ET can be designed with elements of landscape architecture and can create an attractive place for the population.
- Education: ET are a good and tangible example of a measure aimed to achieve sustainable development. They can be used to present the problems of pollution and its remediation in a natural way to different target groups (e.g. how a waste product can be transformed into something valuable).

4. Overview of pollutant pathways and the efficiency of ecosystem technologies

Nutrients and pollutants in natural systems as well as in ET are subdued to numerous processes that enable pollutant and nutrient transformations, degradation or stabilization. The pathways of pollutants in natural ecosystems and ET depend on the characteristics of the system, namely dissolved oxygen concentration, pH, Eh, mineral composition of the media, bacterial, plant and animal communities; besides this, also external influences like climatic conditions and input loads are of significant importance.

4.1 Nitrogen and phosphorous

Regarding the nutrients, nitrogen and phosphorous compounds are of major concern. Increased ammonia concentrations in natural water bodies mainly indicate pollution from wastewater discharge, agricultural runoff or organic waste disposal. The presence of ammonia can significantly reduce the oxygen level in water body. Due to rapid oxidation to nitrate, nitrite is usually present in negligible concentrations in natural water bodies. In contrast to nitrite, the concentrations of nitrate in water surface and groundwater bodies are elevated in many parts of the world. The areas with intense agriculture are the most vulnerable due to nitrate pollution of drinking water.

Nitrogen and phosphorous are also of big interest because they are needed for the production of human food and have an important role in plant growth which further stimulates the wildlife production. Phosphorus in nature does not appear in elementary form but always in different compounds together with other elements. The extraction of P is a damaging process (strip-mining is most common method) having as result toxic by-products (arsenic, cadmium, etc.). Phosphorus is highly bondable and is adsorbed to soil molecules and as such it is not available for the plant use. Plants can uptake phosphorous when the soil is saturated and there is free phosphorous available. The result of this fact is high demand on artificial fertilizers to increase nutrient content in less fertile soil.

Nitrogen and phosphorous removal is required in most European countries in order to reduce the input to water bodies or the sea (eutrophication). Removal of nitrogen compounds is also important in order to prevent oxygen depletion and the toxicity on invertebrate and invertebrate species, including humans.

The processes of volatilization, plant uptake, nitrification and denitrification enable efficient removal of nitrogen from wastewaters in ET (Brix, 1993; Šajn-Slak et al., 2005). However also numerous other processes affect nitrogen elimination and retention in ET, i.e. ammonia volatilization, anaerobic ammonia oxidation (ANAMOX), fixation of nitrogen from the atmosphere, incorporation into plant and microbial biomass, fragmentation, burial, degradation of organic nitrogen compounds (mineralization), reduction of nitrate to ammonia, sorption and desorption from different media, filtration and leaching. Only certain pathways enable N elimination, whereas other processes only transform N from one form to another (Vymazal, 2007). A crucial condition for N elimination in an ET is an exchange of oxic and anoxic areas in micro level, which enable a co-existence of nitrifying and denitrifying bacteria. First step in N elimination is degradation of organic N compounds (ammonification) where ammonia is generated. NH_4-N is then oxidized to nitrite and further to nitrate, but in contrast to organic material, NH_4-N is more demanding for oxidation in TW as nitrifying microbes are autotrophic bacteria which have slow respiration and need a significant amount of oxygen to function. Besides oxygen conditions plants can also be an important factor in nitrogen elimination. Plants preferably take up the reduced form of nitrogen (ammonia), but can also take up nitrate. Nitrogen is integrated in plant tissues and can be eliminated from the ET by mowing and removing the plant biomass. This can present a significant nutrient removal in systems that receive low nutrient loads (Langergraber, 2005); according to Vymazal (2006) plant harvesting can significantly contribute to nitrogen removal in wetlands that receive less than 100-200 g N m^{-2} $year^{-1}$. However, in other free water ET denitrification is usually the main mechanism for nitrate removal), which enables reduction of nitrite and nitrate to gas nitrogen. N_2 is released to the atmosphere and thus presents ultimate elimination of N from the ET. Unlike nitrogen removal, the capacity of ET to eliminate phosphorus is of a concern and has not been reasonably resolved. Phosphorous in TW systems is mainly accumulated in the sediment/media and/or accumulated in the plant biomass. Elimination of P from a wetland treatment system is possible only by harvesting of the plant biomass and removing the saturated sediment.

The investigations on TW show that wetlands have a limited capacity to remove phosphorus and are greatly dependent on the characteristics of the media integrated into TW, namely as the media gets saturated with phosphorus, phosphorus removal efficiency decreases. Phosphorous removal mechanisms can be affected also by other factors, such as biofilm growth on the media, which limits the contact between the media and the treated water. In open water ET P is mainly accumulated in the bottom sediment. A crucial condition for trapping P in the sediment is a consistent and high redox potential in the surface sediment layer. P is leached from the sediments during anoxic conditions, however also microbial activity can cause its release. In neutral and acid conditions, microorganisms can use Fe^{3+} as electron acceptor thus releasing bound P. In basic conditions, microorganisms can dissolve insoluble P by increasing the ion exchange of OH^- and PO_4^{3-}, which arise from Fe-P or Al-P (Huang et al., 2008).

As for nitrogen also for elimination of P with plant harvesting is efficient only in the systems with low P loads. Inflow P concentrations depend on the type of water treated, while P content in the plant tissues remains in the same range, i.e. 1-5 g P m^{-2}. Primary treated municipal wastewater usually presents between 100 in 200 (800) g P m^{-2} $year^{-1}$, which means low P elimination with plant harvesting. However when treating secondary treated wastewater with less than 20 g P m^{-2} $year^{-1}$, harvesting plant biomass can contribute to P elimination from the system for more than 20 % yearly. Despite this, plant harvesting demands specific attention in terms of timing the harvest, as plant harvesting in growth season can severely damage the canopy. Removal of plant biomass can also present a problem in temperate and cold climates due to dead plant material acts as an isolation layer during winter months. In tropical climates with long vegetation period plant harvesting can significantly contribute to nutrient elimination due to several harvests per year are feasible (Vymazal, 2004).

To improve the removal of P in ET numerous solutions have been proposed and examined: e.g. a use of chemically enhanced material with high sorption capacity for P (e.g. Filtralite) for construction of the ET system; an integration of a pre-treatment step for chemical precipitation of P; an elimination of P in separate filters with specific media, etc. The tests of finding appropriate media for P elimination suggest the selection of materials, which must at the same time have good hydraulic features and consistent and continual elimination of P from treated waters.

4.2 Heavy metals

Unlike nutrients, metals in natural ecosystems and ET are not subdued to degradation, but can only change the ionic form. During water treatment with an ET metals cannot be eliminated but are accumulated in the systems' sediment, soil or plant tissues. Many heavy metals are micronutrients for animals and plants and are essential in low concentrations (e.g. Zn and Cu). Other heavy metals (e.g. Cd) as well as high concentrations of micronutrients can be toxic to biota. In water, metals take forms of free metal ions or can be bound to or adsorbed onto organic and inorganic particulate matter and complexes. The most bioavailable form is the soluble form, especially when the metal is present as a free ion or is weakly complexed, and can cause bioaccumulation in the food chain.

The removal of metals from water in wetlands can result in accumulation in the sediment, which might be harmful for the organisms that live or feed on these sediments. To avoid this problem pretreatment to reduce inflow metal concentrations, installation of deep water systems or subsurface wetlands can be considered. In deep water systems with free-floating plants the sediment is deposited at great depths and is thus not available to the top feeders and subsurface wetlands minimize the opportunity for ingestion of metals (Kadlec & Wallace, 2009). Depositing sediments have the ability to adsorb significant quantities of trace metals. Especially organic matter, iron and manganese oxyhydroxides act as metal adsorbents in aerated systems. Under anaerobic conditions iron and manganese oxyhydrohydes dissolves and thus release metals into the aqueous phase. This may lead to a repartitioning of metals into the sulphide or carbonate precipitates. In the conditions where metal's concentrations are in excess of sulphides, metals may complex with organic matter. Organic matter can appear as surface coatings or as particulates and significantly affects metal speciation and bioavailability (Kadlec & Wallace, 2009). Besides, at oxic conditions, co-precipitation of heavy metals with iron, manganese, and aluminum hydroxides also relies on considerable supplies of secondary metals in the system, which might not be present. According to this retention of metals in the sediment can be modified by changes in substrate chemistry and redox potential, which is affected by wetland water depth and biological processes. Besides redox potential, also pH affects sorption/desorption of heavy metals at/from the sediment.

Metals are also accumulated in plant tissues. Most of the metals found in plants are stored in the roots and rhizomes, only small amounts may find their way to stems and leaves as plant physiological mechanisms prevent the transport of heavy metals from the underground tissues to the aboveground tissues, where the toxic heavy metals could damage the photosynthetic tissues. Consequently, harvesting aboveground parts does not enable effective removal of metals from the wetland. However, with the root's death some fraction of the metal may be permanently buried (Kadlec & Wallace, 2009).

4.3 Organic micropollutants

Numerous studies investigate the elimination of organic micropollutants such as polychlorinated biphenyls (PCBs), polycyclic aromatic hydrocarbons (PAHs), phthalates, linear alkylbenzene sulphonates (LAS), nonylphenols etc. Besides listed, there is also a long list of emerging pollutants which are gaining lot of attention from the scientist in recent years. Mainly they focus on personal care products and pharmaceuticals (Hijosa-Valsero et al., 2010a, 2010b).

The pathways of micro and emerging pollutants in ET systems are not clearly known and are still under research. As aromatic compounds, PAHs have low water solubility/high hydrophobicity and thus they tend to absorb to solid particles and are thus removed by sedimentation and filtration. Higher organic carbon content in soils and sediments increases sorption of PAHs. It is known that PAHs are decomposed relatively rapidly in many vertebrates, but more slowly and in a different way in certain other life forms; nevertheless, the degradation products of PAHs can be more harmful than the original compound. The PAHs with four or less aromatic rings are subdued to biodegradation by microbes or are metabolized by multicellular organisms. The heavier PAHs (four, five and six rings) are more insistent compared to the lighter (two and three rings) and tend to have greater carcinogenic and other chronic impacts (Mangas et al., 1998). Also many emerging pollutants are rather lipophilic in character, and are consequently likely to bind with the particulate matter. For this reason mechanical treatment can present an important step in elimination of these substances, but nevertheless, the absence of any biological step may have substantial effect on the total removal of more biodegradable compounds (Vogelsang et al., 2006).

Many treatment processes in ET (e.g. nitrification, aerobic degradation of organic matter and P trapping) are oxygen-limited. Studies have shown that the amount of oxygen transferred through the plants is very small compared to the oxygen demand utilized by the wastewater under usual loading rates. Consequently, many recent studies entirely neglect oxygen transfer by plants and include plants mostly as a microorganism carrier and biodiversity factor. The inadequate oxygen transfer of typical subsurface flow wetlands resulted in the progress of improved treatment systems, which are able to assure adequate oxygen levels for nitrification, degradation of organic matter, and prevention of P leaching. The systems include introduction of oxygen to the wetland by means of regular water level oscillations, passive air pumps or powered mechanical aeration of the reed bed (Nivala et al., 2007). Yet, the elimination of nitrogen in ET has been enhanced by the different flows, cascades, open areas, and with the application of recirculation of the treated outflow back to the inflow (Griessler Bulc & Šajn-Slak, 2009; Griessler Bulc et al., 2011).

4.4 Role of plants and algae

Removal processes of pollutants in ET can be controlled by hydraulic load, ET design as well as with macrophytes and algae. They directly and indirectly influence the physical and chemical environment in ET and play an important role in removal processes. Macrophytes e.g. enhanced sedimentation and sorption on biofilm and therefore accelerate removal of suspended solids, settleable solids, organic N, total N, COD and BOD_5. Floating macrophytes shade the water surface and reduce temperature oscillations, algal development and gas exchange with the atmosphere. Wooden plants can play an important role in phytoremediation processes and evapotranspiration. Algae with photosynthetic activity cause a higher pH (and consequently ammonia volatilisation and ortho-P precipitation), P accumulation and a higher DO concentration in water (consequently higher ortho-phosphate retention and more intensive nitrification). On the other hand, algae in the effluent cause a lower treatment efficiency of suspended solids and BOD_5 (Šajn-Slak et. al., 2005).

5. ET types

ET for water management merges vegetated drainage ditches (VDD), waste stabilization ponds (WSP) and stormwater detention ponds (SDP), TW, buffer zones, phytoremediation with dense woodland establishment, river revitalization, and in stream and bank side river techniques. One of the main aims of ET concept is to integrate exchange, combine and use multi-functionality of different kind of "green technologies" to obtain innovative and sustainable solutions for environmental protection and restoration.

5.1 Vegetated drainage ditches (VDD)

Drainage networks of surface and subsurface drains mainly serve to remove and accumulate excess water associated with irrigation and storm events. In agricultural areas they help to reduce surface water retention and low water tables for optimum plant production and therefore representing integral components for sustaining the economic development. Nonetheless, drainage networks affect several hundred thousand hectares of land in Western and Eastern Europe, leading to reduced water's self purification and retention capacity and the loss of biodiversity. Although the amount of new drainage networks declined significantly in Europe during the 90's, the existing drainage systems continue to pose negative impacts on the environment. Therefore, management of VDD to optimize sorption, complexation and sedimentation processes of pollutants is an important issue in drainage pollution control, which complies with AEM at the point of conservation of habitats and their associated biodiversity.

5.2 Waste stabilization ponds (WSP)

WSP are simple man-made basins for primary, secondary and tertiary treatment of variety of wastewaters. They are used worldwide, alone or in combination with other treatment processes. Anaerobic, facultative and maturation ponds are constructed in one or several series. Anaerobic ponds are designed for primary treatment. They remove suspended solids and some of the soluble element of organic matter (BOD). Most of the remaining BOD is removed in facultative pond (secondary stage) by algae and heterotrophic bacteria. Tertiary treatment takes place in maturation pond where pathogens and nutrients (especially nitrogen) are removed. WSP are low cost treatment technology with simple operation and maintenance (Ramadan & Ponce, n.d.).

5.3 Stormwater detention ponds (SDP)

Detention ponds are open water bodies for the retention of stormwater runoff from urban, agricultural and other areas. The stormwater treatment facility must be flexible to manage high flow rates of the runoff followed by dry periods, and high pollutant concentrations in the first flush followed by diluted concentrations in the main flow. Stormwater detention ponds are diverse biological system with a high buffering capacity that enable water detention, minimize the hydraulic peaks and reduce the pollutant input in downstream facilities and/or receiving waters (Hvitved-Jacobsen et al., 1994). Detained and treated water can be used for different purposes or discharged to the environment. Different plant species can appear in detention ponds (usually colonized by natural way): at the shallower marginal areas emergent and in deeper parts floating and submerged species. The treatment processes in wet detention ponds are similar to those occurring in natural smaller lakes and pools: e.g. contaminant accretion in the bottom deposits via sedimentation, adsorption to colloidal and particulate matter, conversions and degradation of biodegradable compounds by microorganisms and uptake of contaminants by plants. Among those, the key mechanism for pollutants removal in detention ponds is sedimentation. Since the main removal mechanism in wet detention ponds is sedimentation, the wet detention ponds generally have high efficiency in particulate matter removal (Terzakis et al., 2008). Organic matter is subdued to microbial and macroinvertebrate decomposition and final transformation to inorganic matter in the sediment, where it is stored. Sediment accretion at the bottom of wet ponds might vary greatly according to inflow and catchment characteristics. Hvitved-Jacobsen et al. (1994) estimated that excavation and removal of the sediments from wet ponds would be needed every 25 years of the operational period.

5.4 Treatment wetlands (TW)

TW are technically and economically feasible solutions for the treatment of different wastewater types. The technology is widespread around the world. Already ten years ago, more than 5,000 TW were operational in Europe (Vymazal et al., 1998). The performance is thoroughly documented and the systems are capable of reducing the concentration of target pollutants by different bacteriological, physical and chemical processes to acceptable levels before discharging to the environment and therefore mitigating the harmful effect that the disposal of untreated wastewater may have (Vymazal, 2007). TW imitate natural wetlands by using an array of natural processes to transform and remove the contaminants and as such they represent important part of ET. Compared to their natural counterpart, these processes are intensified. This is achieved with appropriate design, filling material, planting, and incorporation of technical equipment (pumps, aeration, pre-treatment) which ensures optimal utilization of the TW area and volume. As TW typically require less or no supplemental energy, their operational costs can be approximately two orders of magnitude lower than those of a standard three-stage waste water treatment plants (WWTP) (Grönlund et al., 2004). The TW removal efficiency usually assessed by the decrease in biochemical and chemical oxygen demand (BOD, COD), total suspended solids (TSS) and nutrient (N, P) load, has already been studied widely (Kadlec & Wallace, 2009). TW can also effectively remove a wide array of persistent pollutants such as pathogens, trace organic contaminants and heavy metals, which all have a negative influence if released into the environment. TW have been used as a treatment step before wastewater is reused in agriculture, but with very variable success. Although wetlands are effective for the treatment of wastewaters, the ever-changing reality of more stringent discharge regulations by the local governments imply that the wastewater treatment systems have to meet high water quality standards before discharge.

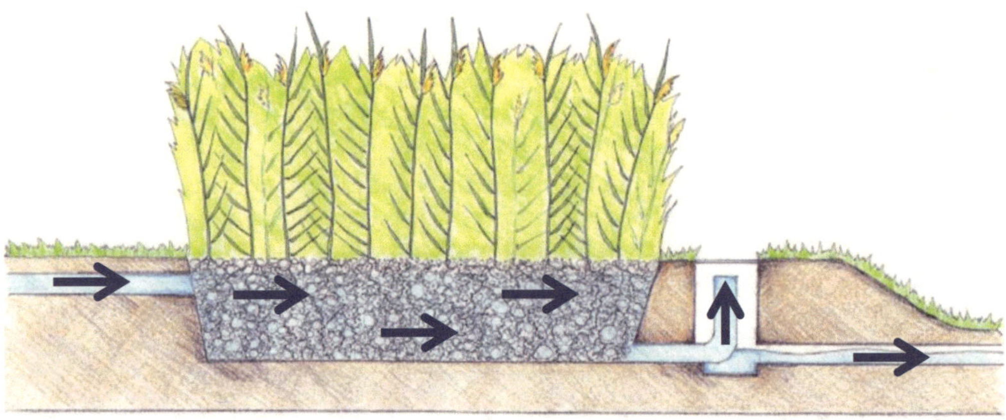

Fig. 1. A simple sketch of horizontal subsurface flow treatment wetland (source: LIMNOS Ltd.)

5.5 Phytoremediation of landfill sites

This ET involves the treatment and recycling of leachate on a vegetative landfill cover in order to avoid additional pollution by treated leachate discharges into the environment, to achieve landfill stabilization and easier public recognition of a reclaimed site. The final aim is to reduce the wastes' impacts on the environment through a closed hydrological and pollution cycle within a landfill site and the utilisation of leachate as a nutrient source. Leachate recycling belongs to a phytoremediation method where the assimilation of plant nutrients from leachate into biomass and faster waste decomposition by enabling leachate infiltration into the landfill body take place. Discharge of treated leachate to vegetation caps can provide an opportunity for closing the nutrient cycling loop and producing an effluent of a suitable quality. A controlled input of leachate results in a better provision of soil with nutrients and organic substances, improved growth of vegetation and intensified microbiological activity in soil. Today, the phytotechnology employing ligneous plants is used for the treatment of various forms of pollution. With a large water uptake from soil pores, plants take up also water pollutants and create a new capacity for water accumulation in soil. Poplars and willows are capable of taking up diverse pollutants and nutrients (nitrate, ammonium, phosphorus), metals, metalloids and petrochemical compounds (fuels, solvents), pesticides and soluble radionuclides (Zupančič et al., 2005). The methods applied for the treatment of leachate are vegetation barriers, filters, vegetation caps and short rotation coppices (SRC) with fast growing woody species. In addition to landfill sites, the planting of trees is used for the remediation of watercourse banks, abandoned and polluted industrial areas, at the margins of intensive agricultural areas and other polluted areas, as well as for the treatment of wastewater and sludge (Griessler Bulc & Zupančič Justin, 2007).

5.6 Watercourse revitalization

Natural watercourses have a great ability of water retention, a great diversity of habitats and biodiversity and high self-cleaning capacity. Regulations or canalizations of watercourses were common in the past but in some places sill appearing in the present with the main goal of flood protection and gaining space for agriculture and urbanization. Canalizations of watercourses do decreased flooding in a local scale but downstream floods were even more severe. Canalized watercourse has trapezium profile; river bed is straightened and often covered with stones or concrete. Habitats for different animal and plant species are destroyed, self-cleaning capacity is scarce and there is no water retention in the riverbed, banks and floodplains. In a canalized watercourse pollutants from surroundings can freely flow into the water. With revitalization of a watercourse ecological balance is restored using appropriate water management interventions. Revitalization of a watercourse enables restoration of habitats for aquatic plants and animals, increases water retention and self-cleaning capacity of a water body. The type of revitalization is chosen according to the scope of revitalization and space abilities in the environment. Where the space around watercourse is limited revitalizations can be implemented inside existing canal. With revitalization measures like stabilization of river banks with vegetation, construction of weirs, pools, rapids, water deflectors, buffer strips along the watercourse, purification beds, creation of meanders and floodplains, backwater etc. habitat and biotic diversity is improved, self-purification capability and water retention are increased (Vrhovšek et al., 2008).

5.7 Additional technologies for integration in ET systems

Additional technologies can be integrated in ET systems in order to enhance the removal of target pollutants. Those technologies mainly target at different soluble pollutants like phosphorous, nitrogen, soluble heavy metals and specific micropollutants. Enhanced removal of dissolved and colloidal pollutants is especially important in case of a discharge into a sensitive recipient and in case of further production of drinking water. Different treatment units can be combined, e.g. coagulation, flocculation and subsequent sedimentation, plant uptake, sorption of dissolved and colloid matter to surfaces, etc. In contrast to sedimentation, the mentioned processes enable higher removal of dissolved and colloidal pollutants. Dissolved and colloidal pollutants are known for its mobile nature in water systems and therefore have the highest risk of causing harmful effects.

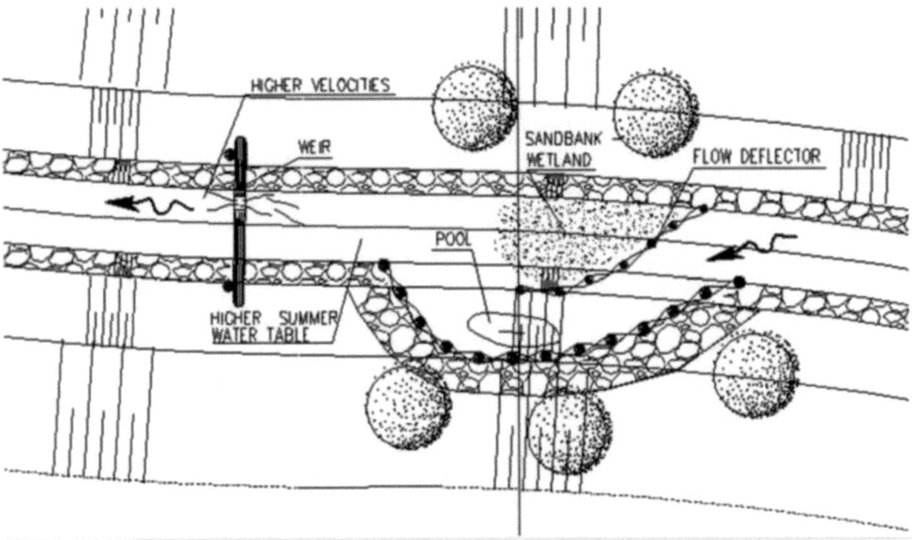

Fig. 2. An example of revitalization measures in a short segment of a watercourse (source: LIMNOS Ltd.)

Flocculation: Aluminum salts form insoluble aluminium hydroxide flocks $Al(OH)_3$ in bulk water. The flocks have good settling properties and high sorption capacity for phosphate, heavy metals, organic micropollutants and algae (El Samrani et al., 2008). Accordingly, these pollutants are removed by sorption to the flocks in bulk water and subsequent sedimentation in the pond. Besides aluminum also lime and iron salts are used, and calcium and iron, which have similar characteristics.

Sediment and media enrichment: Sediment and media in ET can be enriched with minerals that have high sorption capacity for target pollutants. E.g. Ferric iron ($Fe(OH)_3$) binds phosphate and several heavy metals under aerobic conditions (Kadlec & Wallace, 2009). In an aqueous environment, $Fe(OH)_3$ is least soluble at pH between 7 and 10 and provides sorption sites for a number of pollutants. Besides adsorption of pollutants to $Fe(OH)_3$, also insoluble precipitates with iron can be formed, e.g. $FePO_4$ and complexes with metals. Using Fe to adsorb pollutants, it is essential that the redox potential of the media is sufficiently high to prevent reduction of ferric iron to ferrous iron.

Sorption filters: One of the possible technologies for upgrading existing ET is an installation of sorption filters after the system. Dissolved and colloidal pollutants as heavy metals and phosphorous are thus removed by sorption to the filter media. However filter clogging and saturation of the media may be of a concern. Elimination of dissolved pollutants like phosphorous and heavy metals is enabled by the characteristics of the filter materials. Filter materials that comprise a lot of dolomite ($CaMg(CO_3)_2$) or calcite ($CaCO_3$) minerals are effective in P adsorption (Brix et al., 2001) and materials containing iron or alumina are shown to have good sorption capacities for heavy metals (Genc-Fuhrman et al., 2007).

Ultrasound: Sonication is mainly used for algae control and disinfection of water in different systems. Ultrasound breaks up large suspended particles in treated water. The effect of ultrasound to algal cells is not clearly known; however, it is known that ultrasound suppresses algae growth and causes their sedimentation in the open water (Griessler Bulc et al., 2010; Krivograd Klemenčič & Griessler Bulc, 2010).

UV: UV is commonly used for disinfection of treated water, where it usually presents a final stage of the treatment train. The UV lamps produce ultraviolet light that enters cells and damages proteins and genetic material. An ideal wavelength for efficient disinfection is believed to be of approximately 254 nm (Modak, 2008).

6. ET systems in Slovenia

The use of ET, as a new, wider concept of understanding of natural treatment systems has started in Slovenia in the late eighties. The idea of ET was introduced in Slovenia first by floating macrophytes and later by subsurface flow TW for water treatment. An experimental period of treating wastewater with plants, mostly as different types of TW followed. During this period experiences were based on certain European researchers such us Kickuth (1984) and Clayton (1988). The basic design was developed in a project started in 1991 in Austria (Perfler & Haberl, 1992) which was modified to select, apply, and compare various options in situ. After 1995, innovative ET were developed for different applications (*e.g.* protection of lakes and watercourses from non-point pollution) based on design and experiences of TW, primarily regarding geographical, demographical and water management characteristics of Slovenia. The introduction of ET was not systematic, since this alternative way of wastewater treatment was not accepted by the government as a state of the art before the nineties. Most installed systems were pilot-systems, destined above all for experimental work. Nevertheless, from 1989 to 2011, several ET systems were installed in Slovenia; 73 TW, 12 sections of river revitalization, 2 VDD, 2 ET for landfill restoration and 1 WSP were constructed in the Karst, coastal, mountain and agricultural lowland regions of Slovenia. The Karst region, covering about 44 % of the surface, is marked by expressive shortage of surface water and soil, and by scattered communities. All this is reflecting in pollution, which is a serious threat for the extremely sensible underground sources of drinking water, based on the complex underground systems with numerous caves (under UNESCO protection). Similar difficulties are recognized also in the coastal region at the Adriatic Sea, where treated wastewaters are discharged into the sea or in its catchment area in the mountain region, which is conserved because of its ecological and scenic values, and in agricultural lowlands characterized by a high contamination with pesticides and other agricultural contaminants. The majority of inhabitants (60 %) live in the settlements with less than 5,000, most of them even 200 to 500 inhabitants, so usually the only way of treatment is the septic tank. Particular problems are tourist centres with large quantities of wastewater in the high seasons. Nowadays, the ET development in Slovenia is mainly focused on the reduction of dispersed pollution, protection of drinking water sources, revitalization of watercourses, and wastewater separation and reuse.

6.1 Design and performance of different ET types

6.1.1 Vegetated drainage ditches

Design: Two pilot VDD (Figure 3) were constructed in agricultural area to reduce watercourse pollution, draught threat, to mitigate agricultural contaminants, and to develop new wetland habitats in order to improve biodiversity. Ditches approx. 20 m long, 5 m top width, 1.40 m bottom width and 1.5 m deep were filled with selected substrata of 0.4 m in height and planted with macrophytes (*Phragmites australis*) (Griessler Bulc & Šajn-Slak, 2007; Griessler Bulc et al., 2011). In one of the VDD, the treated water flows into a meandering stream of an overall length of 70 m where the revitalization principle was followed to further increase the water quality and biodiversity.

Monitoring: From April 2008 until March 2009, physical and chemical parameters and pesticides in water were sampled and analyzed according to Standard Methods (APHA, 2005). The treatment performance was also monitored by localization of the principal denitrification processes within the VDD. The location and relative abundance of denitrifying microorganisms was determined by real time PCR (rtPCR) of the narG gene.

Results and Discussion: With the exception of SS, pollutant concentrations met the outflow permitted levels (OG RS 47/2005). The comparison of our results with the results of monitoring of the same system in previous years showed that the VDD's efficiency for nitrite and ammonia increased due the maturity of the system. The analyses showed also 91 % removal efficiency for metholaclor pesticide. A relatively even distribution of the narG gene showed the flexibility of the VDD system. The results indicate that the facultative anaerobic denitrifiers were present throughout the system, and when the conditions were suitable, denitrification was performed. The research showed that the regularly maintained VDD efficiently decreased pollutants and is an adequate and promising technology that can be further developed. Start-up period with non-consistent treatment performance could be significantly decreased with bioaugmentation with a proliferous and well adapted microbial community.

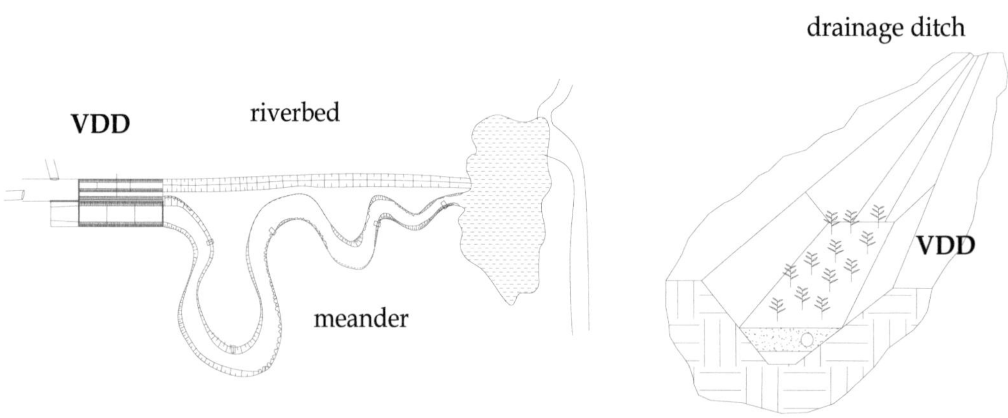

VDD Glinščica **VDD Lešnica**

Fig. 3. Design of the two VDD Glinščica and Lešnica (source: LIMNOS Ltd., CGS plus Ltd.)

6.1.2 Waste stabilization ponds (WSP) / surface flow wetlands (SFW)

Design: In the period between 2000 and 2003, a pilot SFW and a pilot WSP were constructed at the outlet of wastewater treatment plant (WWTP). The SFW was planted with *Phragmites australis* and *Eichhornia crassipes,* while in the WSP development of algae was spontaneous.

Monitoring: The systems were monitored under the same operating conditions. The efficiency was evaluated by means of physical and chemical parameters in the inflow and outflow water, by plant productivity and by the analysis of N and P contents in biomass.

Results and Discussion: The SFW proved more efficient in the elimination of suspended solids (64.6 %), settleable solids (91.8 %), organic N (59.3 %), total N (38 %), COD (67.2 %) and BOD5 (72.1 %) than the WSP. The WSP was more efficient in the treatment of ammonia nitrogen (48.9 %) and orthophosphate (43.9 %). The difference in treatment efficiency between the systems most probably originates from different primary producers (macrophytes vs. algae) and consequent food webs established. The results of this study provide data of help in optimising combinations of SFW and WSP (Šajn-Slak et. al., 2005).

6.1.3 Treatment wetlands (TW)

Design: From 1989 to 2011, over 73 TW were constructed in different regions of Slovenia. Most TW are horizontal or/and vertical systems (VF, HSF), operating in combination or integrated in zero foot print unite. Most of them consist of several interconnected beds. Most TW were installed to treat sewage, industrial wastewater, highway run-off, gray water for toilet flushing, drinking water, water from fish farms and landfill leachate. Pre-treatment mostly comprised septic tanks or sedimentation basins. Excavations were sealed with PVC or HDPE membranes, clay or the combination of both. The medium was mostly a mixture of different material (peat, soil, sand, gravel, expended clay), varying in grain size and proportion. The depths of the TW varied from 0.5 to 0.8 m, and the bottom slope from 0 to 3 %. Most systems were between 20 and 1500 m^2 in area (Table 1). Theoretical hydraulic loading of media was in each case at least 10^{-3} m/s. The TW for sewage vary in size with 2-2.5 m^2 per people equivalent on average. Wide adaptability to different environmental conditions, tolerance to stress, high productivity is evident characteristic of *P. australis* that favoured the use of this species in TW. Different parts of reed were used for planting, most frequently clumps. In shallow beds of integrated systems, where the depth was 0.4 m, other species, such as *Juncus effusus, J. inflexus, Carex gracilis, Schoenoplectus lacustris,* and *Thyphoides arundinacea,* were successfully tested. Systems were planted generally in spring or autumn when the environmental conditions were optimal.

Monitoring: The efficiency of TW was monitored by sampling at the inlets and outlets in different periods between 1989 and 2011. TW for landfill leachate were monitored regularly on a long-term basis, from 1992 till 2003, while other systems were monitored monthly for one year or occasionally for one up to 5 years. The efficiency of TWs was evaluated by analyzing suspended and settlements solids, COD, BOD_5, total phosphorus and ammonia nitrogen. Grab samples were taken mostly according to the measured retention time and analyzed by independent laboratories. Analyses were done according to Standard Methods (APHA, 2005). At sampling sites, flow, temperature, pH, dissolved oxygen and electric conductivity were measured. More extensive chemical and microbiological analyses were done occasionally.

Results and Discussion: Most TW were satisfactory efficient in BOD_5 and COD removal, and only partly efficient in N and P removal. The TW for industry were constructed for the treatment of food processing wastewater, characterized by high COD, BOD_5, and ammonia nitrogen and for dye-rich textile wastewater. The results indicated that TW can be an appropriate technology for the treatment of wastewaters from those industries because the outflow parameters reached the prescribed legislation standards (Griessler Bulc & Ojsteršek, 2008; Zupančič Justin et al., 2009). Two pilot TW were constructed at the end of 2005 for the treatment of water from drinking water wells, polluted with pesticides (atrazine, metholaclor), and pathogens. The results showed the removal efficiency of *E.coli* from 130 to 500 bacteria/100 mL at the inflow to 0 to 3 bacteria/100 mL at the outflow from TW (Istenič et al., 2009). Regarding the pesticides removal, bentazon was reduced from 1.8 µg/L at the inflow to 0.06 µg/L at the outflow, metholaclor from 0.73 µg/L to <0.05 µg/L, and terbutylazine from 0.53 µg/L to <0.03 µg/L (LIMNOS Ltd., results not published). TW for highway run-off treatment showed 69 % removal efficiency for suspended solids, 97 % for settleable solids, 51 % for COD, 11 % for BOD5 and 80 % for Fe. Heavy metals (Cu, Zn, Cd, Ni and Pb) were below the legislation limits at the inflow with the reduction efficiency in the system of over 90 % while the concentrations of N and P showed a low level of nutrients for biological processes (Bulc and Sajn Slak, 2003). The TW for gray water was constructed in 2011. Preliminary data showed that gray water was mostly lost due to evapotranspiration. The TW for landfill leachate were constructed for the landfill sites that cover approximately 0.5 to 2 ha. With regard to the studied parameters, the performance of TW was not influenced by annual seasons, but primarily by precipitation. The reduction efficiency reached on average 50 % for NH_4-N, BOD_5 and COD (Griessler Bulc, 2006; Griessler Bulc & Zupančič Justin, 2007). The results proved that a TW can be considered a method appropriate for the leachate treatments of old waste dumps (Zupančič Justin et al., 2005).

6.1.4 Restorations of landfill sites

Case 1; Design: The ET approach for the reclamation of a 1.5 ha landfill in south-eastern part of Slovenia consists of a landfill soil cover, which is densely planted with grasses and fast growing trees (poplars, willows) as a phytoremediation layer, a TW of 1,000 m^2 with an average hydraulic load of 12 m^3/d and an irrigation system. Landfill leachate is treated in the TW from where it is recycled on the landfill cover without outflow into the environment. Closed loop leachate circulation enables additional leachate treatment by assimilation of nutrients into the plant biomass and by mineralization processes of microbes in the soil layer. Fast growing trees allow the evapotranspiration of a considerable amount of leachate, while the excess percolates back into the landfill body and enables further biodegradation of deposited wastes. The results provided an environmentally and economically-viable solution, with large buffering capacity and simple in concept. The presented methods won international awards (2001 Lillehammer Award; 2008 Global Energy Award, Griessler Bulc & Zupančič Justin, 2007).

Monitoring: To evaluate plant response on leachate irrigation, a remote sensing of the canopy reflectance was performed by ground-based monitoring of vegetation indices of the phytoremediation system with a multispectral camera (Tetracam, USA). Images were taken in regular monthly intervals during one vegetation season, from April to October, after two years of leachate irrigation.

Results and Discussion: The obtained results confirmed the findings that leachate can be a good fertilizer for short rotation coppice produced elsewhere for energy crops. The macronutrient requirements of willow, in relation to nitrogen set to 100, were found to be for N:P:K in the relation of 100:14:72. The N:P:K ratio usually found in leachate ranges from 100:0:54, respectively (Duggan, 2005) to 100:1.5:103 (Dimitriou et al., 2006). The N:P:K ratio calculated from the average concentration of nitrogen, phosphorous and potassium in leachate analyzed during 18-month period was 100:0.5:246, respectively. The potassium concentration was found in excess and phosphorous concentration was low as it is common for leachate. The lack of phosphorous for plant growth is usually expressed in a long-term period (>10 years) and its deficiency could therefore not have been expressed during our observation.

Case 2; Design: A phytoremediation method for the treatment of tannery substrate on the industrial waste dump in Slovenia was used in 2006 to research the potential of various plant species to reduce Cr pollution. Several herbaceous and woody species were planted in the tannery substrate in a greenhouse and on the four testing polygons on the waste dump.

Monitoring: Prior to planting, hardly degradable and toxic substances in the substrate and leachate were analyzed. Moreover, the substrate was examined for its inhibition and chronic toxicity to higher plants according to the ISO standard (ISO/DIS 22030). Several growth parameters were measured in the glasshouse and on the polygons. The growth parameters were measured on a monthly basis in the growing season. Prompt fluorescence, i.e. potential photochemical efficiency (parameter Fv/Fm), was measured on the plants in the glasshouse and on polygons.

Results and Discussion: The preliminary substrate analyses had shown that the most crucial pollutant in the tannery landfill site was Cr (III). The biological accessibility of Cr in roots and shoots of herbaceous and woody plant species showed that beet and sunflower were the most suitable species for phytoremediation, although Cr bioavailability was low. The results revealed that the growth of plants was inhibited and, their health worsened. The results also showed that phytoremediation could be a very delicate method that needs a careful insight into the processes of specific pollutants removal.

6.1.5 River revitalization

Design: The majority of watercourse revitalizations in Slovenia were carried out in the north-east part of the country. Revitalizations of short stretches of rivers and streams were designed in order to increase self-cleaning capacities of the streams in an intensive agricultural landscape and also in order to protect the natural population of otter. According to the data, the most continuous and viable population of *Eurasian otter* in Slovenia lives in the north-east of the country. Threats for otter in this area are degraded habitats due to the agro-operational works, including ameliorations and canalization of watercourses which took place in last decades. The results of non-sustainable management were among others the opening of corridors by removing tree canopies as well as riparian vegetation on watercourse banks (the shelter for otters disappeared, the living conditions for pray species worsened and consequently the food supply was reduced), the permeability of corridors had lowered and the risk for population fragmentation was higher. In order to protect and restore the otter population, ETs were implemented in

Goričko on short sections of watercourses and lakes of eight local communities. Different in-stream and stream bank features were implemented to attain a successful revitalization. Weirs with small pools were constructed in the channels to improve streambed substrate, slow down the flow velocity, retain water and provide proper fish passages. Artificial indentations as well as restored and protected indentations contribute to better water habitats diversity. The passages for otters under the bridges were implemented which enabled otters to pass the roads safely. New vegetation zones (riparian wetlands and constructed wetlands) prevent erosion, provide buffer and better connectivity between terrestrial and aquatic ecosystems (Griessler Bulc & Šajn-Slak, 2009). The revitalization measures were also widely accepted by the local population, satisfied by the re-gained natural appearance of the streams and water murmur.

Fig. 4. Schematic evolution of river revitalization (source: LIMNOS Ltd.)

7. Conclusions

Despite numerous measures to improve the quality of the environment in the last decades, water quality is still seriously threatened by point and non-point sources of pollution. The application and development of ET for water protection, treatment and reuse in Slovenia have shown that they are appropriate measures for reaching EU regulations in terms of good water quality; moreover, they consider also the recycling of nutrients, reuse of water and potentially the production of biomass.

ET or ecoremediations mimic healthy natural ecosystems, have high buffer and self-cleaning capacity and contribute to habitat diversity. Moreover, these systems have the remediation ability, ensure high biodiversity and contribute to the stability of ecosystems. Nutrients and pollutants in ET are subdued to numerous processes that enable pollutant and nutrient transformations, degradation or stabilization. Through these processes, ET enable the reuse or recycling of nutrients as phosphorous and nitrogen, which nowadays have one way flow from consumption to waste, which will result in a shortage of plant fertilizers in a near future. Closing the loops of wastewater treatment is therefore crucial.

In Slovenia ET are in a rise since 1989. Different ET have been applied, namely treatment wetlands, watercourse revitalization, vegetated drainage ditches, waste stabilization ponds, phytoremediation of landfill sites. Most common ET in Slovenia are treatment wetlands for municipal sewage followed by watercourses revitalization. A high number of treatment wetlands indicate the priority of local communities and the authorities to solve deficient wastewater treatment systems in the country. An important part of the research and development of ET in Slovenia was focused also on the restoration of landfill sites, where a closed water and pollutant loop was investigated and successfully implemented; however, the system is not yet successful in the market because the local governance and environmental managers are still focused on wastewater treatment. Due to gradual acceptance and implementation of ET in local environments there is still a long way to walk in order to achieve sustainable society in terms of closing the loops of water and nutrient usage.

Nr.	Type	Area m2	Year of construction	Operation period	Treated water	Average inflow concentration	Efficiency	Reference
55*	Treatment wetland	7.5-1500	1989-2011	1989-2011	Municipal sewage	COD: 239 NH4-N: 496 TP: 8.9	50 51 53	Urbanc-Bercic et al., 1998, Sajn-Slak et al., 2005
9	Treatment wetland	125-750	1991-2008		Industrial	-	-	Vrhovsek et al., 1996, Griessler-Bulc and Ojstersek, 2008; Zupancic-Justin et al., 2009)
8	Treatment wetland	311-1000	1995-2004	1995-2004	Landfill leachate	COD:979 NH4-N:281 TP:2.5	54 51 58	Griessler-Bulc, 2006; Griessler-Bulc and Zupancic-Justin, 2007
1	Treatment wetland	85	2001	2001	Highway runoff	COD:29 NH4-N:- TP: 0.4	51 - 79	Griessler-Bulc and Sajn-Slak, 2003
1	Treatment wetland	20	2011	2011	Grey water	COD: 411 NH4-N: 128 TP: 56	-	Griessler-Bulc et al., 2010
2	Treatment wetland	0.5-1	2009	2009-2011	Drinking water	-	-	Data not published
2	Vegetated drainage ditch	20-90	2006	2006-2011	Surface water, agricultural run off	COD: 25 NH4-N: 0.3 TP: 0.3	neg 7 3	Griessler-Bulc and Sajn-Slak, 2007; Bulc et al., 2011, Griessler-Bulc and Krivograd-Klemenčič, 2011
12*	Watercourse revitalization	10-100*	2006-2008	2006-	Surface water	-	-	Vrhovsek et al., 2007; Griessler-Bulc and Sajn-Slak, 2009
	Waste stabilization pond	36	2000-2003	2001	Sewage	COD:391 NH4-N: 3.3 TP:2.07	67.2 neg-49 33	Sajn-Slak et al., 2005
2	Phytoremediation	15000	2003-2005	2003-2011	Leachate, soil	COD: 1257 NH4-N: 217 TP: 2.5	35 37 49	Griessler-Bulc and Zupancic-Justin, 2007

*estimated numbers

Table 1. Data about ET in Slovenia

8. Acknowledgment

We acknowledge with great appreciation the support of the European Eureka Programme, Life-Environment, Life-Nature, 6FP, 7FP, the University of Ljubljana, LIMNOS Company for Applied Ecology Ltd. and Lutra - Institute for Conservation of Natural Heritage that made it possible to obtain the presented results.

9. References

APHA. (2005). *Standard Methods for the Examination of Water and Wastewater*, 21st Ed., Washington (DC), American Public Health Association (APHA), American Water Works Association (AWWA) & Water Environment Federation (WEF)

Appleyard, S & Schmoll, O. (2006). Agriculture: Potential hazards and information needs. In: *Groundwater for Health: Managing the Quality of Drinking-water Sources*, O. Schmoll, G. Howard, J. Chilton and I. Chorus (Eds.); World Health Organization. IWA Publishing, London, UK

Arias-Estevez, M.; Lopez-Periago, E.; Martinez-Carballo, E.; Simal-Gandara, J.; Mejuto, J.-C. & Garcia-Rio, L. (2008). The mobility and degradation of pesticides in soils and the pollution of groundwater resources. *Agriculture, Ecosystems and Environment*, 123, pp. 247–260

Brix, H. (1993). Wastewater treatment in constructed wetlands: System design, removal processes, and treatment performance. In: Moshiri GA, editor. *Constructed wetlands for water quality improvement*. CRC Press; pp. 9–22

Brix, H.; Arias, C.A. & del Bubba, M. (2001). Media selection for sustainable phosphorous removal in subsurface flow constructed wetlands. *Water Science and Technology*, Vol. 44, No. 11-12, pp. 47-54

Clayton, R.C. (1988). Report on the ATV (Technical Association for Wastewater). Seminar on Reedbed Treatment Systems, 29 - 30 Sept., Nüremberg, FRG

Cooper, P.F. (1990). European Design and Opertaion Guidelines for Reed Bed Treatment Systems. *2nd Int. Conf. on "Use of Constructed Wetlands in Water Pollution Control"*, Cambridge, UK, pp. 33

Cordell, D.; Dranger, J. O.& White S. (2009). The story of phosphorus: global food security and food thought. *Global Environmental Change* 19, pp. 292-305

El Samrani, A.G.; Lartiges, B.S. & Villieras, F. (2008). Chemical coagulation of combined sewer overflow: Heavy metal removal and treatment optimization. *Water Research*, Vol. 42, No. 4-5, pp. 951-960

Erisman, J. W.; de Vries, W.; Kros, H.; Oenema, O.; van der Eerden, L.; van Zeijts, H. & Smeulders, S. (2001). An outlook for a national integrated nitrogen policy. *Environmental Science & Policy*, 4, pp. 87–95

Esrey, S.A. (2000). Towards a recycling society ecological sanitation –closing the loop to food security. *Ecosan- closing the loop in wastewater management and sanitation*. Proceedings of the International Symposium, 30-31 October 2000, Bonn, Germany

Genc-Fuhrman, H.; Mikkelsen, P.S. & Ledin, A. (2007). Simultaneous removal of As, Cd, Cr, Cu, Ni and Zn from stormwater: Experimental comparison of 11 different sorbents. *Water Research*, Vol. 41, No. 3, pp. 591-602.

Griessler Bulc, T. & Ojstrešek, A. (2008). The use of constructed wetland for dye-rich textile wastewater treatment. *J. hazard. mater.*, June 2008, vol. 155, iss. 1/2, pp.76-82

Griessler Bulc, T. & Šajn-Slak, A. (2007). Agricultural run-off treatment with vegetated drainage ditches. V: BORIN, Maurizio (ur.), BACELLE, Sara (ur.). *Proceedings of the International Conference on Multi Functions of Wetland Systems*, Legnaro (Pd), Italy, 26th-29th June 2007. Padova: P.A.N., 2007, pp. 86-87

Griessler Bulc, T. & Šajn-Slak, A. (2009). Ecoremediations - a new concept in multifunctional ecosystem technologies for environment protection. *Desalination*. 2-10

Griessler Bulc, T. (2006). Long term performance of a constructed wetland for landfill leachate treatment. *Ecol. eng.* 26, pp. 365-374

Griessler Bulc, T.; Krivograd-Klemenčič, A. & Razinger, J. (2011). Vegetated ditches for treatment of surface water with highly fluctuating water regime. *Water sci. technol.*, 2011, vol. 63, no. 10, pp. 2353-2359

Griessler Bulc, T.; Zupančič-Justin, M. (2007). Sustainable solution for landfill leachate with a use of phytoremediation. V: VELINNI, Albert A. (ur.). *Landfill research trends.* New York: Nova Science Publishers, pp. 103-139

Griessler Bulc; T., Istenič, D. & Krivograd-Klemenčič, A. (2010). The efficiency of a closed-loop chemical-free water treatment system for cyprinid fish farms. *Ecological Engineering*, Vol. 37, No. 6, pp. 873-882

Grönlund, E.; Klang, A.; Falk, S. & Hanæus, J.(2004). Sustainability of wastewater treatment with microalgae in cold climate, evaluated with emergy and socio-ecological principles. *Ecol. Eng.*, 22, pp.155-174

Herzog, F.; Prasuhn, V.; Spiess, E.; Richner, W. (2008). Environmental cross-compliance mitigates nitrogen and phosphorus pollution from Swiss agriculture. *Env. Science & Policy*, 11, pp. 655-668

Hijosa-Valsero, M.; Matamoros, V.; Martín-Villacorta, J.; Bécares, E. & Bayona, J.M. (2010b). Assessment of full-scale natural systems for the removal of PPCPs from wastewater in small communities. *Water Research*, Vol. 44, No. 5, pp. 1429-1439

Hijosa-Valsero, M.; Matamoros, V.; Sidrach-Cardona, R.; Martín-Villacorta, J.; Bécares, E. & Bayona, J.M. (2010a). Comprehensive assessment of the design configuration of constructed wetlands for the removal of pharmaceuticals and personal care products from urban wastewaters. *Water Research*, Vol. 44, No. pp. 3669-3678

Huang, T.L.; Ma, X.C.; Cong, H.B. & Chai, B.B. (2008). Microbial effects on phosphorous release in aquatic sediments. *Water Science and Technology*, 58, pp. 1285-1289

Hvitved-Jacobsen, T.; Johansen, N.B. & Yousef, Y.A. (1994). Treatment systems for urban and highway run-off in Denmark. *The Science of Total Environment*, 146/147, pp. 499-506

Iital, A.; Pachel, K.& Deelstra, J. (2008). Monitoring of diffuse pollution from agriculture to support implementation of the WFD and the Nitrate Directive in Estonia, *Env. Science & Policy*, 11, pp. 185-193

Istenič, D.; Oblak, L. & Vrhovšek, D. (2009). Conditioning of drinking water on constructed wetland : elimination of Escherichia coli. *Ekológia (Bratisl.)*, 28, pp. 300-311

Jenssen, D. P; Heeb, J.; Huba, M. E; Gnanakan, K.; Warner, S. W.; Refsgaard, K.; Stenström, T-A.; Guterstam, B.& Alsen, K. W. (2004). Ecological sanitation and reuse of wastewater, ecosa. *A thinkpiece on ecological sanitation.* The agricultural University of Norway, 18 pp.

Jenssen, P. D.; Bergstrøm, C. & Vatn, A. (2009). From Crap to Food. *The New Economy* (Sommer 2009).

Kadlec, R.H. & Wallace, S.C. (2009). *Treatment Wetlands.* Second Edition. Boca Raton, CRC Press, Taylor & Francis Group: 1016 pp.

Kickuth, R. (1984). The Root Zone Method. Gesamthochschule Kassel-Universitat des Landes Hessen, 12

Krivograd-Klemenčič, A. & Griessler Bulc, T. (2010). The efficiency of ultrasound on algal control in a closed loop water treatment for cyprinid fish farms. *Fresenius environmental bulletin,* 19, pp. 919-931

Langergraber, G. (2005). The role of plant uptake on the removal of organic matter and nutrients in subsurface flow constructed wetlands: a simulation study. *Water Science and Technology,* 51, pp. 213-223

Mangas, E.; Vaquero, M.T.; Comellas, L. & Broto-Puig, F. (1998). Analysis and fate of aliphatic hydrocarbons, linear alkylbenzenes, polychlorinated biphenyls and polycyclic aromatic hydrocarbons in sewage sludge-amended soils. *Chemosphere,* 36, pp. 61-72

Mara, D.D. (2009) Waste stabilization ponds: Past, present and future. *Desalination and Water Treatment,* 4

Modak, N. (2008). Enhanced photo-sono process for disinfection of surface and subsurface water. *Dissertation Abstracts International* 69, 137.

Neal, C.; Davies, H. & Neal, M. (2008). Water quality, nutrients and the water framework directive in an agricultural region: The lower Humber Rivers, northern England. *J. of Hydrology,* 350, pp. 232- 245

Nivala, J.; Hoos, M. B.; Cross, C.; Wallace, S. & Parkin, G. (2007). Treatment of landfill leacahte using an aerated, horizontal subsurface-flow constructed wetland, *Science of the Total Environment,* doi:10.1016/j.scitotenv.2006.12.030 (article in press)

OECD (2006). Water and Agriculture: Sustainability, Markets and Policies. Organisation for Economic Co-operation and Development, Paris, pp. 480

Perfler, R. & Haberl, R. (1992). Constracted Wetlands for Extended Nutrients Removal. *3rd Int. Spec. Conf. on "Wetland Systems in Water Pollution Control",* Sydney, Australia, pp. 17.1-17.9

Ramadan, H. & Ponce, V. M. (n.d.). Design and Performance of Waste Stabilization Ponds. 1. 07. 2011, available from www.stabilizationponds.sdsu.edu

Šajn-Slak, A.; Griessler Bulc, T.; Vrhovšek, D. (2005). Comparison of nutrient cycling in a surface-flow constructed wetland and in a facultative pond treating secondary effluent. *Water sci. technol.,* letn. 51, no. 12, pp. 291-298

Schertenleib, R. (2001). The Bellagio Principles and a household centered approach in environmental sanitation. In: Ecosan – closing the loop in wastewater management and sanitation. *Proceedings of the International Symposium, Deutsche Gesellschaft für*

Technische Zusammenarbeit (GTZ). October 30-31 2000, Bonn, Germany. 327 pp. www.gtz.de

Simon, R. D. & Makarewicz, J. C. (2009). Storm water events in a small agricultural watershed: Characterization and evaluation of improvements in stream water microbiology following implementation of Best Management Practices. *Journal of Great Lakes Research*, 35, pp. 76-82

Terzakis, S.; Fountoulakisa, M.S.; Georgaki, I.; Albantakis, D.; Sabathianakis, I.; Karathanasis, A.D.; Kalogerakis N. & Manios, T. (2008). Constructed wetlands treating highway runoff in the central Mediterranean region. *Chemosphere*, 72, pp. 141-149

Vinnerås, B. (2002). *Possibilities for sustainable nutrient recycling by faecal separation combined with urine diversion*. Doctoral Thesis. Swedish Univiversity of agricultural Sciences, Uppsala

Vogelsang, C.; Grung, M.; Jantsch, T.G.; Tollefsen, K.E. & Liltved, H. (2006). Occurence and removal of selected organic micropollutants at mechanical, chemical and advanced wastewater treatment plants in Norway. *Water Research*, 40, pp. 3559-3570

Vrhovšek, D.; Vovk Korže, A.; Lovka, M.; Kryštufek, B.; Sovinc, A.; Bertok, M.; Vrhovšek, M. & Kovač, M. (2008). *Ekoremediacije kanaliziranih vodotokov*. Vrhovšek, D. and Vovk Korže, A. (Eds.). Ljubljana: Limnos; Maribor: Filozofska fakulteta, Mednarodni center za ekoremediacije. pp. 219

Vymazal, J. (2004). Removal of phosphorus via harvesting of emergent vegetation in constructed wetlands for wastewater treatment. *9th International Conference in Wetland Systems for Water Pollution Control*, Avignon (France), 26-30 September 2004. Volume 2

Vymazal, J. (2006). Removal of nutrients in various types of constructed wetlands. *Science of the Total Environment*, 380, pp. 48-65

Vymazal, J. (2007). Removal of nutrients in various types of constructed wetlands. *Science of the Total Environment*, 380, pp. 48-65

Vymazal, J.; Brix, H.; Cooper, P.F.; Green, M.B. & Haberl R. (1998). Constructed wetlands for wastewater treatment in Europe. Backhuys Publishers, Leiden, 366 pp.

Werner, C.; Schlick, J.; Witte, G. & Hildebrandt, A. (2000). Ecosan - closing the loop in wastewater management and sanitation. *Proceedings of the International Symposium, Deutsche Gesellschaft für Technische Zusammenarbeit (GTZ)*, October 30-31, Bonn, Germany. 327 pp. www.gtz.de

Zupančič Justin, M.; Bulc, G. T.; Vrhovšek, D.; Bukovec, P.; Zupančič, M.; Zrimec, A. & Berden, Z. M. (2005). Slovenian Experience: MSW landfill leachate treatment in constructed wetland and leachate recycling on landfill cover vegetated with trees. *Proceeding: 10th International Waste management and Landfill Symposium*, Cossu, R. and Stegmann, R. (Eds.), Environmental Sanitary Engineering Centre, Sardinia, Italy, pp. 725-726

Zupančič Justin, M.; Vrhovšek, D.; Stuhlbacher, A. & Griessler Bulc, T. (2009). Treatment of wastewater in hybrid constructed wetland from the production of vinegar and packaging of detergents. *Desalination*. 2009, pp.100-109

Žegura, B.; Heath, E.; Černoša, A.; Filipič, M. (2009). Combination of in vitro bioassays for the determination of cytotoxic and genotoxic potential of wastewater, surface water and drinking water samples. *Chemosphere*, 75, pp. 1453–1460

9

Heterogeneous Photocatalytic Oxidation an Effective Tool for Wastewater Treatment – A Review

Cheng Chee Kaan[1], Azrina Abd Aziz[1],
Shaliza Ibrahim[1], Manickam Matheswaran[2] and Pichiah Saravanan[1]
[1]Department of Civil Engineering, Faculty of Engineering, University of Malaya,
[2]Department of Chemical Engineering, National Institute of Technology Tiruchirappalli,
[1]Malaysia
[2]India

1. Introduction

Heterogeneous photocatalysis has been intensively studied since the discovery of photo-activated water splitting process using titanium dioxide (TiO_2) as electrode (Fujishima & Honda, 1972) in 1972. Heterogeneous photocatalysis can be defined as a reaction in which a catalytic process is initiated by the action of light.

Fujishima and co. discovered that water can be split into hydrogen and oxygen through this process in 1972. Hence early studies were focused on the production of hydrogen using solar energy as a clean fuel from water (Kawai & Sakata, 1980; Sato & White, 1980). Further studies found that irradiated semiconductors particles could catalyze a lot of interesting and useful reduction-oxidation reactions of organic and inorganic compounds (Fox & Dulay, 1993). Some of the semiconductor particles were found to be able to completely mineralize various organic and inorganic substances which are known as environmental pollutants (Fujishima et al., 2007). Since then, many researches were carried out based on the environmental applications of heterogeneous photocatalysis (Herrmann, 1999; Hoffman et al., 1995; Rajeshwar et al., 2001; Saravanan et al., 2009).

Various studies had been carried out to search for an ideal semiconductor photocatalyst, but titanium dioxide (TiO_2) remains as a benchmark among other semiconductors. CdS, SnO_2, WO_3, SiO_2, ZrO_2, ZnO, Nb_2O_3, Fe_2O_3, $SrTiO_3$ etc. were among the semiconductor materials that were being studied but titanium dioxide (TiO_2) remained an excellent photocatalyst for its high resistance to photocorrosion and desirable band-gap energy (Ye & Ohmori, 2002). It can be used to degrade a variety of organic and inorganic pollutants (Fox & Dulay, 1993; Herrmann et al., 2007). Besides, titanium dioxide (TiO_2) is easily available in the market, chemically inert and durable (Saravanan et al., 2009) and non-toxic.

In contrast with other conventional methods in environmental cleanup, heterogeneous photocatalysis involved the breakdown of the pollutants from complex molecules into simple and non-hazardous substances. Hence no residue is left and no sludge is produced from the process. Furthermore, no secondary treatment is needed to process the sludge.

Besides, the catalyst remains unchanged throughout the process and thus can be reuse; therefore no consumable chemical is required. All these result in a significant reduction in overall operating cost. In addition, this process can be carried out at extremely low concentrations because the pollutants were strongly adsorbed on the surface of the catalyst, allowing sub part-per-million condition. Summing up all these benefits and advantages, heterogeneous photocatalysis provides a cheap and effective alternative to clean water production and environmental remediation.

In this study, various issues with respect to the attributes of the photocatalyst and the mechanism behind titania-based photocatalysis will be discussed. The following discussion may be relevant to environmental cleanup context, given that the process is subjected to both contaminant reduction and oxidation relying on the tendency of the former to either accept or give up electrons respectively (Rajeshwar & Ibanez, 1995).

1.1 Advanced oxidation processes

The phrase *advanced oxidation processes* (AOP) refer specifically to processes in which oxidation of organic contaminants occurs primarily through reactions with hydroxyl radicals (Glaze et al., 1995). It involves two stages of oxidation: (1) the formation of strong oxidants (*e.g.*, hydroxyl radicals) and (2) the reaction of these oxidants with organic contaminants in water (Alnaizy & Akgerman, 2000). In water treatment applications, AOPs usually refer to a specific subset of processes that involve O_3, H_2O_2, and/or UV light. However, often AOPs are also referred to a more general group of processes that also involve semiconductor catalysis, cavitation, E-beam irradiation, and Fenton's reaction (Fox & Dulay, 1993; Legrini et al., 1993). All these processes can produce hydroxyl radicals, which can react with and destroy a wide range of organic contaminants, including phenolics. Although many of the processes noted above have different mechanisms for destroying organic contaminants, in general, the effectiveness of an AOP is proportional to its ability to generate hydroxyl radicals (Fox & Dulay, 1993; Legrini et al., 1993).

1.2 Ozonation / UV

The O_3 system is one of the AOP for the destruction of organic compounds in wastewater. Basically, aqueous systems saturated with ozone are irradiated with UV light of 253.7 nm. The extinction coefficient of O_3 at 253.7 nm is 3300 L.mol/cm, much higher than that of H_2O_2 (18.6 L.mol/cm). The decay rate of ozone is about a factor of 1000 higher than that of H_2O_2 (Guittonneau et al., 1991). The AOP with UV radiation and ozone is initiated by the photolysis of ozone. The photodecomposition of ozone leads to two hydroxyl radicals, which do not act as they recombine producing hydrogen peroxide, as shown in the following Eqns. (1) and (2) (Peyton & Glaze, 1988):

$$H_2O_2 + O_3 \xrightarrow{hv} 2OH^{\bullet} + O_2 \quad (1)$$

$$2OH^{\bullet} \rightarrow H_2O_2 \quad (2)$$

1.3 Ultrasonication

Implosion of cavity bubbles in sonicated water containing dissolved gases results in formation of hydrogen and hydroxyl radicals by fragmentation of water molecules. These radicals in turn combine and generate other oxidative species such as peroxy and super oxide radicals ($^{\bullet}OH$) as well as hydrogen peroxide; the quantities of each depend on the ambient conditions and the operating parameters. Such $^{\bullet}OH$ radicals are used for the degradation of the organic compounds (Kidak & Ince, 2006).

1.4 Solar photocatalytic oxidation

In the past years, there have been a number of studies and reviews about this process (Bahnemann., 2004; Fox & Dulay, 1993; Herrmann et al., 2007; Hoffmann et al., 1995; Legrini et al., 1993). Photocatalytic oxidation is based on the use of UV light and a semiconductor. Many catalysts have been tested, although titanium dioxide (TiO_2) in the anatase form seems to possess the most interesting features, such as high stability, good performance and low cost (Bahnemann, 2004; Fox & Dulay, 1993; Hoffmann et al., 1995; Legrini et al., 1993).

Matthews (1990) reported that more than 90% of nitro benzene (NB) mineralization was achieved with TiO_2 and sunlight. Minero et al. (1994) studied the photocatalytic degradation of NB on TiO_2 and ZnO and reported complete mineralization with TiO_2. Titanium dioxide has become the most studied and used photocatalyst, because it is easily available, chemically robust and durable. It can be used to degrade, *via* photocatalysis, a wide range of organic compounds (Herrmann et al., 2007; Hincapié et al., 2005; Leyva et al., 1998; Robert & Malato, 2002). Photocatalytic degradation of phenolic compounds by employing Degussa P-25® in presence of sunlight has been successfully studied by many researchers (Curcó et al., 1996; Minero et al., 1994).

2. Background of photocatalysis

2.1 Photocatalytic process

The presence of non-biodegradable and toxic organic compounds in wastewater is one of the major problems in wastewater treatment. Organic compounds like phenol and its derivatives are known for their toxicity and are classified as persistent organic chemicals (POC) which is a major threat to human health. Phenol in particular, which is carcinogenic, is introduced to the water bodies by various means. Industrial manufacturers, normal households, and landfill leachate contribute these organic compounds (Bahnemann, 2004) into the water bodies and makes wastewater treatment more difficult. All these pollutants need to be removed from wastewater before it can be discharged to the environment. Such contaminants may also be found in surface and subsurface water which require treatment to achieve desirable drinking water quality (Lindner et al., 1995). Conventional water treatment process like activated carbon adsorption, membrane filter, ion exchange etc. generate and produce extra waste during the purification system, which will further increase the cost and time. As a result, many studies and researches have been carried out to develop a sustainable and cost-efficient treatment process that can effectively remove or degrade these organic and inorganic chemicals in wastewater (Ahmed et al., 2010; Zeltner et al., 1996) with photocatalysis gaining much attention in the field of contaminant mineralization.

Majority of the natural purification of aqueous systems such as aerated lagoons or ponds, rivers and streams, lakes etc. are caused by the action of sunlight. Organic molecules were breakdown by the action of sunlight to simpler molecules and finally to carbon dioxide and other mineral compounds. There are many natural accelerators which can be used to accelerate this natural process. The introduction of 'colloidal semiconductor' and catalyst to catalyze distinct redox reactions on semiconductors could boost this sunlight driven natural purification process (Matthews, 1993).

Wastewater treatment using photocatalysis involves the combination of heterogeneous photocatalysis with solar technologies (Zhang et al., 1994). Semiconductor photocatalysis, especially titania-based photocatalysis has been applied to various environmental problems other than water and air purification. Different studies have been carried out from fundamental to practical aspects to improve the process and the properties of the photocatalyst in recent years (Rajeshwar & Ibanez, 1997; Schiavello, 1997; Serpone & Pelizzetti, 1989). Hoffman et al. (1995) reported in that the utilizations of irradiated semiconductors for the degradation of organic pollutants were well documented and have shown positive and encouraging results for various organic pollutants. Various studies have also been carried out from fundamental to practical aspects to improve the process and the properties of the photocatalyst in recent years.

2.2 Mechanisms of generating oxidizing species

The heterogeneous photocatalysis process is very complex. The oxidizing pathway is not very clear yet. Jean-Marie Herrmann (1999) suggested in that the overall of classic heterogeneous photocatalysis process can be divided into five steps (Herrmann, 1999):-

1. Transfer of reactants to the surface
2. Adsorption of one of the reactants
3. Reactions of the reactants in the adsorbed phase
4. Desorption of the product(s)
5. Diffusion of the product(s) from the surface

There are two pathways where the OH radicals can be formed. The valence band hole, h^+_{vb} can either react with the adsorbed water or the surface OH^- groups on the titanium dioxide (TiO_2) particle (Ekabi & Serpone, 1988). Equations 3 and 4 show the two reactions.

$$TiO_2(h^+_{vb}) + H_2O_{ads} \rightarrow TiO_2 + OH^{\bullet}_{ads} + H^+ \tag{3}$$

$$TiO_2(h^+_{vb}) + OH^-_{ads} \rightarrow TiO_2 + OH^{\bullet}_{ads} \tag{4}$$

Generally, an acceptor molecules (A) such as O_2 will be adsorbed and react with an electron in the conduction band while a donor molecules (D) such as H_2O will be adsorbed as well and react with a hole in the valence band. The above reactions are presented in reactions 5 and 6.

$$TiO_2(e^-_{cb}) + A_{asd} \rightarrow TiO_2 + A^-_{asd} \tag{5}$$

$$TiO_2(h^+_{vb}) + D_{asd} \rightarrow TiO_2 + D^+_{asd} \tag{6}$$

It is widely accepted that O_2 plays an important role in these reactions. Oxygen can trap conduction band electrons to form superoxide ions ($O_2^{\bullet -}$) according to reaction 7. These $O_2^{\bullet -}$ can then react with hydrogen ions (H^+) from the water splitting process to form $HO_2^{\bullet -}$.

$$TiO_2(e^-_{cb}) + O_{2ads} + H^+ \rightarrow TiO_2 + HO_2^{\bullet -} \rightarrow O_2^{\bullet -} + H^+ \quad (7)$$

H_2O_2 could also be formed from the $HO_2{\bullet}$- species by reaction 8.

$$TiO_2(e^-_{cb}) + HO_2^{\bullet -} + H^+ \rightarrow H_2O_2 \quad (8)$$

OH radical ($OH^{\bullet}$) may be formed from the cleavage of H_2O_2 via one of the following reactions 9, 10, 11.

$$H_2O_2 + h\upsilon \rightarrow 2OH^{\bullet} \quad (9)$$

$$H_2O_2 + O_2^{\bullet -} \rightarrow OH^{\bullet} + O_2 + HO^- \quad (10)$$

$$H_2O_2 + TiO_2(e^-_{cb}) \rightarrow HO^{\bullet} + HO^- + TiO_2 \quad (11)$$

The OH radical from reaction 4 is the most important oxidant formed in a photocatalytic process. It is the primary oxidant in the degradation of organic compounds (Ahmed et al., 2010). The degradation reaction is expressed in reaction 12.

$$OH^{\bullet} + Organic \rightarrow CO_2 \quad (12)$$

2.3 Photocatalytic materials

2.3.1 Photocatalysts

There are many semiconductor materials can be used as a photocatalyst. Semiconductors like TiO_2, ZnO, Fe_2O_3, CdS, ZnS etc. are all suitable materials to initiate a photocatalytic process. Extensive studies and research concludes that an 'ideal photocatalyst' should possess the attributes shown in Table 1.

Attributes of an ideal photocatalyst for heterogeneous photocatalysis
• Stability and sustained photoactivity • Biologically and chemically inert, non toxic • Low cost • Suitability towards visible or near UV light • High conversion efficiency and high quantum yield • Can be react with wide range of substrate and high adaptability to various environment • Good adsorption in solar spectrum

Table 1. Attributes of an ideal photocatalyst for heterogenous photocatalytic process (Bhatkhande et al., 2001)

Titanium dioxide (TiO_2) has been widely recognized as an excellent photocatalyst. It is known to have superb pigmentary properties, high adsorption in the ultraviolet region, and high stability which allows it to be used in various applications such as electroceramics, glass, and photocatalytic purification of chemical in air and water. Two types of reactors have been developed which are suspension/slurry and thin film/fixed in wastewater treatment (Chang et al., 2000; Huang et al., 1999; Matthews et al., 1990). The details of these two reactors will be discussed later.

Titanium dioxide (TiO_2) exist as many crystalline forms. The most common forms of crystalline structures are anatase and rutile. Brookite is the most uncommon form due to its instability in terms of the enthalpy of formation. Anatase is the most stable among all the different crystalline forms with 8-12kJmol^{-1} (Cotton et al., 1999). It can be converted to rutile when it is heated to approximately 700°C (Bickley et al., 1991). Anatase is less dense compared to rutile, has a density of 3900kg/m^3 while rutile has a density of 4260kg/m^3. In the application of photocatalysis, anatase is a more efficient photocatalyst compared to rutile due to its open crystalline structure.

2.3.2 Titanium dioxide (Degussa P25)

It was been used extensively in many studies regarding photocatalytic degradation. Photocatalytic degradation studies utilizing Degussa P25 have been well documented due to its chemical stability, readily availability, reproductive ability, and activity as a catalyst for oxidation process (Bekbolet et al., 1998; Bekbolet & Balcioghu, 1996; Saravanan et al., 2009). Vigorous activities and researches are in process to further develop the existing Degussa P25 or synthesizing new materials, which can initiate photocatalysis using solar energy and hence reduce the cost and shortening the total time needed for the degradation. These developments include increasing the effective surface area, increasing the photoactivity, increasing the active sites, enhancing the absorption of photon energy and reducing the band-gap energy.

2.4 Photocatalytic reactors

There are many types of reactors have been developed and can be used in photocatalytic studies. These reactors were developed based on the different needs of applications. The selection of these reactors was according to the experiment conditions and applications. Generally the reactors can be briefly categorized into two groups, a suspension/slurry type and a thin film type. A slurry type reactor uses the catalyst in a suspension form whereas a thin film type reactor uses a thin film catalyst. Both types of reactors can be designed to be an immersion well reactor or flat wall reactor. Immersion well reactors were generally used in laboratory scale works for evaluation purposes. It can be run on either batch or continuous mode. The flow of oxidant and the temperature can be easily controlled and monitored. The source of the light can either be single or multiple with or without any reflectors. A suspension form is preferred because it is normally more efficient compared to the thin film reactor. It is because in a suspension type reactor, the catalyst has a higher effective surface area and hence larger surface area in contact with the substrate. This allows a larger amount of photon to hit the surface and results in large adsorption capacity.

Other types of reactors are flat wall and tubular photoreactors. These types of reactors are simple and easy to design. Air can be use as an oxidant option for these reactors. Besides, solar energy can be utilized by these types of reactors. Moreover, reflectors are used in a tubular reactor to concentrate sunlight so that it can enhance the photoreaction.

For the past 20 years, several photocatalytic water treatment reactors have been developed and tested. Different rectors were developed to find the best way of conducting solar wastewater treatment process. Different principles and technologies were adopted including sunlight concentrating system. Four most frequent used photoreactors will be presented in the following text.

2.4.1 Parabolic through reactor (PTR)

A parabolic through reactor adopted the principle of parabolic through solar concentrating system' to concentrate the sunlight on the focal point using Dewar tube. The schematic presentation is given in Fig. 1. The PTR concentrates the parallel (direct) rays of the photocatalytically active ultra-violet part of the solar spectrum and can be characterized as a typical plug flow reactor. Borosilicate glass tube which positioned along the focal line was filled with contaminant with titanium dioxide (TiO_2) in suspension with a flowrate ranges between 250–3500Lh^{-1}. This type of reactor had been selected as the first solar detoxification loops in Albuquerque and California in USA and Almeria in Spain (Bahnemann, 2004). Several research groups from the European continent have tested the PTR which installed at the Plataforma Solar de Almeria (PSA) in Spain for solar wastewater purification in the early 1990s (Bahnemann, 2004).

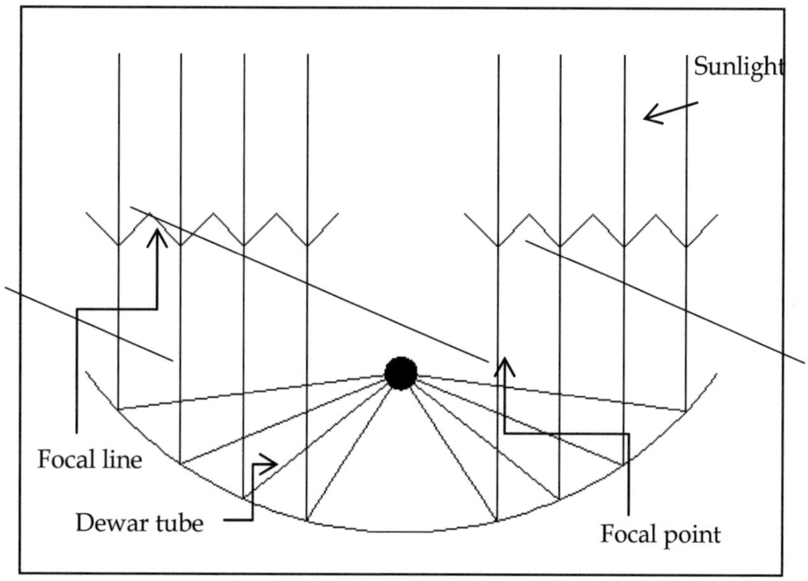

Fig. 1. A schematic view of a parabolic through solar concentrator

2.4.2 Thin film fixed bed reactor (TFFBR)

Thin film fixed bed reactor (TFFBR) is one of the very first solar reactor which does not utilize a solar concentrating system. It implies that the TFFBR can utilize the diffuse as well as the direct portion of the solar UV-A illumination for the photocatalytic process. A TFFBR installed at PSA is depicted in Fig. 2 (Bockelmann et al., 1995; Goslich et al., 1997; Hilgendorff et al., 1993). The most important aspect in the TFFBR is the slopping plate coated with photocatalyst like Degussa P25 (Bockelmann, 1993) and rinsed with contaminated water in a very thin film (~100µm). The flowrate was controlled by a cassette peristaltic pump and ranges from 1–6.5Lh^{-1} (Bockelmann et al., 1995).

Fig. 2. A TFFBR installed at the PSA in Spain (Bockelmann et al., 1995)

The TFFBR was tested during its operation at the PSA. The efficiency of the performance of the TFFBR was found to be higher than that of the PTR during several test campaigns utilizing both model pollutants dissolved in pure water and real wastewater samples collected from a variety of industrial companies (Bahnemann, 2004).

2.4.3 Compound parabolic collecting reactor (CPCR)

The compound parabolic collecting reactor (CPCR) is a through reactor without any solar concentrating properties. The major difference between the PTR and CPCR is the shape of their reflecting mirrors. The reflector of a PTR has a parabolic profile and the pipe is

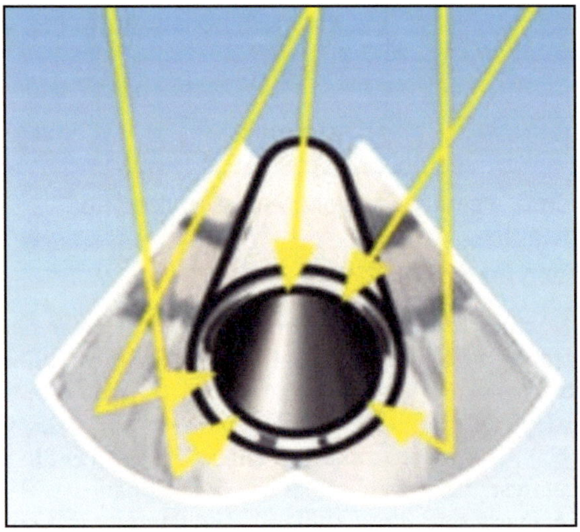

Fig. 3. Schematic view of the reflector of a CPCR

positioned along the focal line. Subsequently only parallel light which enters the parabolic through can be concentrated on the pipe. The reflector of a CPCR generally made up of two half circle profiles side by side. The focal line is situated just above the connections of the two circles. Thus light entering from almost any angle can be reflected to the focal line of the CPCR. Fig. 3 shows the schematic view of the reflector of a CPCR.

2.4.4 Double skin sheet reactor (DSSR)

The double skin sheet reactor is a new kind of reactor which does not have a light concentrating properties. It is a flat and transparent structured box made of PLEXIGLAS® (Van Well et al., 1997). PLEXIGLAS® is a trademark of a commercialized Poly(methyl methacrylate) (PMMA) which is a transparent thermoplastic. The inner structure of the DSSR is depicted in Fig. 4. The suspension of the model pollutant and the photocatalyst is allowed to flow through these channels. The DSSR can use both the diffuse and direct portion of the sunlight.

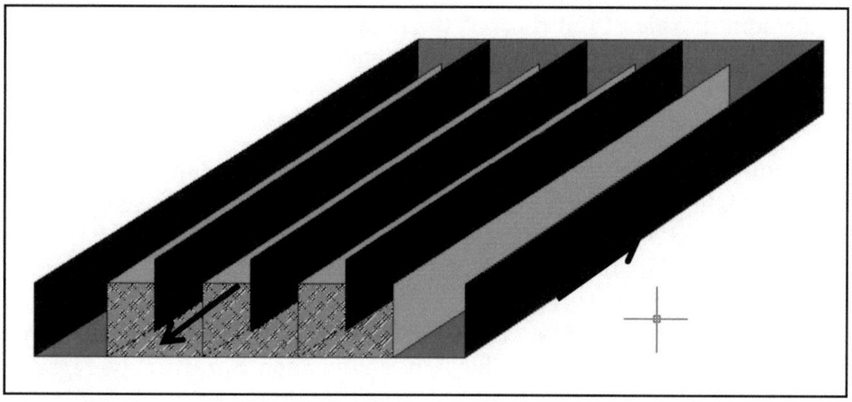

Fig. 4. A schematic view of the inner structure of a DSSR

2.5 Synthesis and doping method

Many studies had been carried out to alter the characteristics of the titanium dioxide (TiO_2) in order to improve the practical and commercial values of titanium dioxide (TiO_2) as a photocatalyst. For most of the cases, doping was carried out to improve the photocatalytic activity, the absorption of visible region of the solar spectrum, and to impart separable property.

Ao et al. (2009) reported the degradation of a dye (Red X-3B) under sunlight using N-doped titania-coated g-Fe_2O_3 magnetic activated carbon (NT-MAC). The titanium dioxide (TiO_2) was doped with nitrogen to improve the visible light absorption while the g-Fe_2O_3 magnetic activated carbon was coated to impart the magnetic properties. The preparations were carried out under low temperature and ambient pressure. It is reported that the photocatalytic of the NT-MAC was approximately three times than that of Degussa P25. The separation can be done easily using an external magnetic field. Furthermore, the prepared NT-MAC can be recycled and reused without any mass losing and the degradation of the X-3B remains higher than 85% after six cycles (Ao et al., 2009).

Han et al. (2009) studied the degradation of organic dyes using various modified titanium dioxide (TiO_2) photocatalysts. The modifications include doping with metals (noble metals, transition metals, lanthanide metals, alkaline and alkaline earth metals, cadmium sulphide etc.) and non-metals (nitrogen, fluorine, sulphur, carbon etc.). The purposes of these modifications and doping were to improve photocatalytic efficiency, complete degradation of organic dyes, improve visible light absorption, improve stability and reproducibility, and to improve recycle and reuse abilities of titanium dioxide (TiO_2). The modified titanium dioxide (TiO_2) showed considerably improved photocatalytic activity. For example, a complete degradation of Rhodamine (RB) in 105 minutes was observed using silver doped indium (III) oxide-coated TiO_2 (Ag/In_2O_3-TiO_2) as photocatalyst in 2008. It is more efficient than degradation using Degussa P25 which is 85.9% (Han et al., 2009).

Narayana et al. (2011) studied the photocatalytic decolourization of basic green dye using pure and ferum (Fe) and cobalt (Co) doped titanium dioxide (TiO_2) under sunlight irradiation. The purpose of doping was to improve the visible light absorption of the photocatalyst. The doped titanium dioxide (TiO_2) was prepared using sol-gel method. The Fe-doped titanium dioxide (TiO_2) showed the highest photoactivity among the other two with a 98% degradation of dye under sunlight illumination. Hence, doped titanium dioxide (TiO_2) can have very high commercial value in wastewater treatment since it utilizes only sunlight, which is a natural resource for reaction activation (Narayana et al., 2011).

Wang et al. (2010) doped titanium dioxide (TiO_2) with tin (Sn) and nitrogen (N) intended to improve the visible light absorption of titanium dioxide (TiO_2) photocatalyst. The doping was successfully carried out via simple sol-gel method. Pure TiO_2, N-doped TiO_2, Sn-doped TiO_2, and co-doped N/Sn-TiO_2 were tested separately to compare their characteristics. N/Sn-TiO_2 recorded the highest absorption in the visible region of solar spectrum. Besides, N/Sn-TiO_2 also recorded the highest visible-light activity among the other three by using 4-chlorophenol (4-CP) in water under visible light illumination. Surprisingly, N/Sn-TiO_2 also had the highest photoactivity under UV irradiation. This implies that the co-doping of two foreign ions is more efficient in improving photoactivity of titanium dioxide (TiO_2) compared to doping of one ion (Wang et al., 2010).

A simple sol-gel method to prepare titania-coated magnetic porous silica (TMS) photocatalyst was reported by Wang et al. (2010). The TMS was then employed in the degradation of red X-3B dye under UV and visible light irradiation to determine its photocatalytic activity. The same was done using commercialized titanium dioxide (TiO_2), Degussa P25 for comparison purpose. They recorded that the TMS had considerably higher photoactivity compared to that of Degussa P25, under either UV or visible light illumination. The TMS can be separated by applying external magnetic and thus can be reused without any mass loss. Hence, TMS can be a suitable photocatalyst for practical water purification system due to its high photocatalytic activity and separability (Wang et al., 2010).

2.6 Applications of photocatalysis on water and wastewater treatment

Since the discovery of water splitting phenomenon via photocatalysis by Fujishima and Honda in 1972, the research and development of the heterogeneous photocatalytic process has never been stop and has been growing rapidly (Linsebigler et al., 1995). Though early studies and researches were focused on the energy production i.e. the production of clean fuel using hydrogen, heterogeneous photocatalysis has taken a new step since its breakthrough in the environmental remediation field. Today, this technique has been implemented in various applications, from water and air treatment to health applications.

2.6.1 Degradation of organics

The degradation of organics perhaps is the most important applications of photocatalysis. The degradations of organic compounds such as alcohols, carboxylic acids, phenolic derivatives or chlorinated aromatics into non-hazardous and harmless products or residues such as carbon dioxide, water or other minerals had been well documented (Bhatkande et al., 2001; Chen & Ray, 2001; Michael et al., 1995; Mills et al., 1993; Pirkanniemi & Sillanpaa, 2002). Joanna et al. (2000) reported that oily water can also be treated effectively by photocatalysis. Herbicides and pesticides like 2, 4, 5, trichlorophenoxyacetic acid, 2, 4, 5, trichlorophenol, s-triazine herbicides and DDT which generally considered as hazardous pollutants can also be completely mineralized (Olis et al., 1991).

2.6.2 Elimination of inorganics

Other than the organic chemical compounds mentioned above, a variety of inorganic compounds are sensitive to photochemical conversion on the catalyst surfaces. Inorganics such as chlorate and bromate (Mills et al., 1996), azide, halide ions, nitric oxide (NO), palladium (Pd) and rhodium (Rh) species, and sulphur species can be broken down (Michael et al., 1995). Metal salts like silver nitrate ($AgNO_3$), mercury (II) chloride (HgCl) and organometallic materials can be eliminated from water (Bhatkande et al., 2001), as well as cyanide (CN), thiocyanate (SCN^-), ammonia (NH_3), nitrates (NO^{3-}) and nitrites (NO^{2-}) (Blake, 2001).

2.6.3 Elimination of natural organic matter

Humic substances (HS) can be generally defined as a class of naturally existing biogenic heterogeneous organic substances that can be further classified as being yellow-brown and having high molecular weights (MacCarthy, 2001). HS can also be defined as the fraction of filtered water that adsorb on XAD-8 resin (a non-ionic polymeric adsorbent) at pH 2 (Obernosterer & Herndl, 2000). They are the major components of the dissolved organic carbon (DOC) pool in surface water (marine waters and fresh waters) and sub-surface or ground waters. They are often said to be a main factor that lead to yellowish-brown colour in the water bodies (Schmitt-Kopplin, 1998). The concentration of the HS differs from place to place; seawater normally contains 2-3mg/L of HS. According to Gaffney et al. (1996), their physical properties like size and their chemical properties like the structure and the number and position of the functional group differ, relying on the origin and the age of the substance (Gaffney et al., 1996).

HS are known to have the ability to change the behaviour of certain pollutants considerably, such as trace metal speciation and toxicity, (Bekbolet & Balcioghu, 1996; Shin et al., 1996), solubilisation and adsorption of hydrophobic contaminants (Chiou et al., 1986; Tanaka et al., 1997) and aqueous photochemistry (Fukushima et al., 2000). HS can act as substrates for bacterial growth, hinder the bacterial degradation of impurities (colours), interact with heavy metals such as Fe, Mn and Pb and thus making them difficult to remove, help to transport the metals into the environment and also contribute to pipe corrosions (Motheo & Pinhedo, 2000). Besides, HS can also act as a source of methyl groups and hence react with hypochlorite ions which are being used as biocide in water treatment plants to produce disinfectant by-products. Examples of these by-products are trihalomethanes, haloacetic acids, other chlorinated compounds and nitriles. Some of these by-products are suspected to be carcinogenic. Till date, more than 150 products have been recognized as the products of the reaction between HS and chlorine.

The advance oxidation process has been implemented to decrease the organic matter in water including the HS. Its major advantage is that the advance oxidation process does not produce any toxic by-products or residues which required further treatment or disposal. Till date, the degradation of HS using photocatalytic process has not been studied well. The very first study based on this was carried out by Bekbolet in 1996, who studies the effectiveness of photocatalytic treatment on the degradation of model humic substances or humic acid (Bekbolet & Ozkosemen, 1996).

Bekbolet and Ozkosemen studied about the degradation via photocatalytic process using humic acid as a model pollutant. Through the experiment, they found out that after one hour of illumination in the presence of 1.0g of Degussa P25, 40% of the TOC and 75% of the colour (400nm) were removed (Bekbolet & Ozkosemen, 1996).Bekbolet and co. again studied the removal of colour caused by humic acid in the presence of common inorganic ions (e.g. chloride, nitrate, phosphate and sulphate ions) at pH 6.8, and they found some removal (Bekbolet et al., 1998). In other researches where humic acid was used as an additional matrix for the degradation of some organic pollutants, a 80% removal of commercialized humic acid was recorded by using irradiation in the presence of Degussa P25 (Minero et al., 1997). Another similar study showed a reduction around 50% of the concentration of humic acid in just 12 minutes using suspension of Degussa P25 irradiated by a mercury lamp (Eggins et al., 1997).

2.6.4 Removing trace metals

Trace metals especially mercury (Hg), lead (Pb), chromium (Cr) and many others are considered extremely hazardous to human being. The presence of these metals in water bodies should be removed. Photocatalysis can be used to remove heavy metals like mercury (Hg), chromium (Cr), lead (Pb), cadmium (Cd), arsenic (As), nickel (Ni), and copper (Cu) (Blake, 2001; Olis et al., 1991). Other than that, the photochemical ability of the photocatalysis enables it to recover costly metals from industrial waste discharge such as gold (Au), platinum (Pt) and silver (Ag) (Olis et al., 1991).

2.6.5 Water disinfections

Photocatalysis can be used in water disinfections because it can kill or destroy various bacteria and viruses. In 1997, a study by Mills and LeHunte reported that *Streptococcus mutans, streptococcus natuss, streptococcus cricetus, escherichia coli, scaccharomycescerevisisas, lactobacillus acidophilus,* poliovirus 1 were destructed effectively using heterogeneous photocatalysis (Mills & LeHunte, 1997). With algae blooming in fresh water supplies becoming more and more common, the subsequent possibility of cyanobacterialmicrocystin pollution of portable water caused by *Microcystin* toxins. In 2002, Shephard et al. reported that Microcystin toxins can be degraded on immobilized titanium dioxide (TiO_2) photocatalyst (Shephard et al., 2002). Matsunaga and Tomoda in 1985 first reported on photocatalytic disinfection.They found that in the presence of high concentrations of microorganisms, disinfection process was effective and more efficient. Similarly, Belapurkar et al., 2006 prepared high surface area TiO_2 by hydrothermal method using titanium isopropoxide. They applied for disinfection of water and found to be effective when 1 L of water was photolysed by solar light in a plastic tray containing TiO_2 photocatalyst coated on a stainless plate. Their study also proved that the technique can be used for disinfection of ~20 L water daily using solar light. Finally concluded that the photocatalytic technique using solar light, a viable,simple and easy-to-use device for disinfection of drinking water on liter scale (Belapurkar et al., 2006)

2.6.6 Seawater treatment

Lately, the decomposition of humic substances in artificial seawater (highly saline water) and natural seawater were studied by Al-Rasheed and Cardin (Al-Rasheed & Cardin, 2003). Although the decompositions were found to be slower compared with a fresh water media usually employ by other researches, no toxic or hazardous by-products were found throughout the decomposition process.

The degradation of some crude oil components (dodecane and toluene) via photocatalysis using seawater media was carried out in 1997 (Minero et al., 1997). No chlorinated compounds were found over the course of irradiation. 100% degradation was recorded after just few hours of illumination. Ziolli and Jardim reported in 2002 that seawater-soluble crude oil fractions can be decomposed under the irradiation of nanoparticles of titania using artificial light (Ziolli & Jardim, 2002).

2.7 Current and future scope

The number of new publications regarding photocatalysis for water and wastewater treatment has been increasing significantly since the last decade.

The photocatalytic oxidation of towards water treatment has caught up most of the attention. Recently, the attention started to shift onto the oxidation of volatile organic or inorganic compounds present in ground water for an efficient treatment. Photocatalytic reduction organic compounds and metal-containing ions and researches on cell destroying and disinfection by irradiated titania has also caught up some attention (Zaleska, 2008).

Subsequently, titania-based photocatalysts has been commercialized in various fields firstly in Japan, followed by the United States and then China. This commercialization of TiO_2-based photocatalysts products was started during the mid 90s in Japan. Among commercialisation the purification equipment (e.g. air purifiers, air conditioners, portable water purification system, purification system for pools) and household equipment is more promising achievements in field of water treatment. (Zaleska, 2008).

Though a number of commercial TiO_2 is available in market they lag in low sensitivity of photocatalyst towards visible light, which cannot take up the visible spectrum (largest part) in the solar radiation for waste treatment. Hence scientific community is eager to increase the sensitivity of photocatalyst to visible light so that sunlight could be used for excitation for a sustainable waste treatment.

In the present scenario, the major difficulty regarding the doped TiO_2 photocatalyst is the possible loss of photoactivity due to recycling of photocatalyst and long-term storage. It is believed that the efficiency of the metal-doped TiO_2 under visible light wholly depends on the synthesizing and doping method adopted. In some cases, such doped photocatalysts showed zero activity under visible light or considerably lower activity in the ultraviolet spectral range compared to the non-doped TiO_2 because of high carrier recombination rates through the metal ion levels. The problem is that the non-metal-doped TiO_2 catalyst has very low photoactivity under visible light compared to that under UV light (Zaleska, 2008).TiO_2 with visible light absorption can be employed to purify and disinfect the water and make it more suitable for consumption. Beside these limitations the major edge of the photocatalytic oxidation process over other process is because it's an only green and sustainable process towards waste treatment.

3. Conclusion

Strictly speaking, contaminant treatment can be defined as the complete degradation or mineralization of the contaminants. However, the photocatalytic degradation is suitable for treating hazardous organic pollutants. Generally, biological treatment is the most economical treatment option and the most compatible with the environment when feasible. Though the feasibility of adopting the photocatalytic oxidation is much explored they are not adopted practically, due to the problem of surplus power needed for generation of UV radiation. But in the present years ferrite doped titanium dioxide (TiO_2) addresses this issue and, it also enhances the reusability of the catalyst. Hence it could be an ideal treatment to transform the bio-persistent compounds for an effective treatment system with sustainability. They also address the green technology, by utilizing sunlight as their source of excitation. This could be further eliminates or reduces the production of sludge's, a secondary pollutant. In near future the practise of biological methods could be replaced by the heterogeneous oxidation process. Such replacement will lead pathway to a green technology for the sustainable development.

4. Acknowledgement

The authors are grateful to IPPP, Department of Civil Engineering and Faculty of Engineering University of Malaya for their grant (RG091/10SUS) and moral support respectively.

5. References

Ahmed, S., Rasul, M., Martens, W., Brown, R., & Hashib, M. (2010). Heterogeneous photocatalytic degradation of phenols in wastewater: A review on current status and developments. *Desalination,* 261, pp. 3-18

Alnaizy, R. & Akgerman, A. (2000). Advanced oxidation of phenolic compounds. *Advances in Environmental Research,* 4,pp. 233-244

Al-Rasheed, R., & Cardin, D. (2003). Photocatalytic degradation of humic acid in saline waters. Part 1, Artificial seawater: Influence of TiO_2, temperature, pH, and air-flow. *Aldrich. Chemosphere,* 51, pp. 925-933

Al-Rasheed, R., & Cardin, D. (2003). Photocatalytic degradation of humic acid in saline waters. Part 2, Effect of various photocatalytic materials. *Applied Catalysis A:General,* 246, pp. 39-48

Ao, Y., Xu, J., Zhang, S., & Fu, D. (2009). Synthesis of a magnetically separable composite photocatlyst with high photocatalytic activity under sunlight. *Journal of Physics and Chemistry of Solids,* 70, pp. 1042-1047

Asahi, R., Morikawa, T., Ohwaki, T., Aoki, K., & Taga, Y. (2001). Visible-light photocatalysis in nitrogen-doped titanium oxides. *Science,* 293, pp. 269-271

Bahnemann, D. (2004). Photocatalytic water treatment: Solar energy applications. *Solar Energy,* 77, pp. 445-459

Bekbolet, M., & Balcioglu, I. (1996). Photocatalytic degradation kinetics of humic acid in aqueous TiO_2 dispersions: The influence of hydrogen peroxide and bicarbonate ion. *Water Science and Technology,* 34, pp. 73-80

Bekbolet, M., & Ozkosemen, G. (1996). A preliminary investigation on the photocatalytic degradation of a model humic acid. *Water Science and Technology,* 33, pp. 189-19

Bekbolet, M., Boyacioglu, Z., & Ozkaraova, B. (1998). The influence of solution matrix on the photocatalytic removal of color from natural waters. *Water Science and Technology,* 38, pp. 155-162

Belapurkar, A. D., Sherkhane, P. & Kale, S. P. (2006). Disinfection of drinking water using photocatalytic technique. *Current Science,* 91,pp.73-76

Bhatkhande, D., Pangarkar, V., & Beenackers, A. (2001). Photocatalytic degradation for environmental applications - A review. *Journal of Chemical Technology, 77,* pp. 102-116

Bickley, R., Gonzalez-Carreno, T., Lees, J., Palmisano, L., & Tilley, R. (1991). A structural investigation of titanium oxide photocatalyst. *Solid State Chemistry,* 92, pp. 178

Blake, D. (2001). *Bibliography of Work on the Heterogeneous Photocatalytic Removal of Hazardous Compounds from Water and Air,* National Renewable Energy Laboratory, pp. 1-158

Bockelmann, D. (1993). *Patent No. P 4 237 390.5.* Germany.

Bockelmann, D., Weichgrebe, D., Goslich, R., & Bahnemann, D. (1995). Concentrating versus non-concentrating reactors for solar water detoxification. *Solar Energy Materials and Solar Cells,* 38, pp. 441-451

Chang, H., Wu, N.-M., & Zhu, F. (2000). A kinetic model for photocatalytic degradation of organic contaminants in a thin-film TiO_2 catalyst. *Water Research,* 34, pp. 407-416

Chen, D., & Ray, A. (2001). Removal of toxic metal ions from wastewater by semiconductor photocatalysis. *Chemical Engineering Science,* 56, pp. 1561-1570

Chiou, C., Malcolm, R., Brinton, T., & Kile, D. (1986). Water solubility enhancement of some organic pollutants and pesticides by dissolved humic and fulvic-acids. *Environmental Science and Technology,* 20, pp. 502-508

Cotton, F., Wilkinson, G., Murillo, C., & Bochmann, M. (1999). *Advance Inorganic Chemistry,* Wiley, New York

Eggins, B., Palmer, F., & Byrne, J. (1997). Photocatalytic treatment of humic substances in drinking water. *Water Research,* 31, pp. 1223-1226

Fox, M., & Dulay, M. (1993). Heterogeneous photocatalysis. *Chemical Reviews,* 93, pp. 341-357

Fujishima, A., & Honda, K. (1972). Electrochemical photolysis of water at a semiconductor electrode. *Nature,* 238, pp. 37-38

Fujishima, A., Zhang, X., & Tryk, D. (2007). Heterogeneous photocatalysis: From water photocatalysis to applications in environmental cleanup. *International Journal of Hydrogen Energy,* 32, pp. 2664-2672

Fukushima, M., Tatsumi, K., & Morimoto, K. (2000). Influence of iron(III) and humic acid on the photodegradation of pentachlorophenol. *Environmental Toxicology and Chemistry,* 19, pp. 1711-1716

Gaffney, J., Marley, N., & Clark, S. (1996). Humic and fulvic acids and organic colloidal material in the environment, in: Humic and fulvic acids: isolation, structure, and environmental role. *Washington: Amer. Chemical. Soc.*, pp. 2-16

Glaze, W., Lay, Y. & Kang, J. (1995). Advanced oxidation processes. A kinetic model forthe oxidation of 1,2-dibromo-3-chloropropane in water by the combination of hydrogen peroxide and UV radiation. *Industrial and Engineering Chemistry Research,* 34,pp. 2314-2323

Goslich, R., Dillert, R., & Bahnemann, D. (1997). Solar water treatment: principles and reactors. *Water Science and Technology,* 35, pp. 137-148

Guittonneau, S., Glaze, W., Duguet, J. & Wable, O. (1991). Characterization of natural waters for potential to oxidize organic pollutants with ozone *Proceedings of 10th Ozone World Congress*, Zürich, Switzerland

Han, F., Kambala, V., Srinivasan, M., Rajarathman, D., & Naidu, R. (2009). Tailored titanium dioxide photocatalysts for the degradation of organic dyes in wastewater treatment: A review. *Applied Catalysis A: General,* 359, pp. 25-40

Herrmann, J. (1999). Heterogeneous photocatalysis: fundamentals and applications to the removal of various types aqueous pollutants. *Catalysis Today,* 53, pp. 115-129

Herrmann, J., Duchamp, C., Karkmaz, M., Hoai, B., Lachheb, H., & Puzenat, E. G. (2007). Environmental green chemistry as defined by photocatalysis. *Journal of Hazardous Materials,* 146, pp. 624-629

Hilgendorff, M., Hilgendorff, M., & Bahnemann, D. (1993). Reductive photocatalytic elimination of tetrachloromethane on platinized titanium dioxide in aqueous suspension. In M. Tomkiewicz, R. Haynes, H. Yoneyama, & Y. Hori (Ed.), *Environmental Aspects of Electrochemistry and Photoelectrochemistry, Proceedings of Electrochemical Society* (pp. 112-121). Pennington: The Electrochemical Society

Hincapié, M., Maldonado, M.I., Oller, I., Gernjak, W., Sánchez-Pérez, J.A., Ballesteros, M.M., & Malato, S. (2005). Solar photocatalytic degradation and detoxification of EU priority substances. *Catalysis Today,* 101, pp. 230-210

Hoffmann, M., Martin, S., Choi, W., & Bahnemann, W. (1995). Environmental applications of semiconductor photocatalysis. *Chemical Reviews,* 95, 1, pp. 69-96

Huang, A. C., Spiess, F.-J., Suib, S., Obee, T., Hay, S., & J.D., F. (1999). Photocatalytic degradation of triethylamine on titanium dioxide thin films. *Journal of Catalysis,* 188, pp. 40-47

Joanna, G., Maciej, H., & Antoni, W. (2000). Photocatalytic decompstion of oil in water. *Water Research,* 34, pp. 1638-1644

Kawai, T., & Sakata, T. (1980). Conversion of carbohydrate into hydrogen fuel by a photocatalytic process. *Nature,* 286, pp. 74-476

Kidak, R. & Ince, N.H. (2006). Ultrasonic destruction of phenol and substituted phenols: A review of current research. *Ultrasonics Sonochemistry,* 13,pp.195–199

Legrini, O., Oliveros, E. & Braun, A.M. (1993). Photochemical processes for water treatment. *Chemical Reviews*, 93,pp. 671- 698

Leyva, E., Moctezuma, E., Ruiz, M.G., & Torez-Martínez, L. (1998). Photo degradation of phenol and 4-chlorophenol by BaO-LiO_2-TiO_2 catalysts. *Catalysis Today,* 40, pp. 367-376

Lindner, M., Bahnemann, D., Hirthe, B., & Griebler, W. (1995). Solar water detoxification: Novel TiO_2 powders as highly active photocatalysts. *Solar Engineering, 1 ASME*, pp. 399- 408

Linsebigler, A., Lu, G., & Yates, J. (1995). Photocatalysis on surfaces: Principles, mechanisms, and selected results. *Chemical Reviews,* 95, pp. 735-758

MacCarthy, P. (2001). The principles of humic substances. *Soil Science,* 166, pp. 738-751

Malato, S., Fernández-Ibáñez, P., Maldonado, M., Blanco, J., & Gernjak, J. (2009). Decontamination and disinfection of water by solar photocatalysis: Recent overview and trends. *Catalysis Today,* 147, pp. 1-59

Matsunaga, T., Tomoda R., Nakajima T. & Wake H., (1985). Photoelectrochemical sterilization of microbial cells by semiconductor powders, *FEMS Microbiology Letters,* 29, 211-214

Matthews, R. (1990). Purification of water with near-UV illuminated suspensions of titanium dioxide. *Water Research,* 24, pp. 653-660

Matthews, R. (1993). Photocatalysis in water purification: Possibilities, problems and prospects. In D. Ollis, H. AL-Ekabi, D. Ollis, & H. AL-Ekabi (Eds.), *Photocatalytic Purification and Treatment of Water and Air,* pp. 121-139, Elsevier Science, Amsterdam

Matthews, R.W., Abdullah,M.,& Low, G. K.C. (1990). Photocatalytic oxidation for total organic carbon analysis. *Analytica Chimica Acta,* 223, pp. 171-179

Michael, R., Scot, T., Wonyong, C., & Detlef, W. (1995). Environmental application of semicondutor photocatalysis. *Chemical Reviews,* 95, pp. 69-96

Mills, A., & Hunte, S. (1997). An overview of semiconductor photocatalysis. *Journal of Photochemistry and Photobiology A: Chemistry,* 108, 1, pp. 1-35

Mills, A., Belghazi, A., & Rodman, D. (1996). Bromate removal from drinking water by semiconductor photocatalysis. *Water Research,* 30, pp. 1973-1978

Minero, C., Maurino, V., & Pelizzetti, E. (1997). Photocatalytic transformations of hydrocarbons at the sea water/air interface under solar radiation. *Marine Chemistry,* 58, pp. 361-372

Motheo, A., & Pinhedo, L. (2000). Electrochemical degradation of humic acid. *Science of The Total Environment,* 256, pp. 67-76

Narayana, R. L., M, M., Aziz, A., & Saravanan, P. (2011). Photocatalytic decolourization of basic green dye by pure and Fe, Co doped TiO_2 under sunlight illumination. *Desalination,* 269, pp. 249-253

Obernosterer, I., & Herndl, G. (2000). Differences in the optical and biological reactivity of the humic and non-humic dissolved organic carbon component in two contrasting coastal marine environments. *Limnology and Oceanography,* 45, pp. 1120-1129

Ohno, T., Sarukawa, K., Tokieda, K., & Matsumura, M. (2001). Morphology of a TiO_2 photocatalyst (Degussa, P-25) consisting of anatase and rutile crystalline phases. *Journal of Catalysis,* 203, pp. 82-86

Ollis, D., Pelizzetti, E., & Serpone, N. (1991). Photocatalyzed destruction of water contaminants. *Environmental Science and Technology,* 25, pp. 1522-1529

Peyton, G. & Glaze, W. (1988). Destruction of pollutants in water with ozone in combination with ultraviolet radiation. 3. Photolysis of aqueous ozone. *Environmental Science and Technology,* 229,pp.761-767

Pirkanniemi, K. & Sillanpaa, M. (2002). Heterogeneous water phase catalysis as an environmental application: A review, *Chemosphere,* 48, pp. 1047-1060

Rajeshwar, K. & Ibanez, J.G.(1995). Electrochemical aspects of photocatalysis: application to detoxification and disinfection scenarios. *Journal of Chemical Education,* 72, pp. 1044-1049

Rajeshwar, K. & Ibanez, J.G. (1997). *Environmental Electrochemistry: Fundamentals and Applications in Pollution Abatement,* Academic Press, San Diego

Rajeshwar, K., Chenthamarakshan, C., Goeringer, S., & Djukic, M. (2001). Titania-based heterogeneous photocatalysis. Materials, mechanistic issues, and implications on environmental remediation. *Pre Applied Chemistry* 73, 12, pp. 1849-1860

Robert, D., & Malato, S. (2002). Solar photocatalysis: A clean process for water detoxification. *The Science of the Total Environment*, 291, pp. 85-97

Saravanan, P., Pakshirajan, K., & Saha, P. (2009). Degradation of phenol by TiO_2-based heterogeneous photocatalysts in presence of sunlight. *Journal of Hydro-environment Research*, 3, pp. 45-50

Sato, S., & White, J. (1980). Photodecomposition of water over Pt/TiO_2 catalyst. *Chemistry Physics Letters*, 72, pp. 83-86

Schiavello, M. (1997). *Heterogeneous Photocatalysis* (M. Schiavello, Ed.), Wiley, Chichester

Schmitt-Kopplin, P., Hertkorn, N., & Kettrup, A. (1998). Structural changes in a dissolved soil humic acid during photochemical degradation processes under O_2 and N_2 atmosphere. *Environmental Science and Technology*, 32, pp. 2531-2541

Serpone, N., & Pelizzetti, E. (1989). *Photocatalysis, Fundamentals and Applications* (N. Serpone, & E. Pelizzetti, Eds.), Wiley, New York

Shephard, G., Stockenstrom, S., de Villiers, D., Engelbrecht, W., & Wessels, G. (2002). Degradation of microcystin toxins in a falling film photocatalytic reactor with immobilized titanium dioxide catalyst. *Water Research*, 36, pp. 140-146

Shin, H., Rhee, S., Lee, B., & Moon, C. (1996). Metal binding sites and partial structures of soil fulvic and humic acids compared: Aided by Eu(III) luminescence spectroscopy and DEPT/QUAT C-13 NMR pulse techniques. *Organic Geochemistry*, 24, pp. 523-529

Tanaka, S., Oba, K., Fukushima, M., Nakayasu, K., & Hasebe, K. (1997). Water solubility enhancement of pyrene in the presence of humic substances. *Analytica Chimica Acta*, 337, pp. 351-357

Van Well, M., Dillert, R., Bahnemann, D., Benz, V., & Müller, M. A. (1997). A novel non-concentrating reactor for solar water detoxification. *Trans. ASME, J. Solar Energy Engineering*, pp. 114-119

Wang, C., Ao, Y., Wang, P., Hou, J., & Qian, J. (2010). A facile method for the preparation of titania-coated magnetic porous silica and its photocatalytic activity under UV or visible light. *Colloids and Surfaces A: Physiochemistry Engineering Aspects*, 360, pp. 184-189

Wang, E., He, T., Zhao, L., Chen, Y., & Gao, Y. (2010). Improved visible light photocatalytic activity of titania doped with tin and nitrogen. *Journal of Materials Chemistry*, 21, pp. 144-150

Ye, F., & Ohmori, A. (2002). The photocatalytic activity and photo-absorption of plasma sprayed TiO_2-Fe_3O_4 binary oxide coatings. *Surface and Coatings Technology*, 160, pp. 62-67

Zaleska, A. (2008). Doped-TiO_2: A review. *Recent Patents on Engineering*, 2, pp. 157-164

Zeltner W.A. & M.A. Anderson (1996). The use of nanoparticles in environmental applications, in: *Pelizzetti E. ed. Fine Particles Science and Technology*, Kluwer Academic Publishers, pp. 643–656

Zhang, Y., Crittenden, J.C., Hand, D.W., & Perram, D.L. (1994). Fixed-bed photocatalysts for solar decontamination of water, *Environmental Science and Technology*, 28, pp. 435

Ziolli, R., & Jardim, W. (2002). Photocatalytic decompsition of seawater-soluble crudeoil fractions using high surface areacolloid nanoparticles of TiO_2. *Journal of Photochemistry and Photobiology A-Chemistry*, 147, pp. 205-212

Part 3

Water Resources Planning and Management

10

Novel SPP Water Management Strategy and Its Applications

Shu-Qing Yang[1], Bo-Qiang Qin[2] and Pengzhi Lin[3]
[1]*School of Civil, Mining & Environmental Engineering, University of Wollongong, NSW2522,*
[2]*Nanjing Institute of Geography and Limnology, Nanjing*
[3]*State Key Lab. of Hydraulics and Mountain River Engineering,*
[1]*Australia*
[2,3]*China*

1. Introduction

Clean freshwater is the most precious resource in the world and the development of water resources has had a very long history, as early as humans changed from being hunters and food collectors to modern civilization. At very early stage, people had to rely on creeks, rivers and lakes for their water demand that was relatively small, and today humans have accumulated the knowledge and techniques for water storage, building artificial lakes or reservoirs to meet their huge water demand due to industrialization and urbanization. The world's earliest large dam was the Sadd-el-kafara Dam built in Egypt between 2950 and 2690 B.C. Up to now, water from lakes and reservoirs is still the main source for people's water supply. However these large water bodies suffer two problems incurred by nature and human being, one is *sedimentation* and the other *water pollution*. Two of them jointly reduce the available amount of clean water and deteriorate the water quality. Consequently, approximate 1.1 billion people lack of safe drinking water and between 2 and 5 million people die annually from water-related disease (Gleick, 2004). It is understandable that with the population growth in the world, it is difficult to provide sufficient clean water to meet the demand; on the other hand, our natural systems are under pressure from drought (too little), floods (too much), pollution (too dirty), climate change, and other stresses. This creates serious challenges for water management.

Within a generation, water demand in many countries is forecast to exceed supply by an estimated 40%. In other parts of the world prone to flooding, catastrophic floods normally expected once a century could occur every 20 years instead.

Currently, there are about 40,000 large reservoirs worldwide used for water supply, power generation, flood control, etc. The total sediment yield in the world is estimated to be 13.5×10^9 tonnes/a or 150tonnes/km^2 and about 25% of this is transported into the seas and oceans and the rest 75% is trapped, retained and stored in the lakes, reservoirs and river systems (Batuca and Jordaan, 2000). Consequently the silting process is reducing the storage capacity of the world's reservoirs by more than 1% per year. As a result of sedimentation,

300-400 new dams would need to be constructed annually to maintain current total storage (White, 2001). On the other hand, due to climate change, the natural erosion rates will be accelerated. UN experts warned that a fifth of the current storage capacity of reservoirs worldwide or 1,500 km^3 will be gradually lost over the coming decades as global warming may increase the severity of storms and rains. Thus, one may conclude that the worst enemy of sustainable water resources management is sedimentation (USBR, 2006).

In 1998 the U.S. Environmental Protection Agency has identified that sediment in waterways is the largest single pollutant in the ecosystem (National Water Quality Inventory Section 305(b) Report to Congress), because sediment transported downstream can fill reservoirs, reduce its capacity and impair aquatic habitats. Detrimental effects to fish and aquatic invertebrate have been directly related to increases in the magnitude and duration of suspended sediment concentrations (Newcombe and Jensen, 1996 and Kuhnle et al., 2001). It is found that eutrophication can result from a high sediment load that has elevated levels of colloidal material, phosphorus and nitrogen that are transported in association with the sediment (Davis and Koop, 2006).

Water pollution, the main threat of water quality, began with the industry revolution, and the word "pollution" is an adaptation of the Latin "pollutionem", meaning defilement from "polluere". To many people, water pollution means the introduction into natural water of foreign substance, but strictly speaking, the water pollution refers to the introduction into water of any substances that makes water hazardous to public health. The impurities or foreign substances could be organic, inorganic, radiological or biological, and its presence in water tends to degrade water's quality or impair the usefulness of the water. The substances in water can be classified into three groups as shown in Fig. 1, i.e., dissolved; suspended and colloidal.

A dissolved substance is one which is truly dispersed in the liquid, and cannot be removed from the liquid without accomplishing a phase change such as: distillation, precipitation, adsorption, extraction or passage through ionic pore sized membranes. Suspended solids are large enough to settle out of solution or be removed by filtration. The lower size range of this class is 0.1 to 1 mm that is about the size of bacteria. The suspended solids can be removed from water by physical methods such as: sedimentation, filtration and centrifugation. Colloidal particles are in the size between dissolved substances and suspended particles. They can be removed from the liquid by physical means such as very high forced centrifugation or filtration through membranes with very small pore spaces. When light passes through a liquid containing colloidal particles, the light is reflected, which is measured by turbidity.

The typical impurities in water can be further divided based on its density (equal to or larger or less than the density of water) or size:

Algae (0.3-100 mm),
Bacteria (0.1-100 mm),
Viruses (0.003-0.3 mm),
Fungi (1-90 mm),
Giardia cysts (5-16 mm),
Colloids (0.01-7 mm),
Suspended solids (0.3-100 mm),

Humic acids (0.01-0.1 mm),
Post filters particles (0.4-10 mm),
Flocculated particles (1-100 mm).

Until 1800s, most materials used in homes and industries were natural products. In 1900s petroleum was used widely, in 1940s explosion in chemical production and in 1930s to 1950s chemicals like fertilizers and pesticides were invented and found very effective. On the other hand, the urbanization and modern agriculture have changed people's living habit and have greatly enhanced the food productivity. Consequently, domestic/agricultural waste water has also increased significantly with the same trend as industrial wastewater. The United Nation and World Bank's statistics shows that the world discharges 400 billion tons of wastewater every year, resulting in that 5,000 billion ton of clean water being polluted. Among them, China releases 60 billion ton of wastewater every year into rivers and lakes, greatly damaging its ecosystem. As a result of the rapid growth in industrial development, urbanization and population, the world's water resources are grossly polluted by human, agricultural and industrial wastes, to the point that vast stretches of rivers are dead and dying and lakes are cesspools of waste.

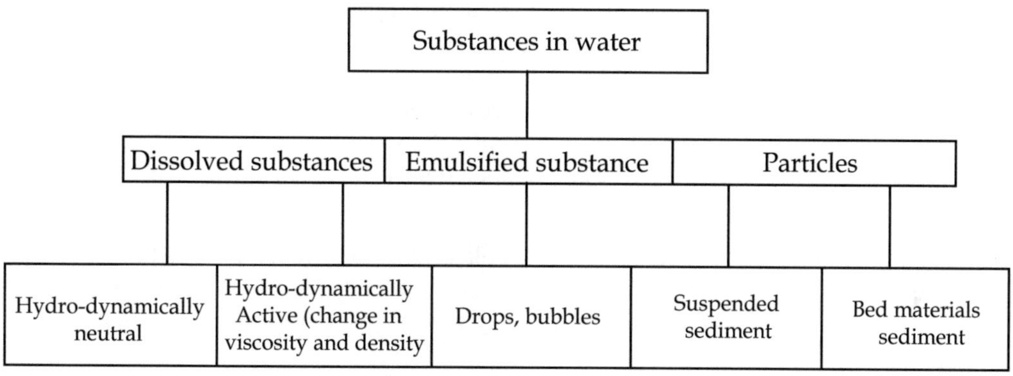

Fig. 1. Classification of substances in water

The world's current strategies in relation to water management are mainly focused on construction of reservoirs, wastewater reuse, desalination technology, alongside non-engineering methods like marketplace allocation etc. Increasingly, water planners have to take into account societal responses to proposed technologies as illustrated by the opposition to new dams, not to mention the hostile public reception met by wastewater and desalination strategies. Thus, it is worthwhile investigating other alternative water management strategy as long as it proves to be cost effective, energy thrifty and environment-friendly. By far, reservoirs and lakes play an important role for the modern society. As mentioned above, the biggest challenge is that the incoming rivers to these large water bodies are severely polluted by contaminants particle or solvable pollutants in water. The almost stagnant water together with high temperature, suitable sun shine and nutrients often leads to the algal blooms on vast scales that have become a major water quality issue for ecosystems throughout the world. It is an urgent threat facing surface water resources today, because of their ecological, aesthetic, and human health impacts. Blooms involving toxin-producing species pose serious threats to animals, plants and humans. Bloom occurrence is visible evidence that humans now strongly effect almost every aspect of our ecosystem. Just like the carbon dioxide emission and climate change, human activities have dramatically increased the emission of nutrients, contributing to increase in phytoplankton biomass and even algal blooms formation in receiving water bodies, e.g. rivers, lakes, coastal waters and reservoirs. With rapid economic/development and population growth worldwide, harmful algal blooms have become more frequent, more extensive, and more severe (Hallegraeff, 1993, Heisler, et al. 2008).

Algal blooms typically appear in slow moving water bodies with excess nutrients e.g. phosphorus, and nitrogen etc., when sunlight, turbulence, transparency, salinity and temperature are suitable. Over the past four to five decades despite extensive and intensive research, many key questions in eutrophication science remain unanswered (Smith and Schindler, 2009). The causes of these blooms are very complex. Never-the-less, in practice there is a pressing need to provide sufficient clean water to our society. The need is urgent to develop reliable methods and strategies to control the algal blooms in our water sources, i.e., lakes or reservoirs. Such a method should be able to restore water quality of these water sources to an acceptable (i.e., no harmful blooms) level in a short of period, e. g. 3-5 years, and also it should be technically feasible, cost-effective and environment-friendly.

Therefore, a big question to ask is how to manage our water resources, and our target is to provide the society enough clean water to sustain their other activities. Any effective strategy of water management should be able to manage or mitigate the disasters caused by floods (too much), droughts (too little), deterioration of water quality (too dirty) and siltation (too turbid). The above mentioned clearly shows that all problems of water supply are directly or indirectly related to the contaminated particle management that affects either water quantity or water quality.

One such method exists (Yang, 2004) and its application is described below. After reviewing the existing problems in water management, Yang and Liu (2010) proposed an effective method to control water quality in lakes; the "separation, protection and prevention" or SPP strategy comes from the following facts:

1. Algal blooms and siltation often appear in slowly moving water bodies. At the same nutrient level, fast moving water has less likelihood to induce massive algal blooms. Large-scale and sustained blooms are not a common occurrence if nutrient loading is very low, or blooms are governed by environmental factors (e.g. temperature, light extinction, nutrients), but equally important are the physical processes such as flow velocity, turbulence, mixing process and dispersion. Similarly, fast moving water has less siltation problems, all reservoirs/lakes are silted by high-sediment laden flows. Hence, it is important to separate clean water from water with too dirty and too turbid waters.
2. For typical watersheds, rivers always play a major role in assimilating or carrying off municipal and industrial wastewater as well as runoff from the catchment. Reservoirs/Lakes receive a major portion of their pollutants/sediment from river inflows; therefore excessive wastewater/sediment inputs can cause serious ecological problems in the ecosystem. However, rivers also constitute the main clean water sources to a lake/reservoir. River water quality is heterogeneous spatially and temporally. In order to manage water quality in a lake, the temporal and spatial variation in water quality must be understood. The typical hydrograph for a catchment is simplified in Fig. 2 where Q is flow rate, Q_r is runoff due to rainwater and

groundwater without being polluted, Q_w is the domestic and industrial wastewater discharge, Q_s is the sediment discharge, Q_a is the rate of wastewater from agriculture or the non-point source pollutant. The year-round rainwater is unsteady, and floods often appears in wet seasons, but wastewater rate from industry and domestic sources is relatively constant year-round even their concentrations in the rivers are not even after mixing. Inflow-river waters are heavily polluted in the dry period while pollutant concentrations are reduced during the wet (flood) period after mixing of clean rainwater with the domestic and industrial wastewater. But this does not mean that the wastewater concentration in flood period is always low as the first flush of storm often drives the non-point source pollutants to the waterways, thus the peak of Q_a appears before the peak of Q_r. Different from the agricultural wastewater, sediment discharge concentration is roughly proportional to the river flow, i.e., very high during wet season, very low in dry seasons.

3. Currently, throughout the world it is rare to find an integrated water resources management plan that can reduce the external nutrients, and at the same time mitigate flood disaster and solve the water shortage problems, i.e., simultaneously solve four problems of flooding (too much), droughts (too little), algal blooms (too dirty) and siltation (too turbid) for a given basin where reservoirs/lakes are located.

Based on these realizations, Yang and Liu (2010) proposed the following SPP strategy to manage different waters in a lake/reservoir:

- Clean water and wastewater should be *separated* spatially and temporally;
- Clean water should be stored and *protected* and wastewater should be discharged as fast as possible, i.e., the detention time of wastewater in the lake should be as short as possible, whilst the residence time of clean water should be as long as possible.
- One of the effective ways to *prevent* water deterioration is to maintain water movement since moving water has the self cleaning capacity (self-degradation, self-decomposition and self-purification), and high turbulence shears dispersed planktonic flocs and limits sessile microbial growth.

The objectives of this chapter will be 1) to show how to manage the water and the particles it carries simultaneously, thus the floodwater is regulated for the use in drought seasons, and the lakes/reservoirs' storage capacity is protected with the damaged ecosystem being restored; 2) to compare the new strategy of water management with existing strategies, and the case studies will demonstrate whenever the SPP strategy is applied, the water related disasters can be mitigated, vice versa.

2. SPP strategy

Normally, a natural or artificial lake may comprise multiple incoming and outgoing rivers that collect the rainwater/wastewater from upstream of the lake and drain the lake water to downstream, respectively. The SPP strategy is achievable if an internal levee with sluice gates is built in the lake. By doing so, flood disasters (too much water) of the watershed will be significantly reduced, water shortage problem (too little water) can be alleviated simultaneously, the siltation rate is reduced (too turbid) and clean water resources are protected against pollution. These two levees around the lake shoreline together would form an artificial canal or by-pass channel (BPC) as shown in Fig. 3. The inner bank of the canal would be built in the lake and could be constructed using dredged sediments to deepen the BPC.

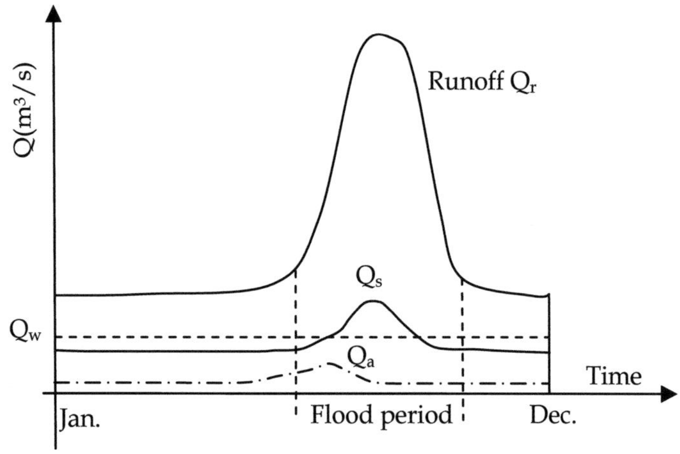

Fig. 2. Simplified hydrograph of runoff and wastewater yielded in a catchment

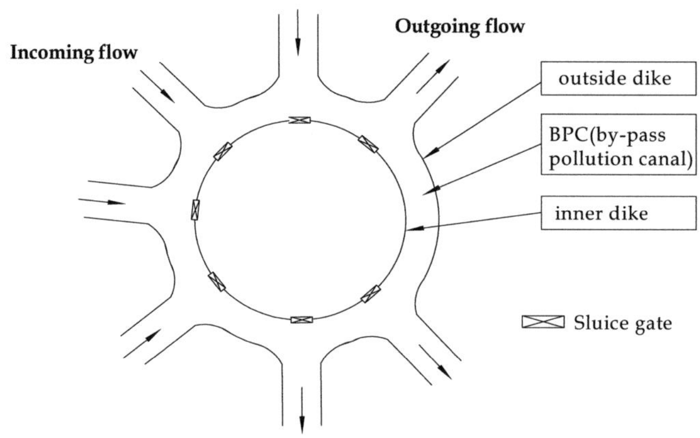

Fig. 3. Simplified water system for a lake and the proposed infrastructure for water resources development and pollution prevention

Separation in space and time: The sluice gates would be used to regulate the unwanted water. If the river water is heavily polluted or sediment discharge is very high, the sluice gates will be closed so that all polluted water by-passes the lake and is discharged to the downstream of the catchment via the lake outlets. During flood periods, nutrient concentration is very low, except first flushes, the water quality from all inflow rivers will be relatively good. In such case, the sluice gates will be opened so that clean floodwater can enter the lake and be stored. Thus the water separation can be conducted based on the quality of river water in time (flood/drought) and space, i.e., good quality water enters to the lake for storage; but unwanted water runs to the downstream via BPC. Without the BPC, the lake receives all sediment particles from the rivers, with the BPC, the sedimentation rate is reduced to the particle volume carried by the stored water that is only a small fraction of all sediment yielded by the catchment.

Protection against external pollution: when the good quality floodwater enters the lake, this water is protected by the inner levee as the sluice gates are closed when the incoming river water quality becomes poor later. Thus the external contaminants in the rivers cannot mix with the clean water in the lake. In this stage (drought periods), the river water and the lake water are separated and the clean water is fully protected.

Prevention of water degradation: As most of unwanted water (heavily polluted and high sediment-laden) will be concentrate in the BPC where the cross-section area is much smaller relative to the lake, thus for a certain discharge at the outlet, the velocity in the BPC will be rather fast. Consequently, the high level of turbulence will disperse all aggregated flocs and to purify the water, and the flow is capable to carry the sediment to the downstream without deposition. It may be hard to imagine that the water quality in the "by-pass canal" will be better than the water without the canal in the same conditions. This seems counter-intuitive as most of unwanted water would flow in the canal surrounding the lake. But this is possible because, as mentioned early, hydrodynamic parameters are also a dominant factor for eutrophication and sedimentation, responsible for the appearance of algal blooms. It is often observed that higher velocities can prevent blooms/siltation from developing, and slow water movements create an almost quiescent environment where phytoplankton grow quickly with favourable nutrient, light and temperature conditions because microbial particles can stick together to organic aggregates.

Too Much: The above analysis shows that the proposed scheme can well solve the "unwanted water" (i.e., too dirty and too turbid) problem. Furthermore, the flood disaster can also be mitigated. Under natural conditions, i.e., without BPC in Fig. 3, the highly contaminated river water with first flushes at the beginning of flood period occupies a lake's storage capacity. Consequently, the active storage is not sufficient to accommodate the following peak flow and flood disaster occurs. SPP approach can considerably reduce the dead storage of a lake as all unwanted water is discharged to the downstream via BPC and does not occupy the lake's capacity, thus it significantly increases the effective flood-control storage of the lake.

The flood disaster can be mitigated, because the sluice gates are normally closed during dry period, so that the water level in the lake is very low due to regulation and water use after last year's flood period. Then just before the onset of the rainy season, water level in the lake is very low, and the low level can be kept until the arrival of peak flow, during which the sluice gates is open, thus the water level in BPC and the rivers can be lowered to acceptable level for disaster mitigation.

In other words, with BPC, the lake can be operated intentionally, and the lake's storage volume can be kept to accommodate the coming floods. The strategy of transforming a lake to a flood retention zone can greatly expand its capacity for flood control. During flood seasons, the sluice gates are opened when the water level in the rivers reaches the designed high level. Once opened, the velocity of flood wave propagation will be increased as very low water level in the lake increases the hydraulic gradient, then the scheme can greatly mitigate flood disasters. At the same time it transforms the floodwater to clean lake water that is protected from pollutants by enclosing levee with the sluice gates.

Too Little: This innovative strategy of water management can also alleviate the water stress in a lake, because under natural conditions, water from the lake keeps losing by gravity via its outlets, and the water level quickly drops after flood period. If the proposed scheme is applied, the clean water will be protected, and sufficient high quality freshwater during flood period will be kept in the lake through the drought period until the next rainy season just like a reservoir. In other words, the water level is still high even after flood period. This water can be used by the human society to sustain their developments. Most importantly, during dry periods, all sluice gates are closed, the inner levee protects the lake water against external pollution, and it also prevents unnecessary loss of lake water.

The core idea of SPP is that the water should be separated based on its quality or turbidity; the unwanted water should be prevented mixing with the clean water, and good quality water must be stored and protected. In other words, the residence time of unwanted water in an ecosystem should be as short as possible; while the detention time of wanted water (or clean water) should be as long as possible.

3. Successful cases by applying SPP strategy (too turbid)

The idea of SPP for lake and reservoir water management is new, but this does not mean that the similar strategy has never been applied by far. In this section, we will discuss two cases to show that if SPP is not applied, the water related problems cannot be solved, while if the strategy is applied, all problems become solvable.

In 1854 Dr John Snow demonstrated the connection between water supply and Cholera as shown in Fig. 4. Several years later, Thomas Hawksley in Nottingham suggested that water supply systems should be piped to prevent external pollution.

Here the suggestion of pipe transport implied the idea of "separation, prevention and protection". Like the enclosed levee shown in Fig. 3, the pipe can *separate* clean water from the polluted water, and only clean water is allowed to enter the pipe network; likewise, the pipe protects the clean water inside the system and prevents the external pollutions. Consequently, the problem of waterborne illness death rate like Cholera, typhoid fever and dysentery had rapidly dropped to a very low level as shown in Fig. 5. This is probably the first time in history using the strategy of SPP for water resources management, and the concept of pipe water has been widely accepted in the world since then.

However, for large water system, so far there is no direct example of SPP strategy in application. Sometimes, people were forced to adapt the strategy of SPP to solve their problems. One of the examples is the Sanmenxia Reservoir in the middle reach of Yellow River, China. Yellow River basin is one of the largest basins in the world, and it is notorious in the world for high sediment transport rate in the middle reach. Between 1919 and 1960, the measured data at Sanmenxia Station showed that the mean annual runoff was 42.3×10^9 m^3 and the long-term average annual sediment load was 1.57×10^9 t with the average concentration of the river water being 49.8 kg/m^3, the highest in the world.

Sanmenxia Reservoir is created by the Sanmenxia dam. It is a large-scale multipurpose project, and the first one constructed on the main stream of the Yellow River. The construction of the dam was started in 1957, and the impoundment of water commenced in September 1960. The dam height is 106m at the pool level 323m with the storage capacity of 3.6×10^9 m^3.

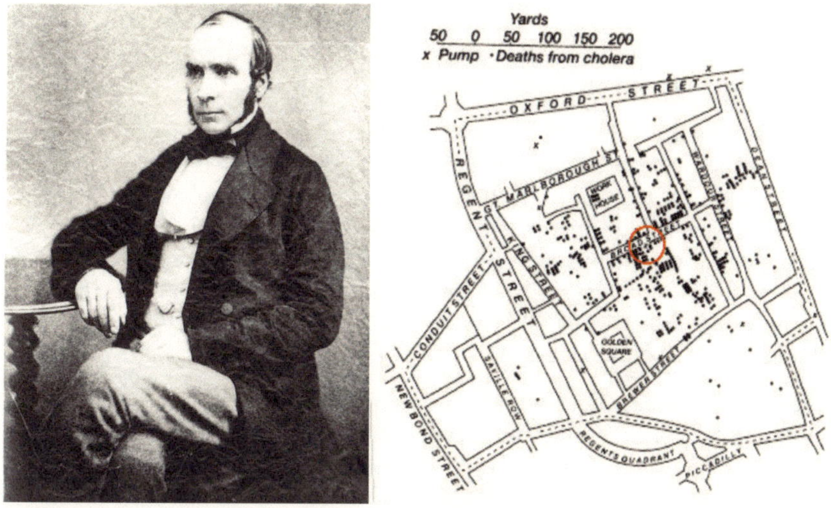

Fig. 4. Dr. John Snow and the map he used in 1854 to identify the source of cholera; this discover leads to the first application of SPP strategy in water management

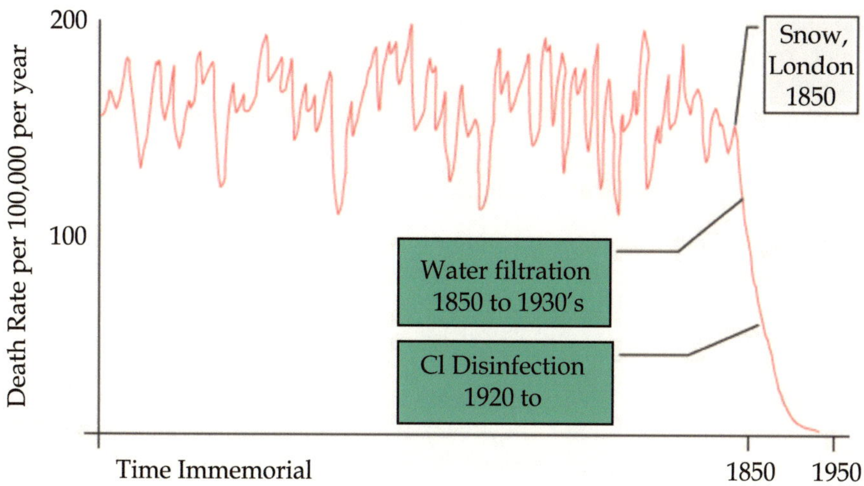

Fig. 5. Death rate of human being in the history, indicating the application of SPP could immediately reduce people's death rate

In its planning stage, engineers and decision makers used the same practice as they did for other reservoirs, i.e., no special measures had been taken to separate incoming water based on its quality, and nothing had been done to protect the reservoir water and to prevent external high turbidity water. Consequently the reservoir storage capacity was decreasing at an astonishing rate in the reservoir, which caused the unacceptable negative impact of rapid development of backwater sediment deposition. In 4 years time, the accumulative sediment silted in the reservoir reached alarming level, i.e., 4.44×10^9 m^3, in other words, the reservoir's storage capacity had almost totally lost. Under such circumstance, engineers were forced to reconsider the strategy of water management, and the dam had to be reconstructed to minimize the adverse effect of reservoir sedimentation.

The engineers and decision makers realized that there was no hope to solve the sedimentation problem without separation, then they divided the dam's operation into two time periods or two different modes of operation:

1. in flood seasons (July–October), turbid water is discharged to the downstream by lowing the water level at 305m, thus the flow velocity is high enough to carry the highly suspended load. In this period, if the flood discharge is greater than 2500 m^3/s, all the outlets of the dam will be opened to flush the sediment as much as possible.
2. in non-flood seasons (November–June), the relatively clear water is stored and the reservoir is operated at a high operational level, with the maximum level not exceeding a value between 322 m and 326 m, to store water for irrigation and hydropower generation in non-flood seasons;

After the separation, the new design was found effective, till now, the dam is still providing the basin with flood control, irrigation, and hydropower generation, even though some benefits are lower than the original design. The engineering experiences and management practices of Sanmenxia Dam are valuable assets to the sustainable management of reservoirs/lakes on sediment-laden or nutrient-rich rivers. It can be seen that the above operation strategy is similar to the SPP introduced in this chapter as the unwanted water (high turbid floodwater) has a short residence time in the reservoir, and the clean water in non-flood season (wanted water) has a relative long residence time for the eco-system. This project not only demonstrated the perils of existing water management, but also demonstrated that a sediment balance may be achieved across an impounded reach even in a large reservoir built on a river with an extremely high sediment load.

4. Flood control in Dong-Ting Lake, China (too much)

Different from the Yellow River basin where droughts (too little) and high turbid water (too turbid) are two key challenges for water resources managers, the Yangtze river basin where the Dong-Ting lake is located often has the threat of flood (too much). Yangtze River is the third longest river in the world, and its catchment area is about $1.91\times10^6 km^2$. The average annual water discharge is $918km^3$ and average annual load of suspended sediment is 500×106 t (1956-2003, Zhang 1995), respectively the fifth and fourth in the world (Milliman and Meade, 1983; Milliman and Syvitski, 1992). Its middle reach is known as the country of a thousand lakes that serve as an efficient flood regulator, among which Dongting is the largest lake. Due to sedimentation and reclamation over the past 50 years only few lakes have been left in this region, and the total capacity has been significantly shrunk. This region plays an important role in China's economic development, and this can be inferred from the historical proverb "If southern and northern regions of Dongting Lake harvest, the whole China's prosperity is ensured"; but if this region suffers from flooding, the whole China's economical development will be badly affected.

Dongting Lake, used to be the largest freshwater lake in China before 1950. Now it has been shrunk to be the secondary largest lake in China due to high sedimentation rate. The lake valley is surrounded by mountains on the east, south, west and Yangtze River on the north. The population in this basin in 1997 was 15 million. Many rivers drain into the lake from mountainous south and west, besides three channels connect the lake to the Yangtze River, which dumps a major proportion of the sediment suspended from Yangtze River to the lake when the lake accommodates floodwater. The lake has only one outlet on the east side.

This lake annually receives 173 million tons of sediment, 83% of which are carried into the lake from the Yangtze River, and 17% from other inflow rivers. Over the past 150 years, the lake area has reduced from about 6000km^2 to about 2600km^2 due to both natural siltation and human activity, consequently flood modulation of the lake capacity has been reduced significantly from about 30km^3 in 1949 to 17km^3 in 1995, about 50% of the lake's capacity has been silted. It is certain that the lake will continue to shrink, and in the near future it may serve only as a river channel during the non-flood reasons (Du et al. 2001). In another 50-100 years, this lake may totally disappear if there is no human's interference.

Similar to other plain lakes, Dongting Lake has suffered many problems, if not solved properly, these problems may cause serious consequences to environment and bring about huge economic and life loss. These problems can be summarized as flood disasters in wet seasons (too much), water shortage for the ecosystem in dry seasons (too little), lake's shrinkage (too turbidity) and water pollution (too dirty).

Apart from too much and too turbid problems, the lake water was polluted by the incoming wastewater due to rapid industrial development and population growth and agriculture fertilization. In this fertile "land of fish and rice", farmers every year use 18k tons of pesticide and 1700k tons excessive fertilizer, which flows into the lake via the rivers. After 1990s eutrophication tendency has been accelerated, according to the monitoring data, 1996-2005 main pollutants are total phosphorus and total nitrogen (Guo et al., 2007).

Li et al (2009) summarized the problems in the Donting Lake area as a chain:

$$\begin{aligned}&\text{Disaster in wet seasons: sedimentation} \rightarrow \text{marshland expansion} \rightarrow \text{water}\\&\text{space shrinkage} \rightarrow \text{flood level raised and flood disaster aggravated.}\\&\text{Disaster in dry seasons: sedimentation} \rightarrow \text{marshland expansion} \rightarrow \text{water}\\&\text{space shrinkage} \rightarrow \text{water pollution and fish resource depletion} \rightarrow \text{wetland and}\\&\text{biodiversity reduction.}\end{aligned}$$

The average reoccurrence interval (ARI) of flood disaster was 83 years from 276 B.C.-1525, and ARI = 20 years from 1525-1853; ARI = 5 years from 1853-1949 and ARI = 4 year from 1949-2000, recently this area suffers the flood disaster almost every year. Obviously this is caused by the sedimentation that keeps reducing the lake's capability for flood retention (Yang, 2004).

All problems that the lake is facing can be well solved if the proposed SPP strategy is applied as shown in Fig. 6, in which the length of inner levee is about 240km. This is feasible because:

1) *Sedimentation*: in the current condition, almost all incoming sediment particles are deposited in the lake, i.e., 1730 million tons of sediment, and the lake functions like a sedimentation tank where the clean water run out of the tank. But with the inner dike, the incoming flow can be separated into wanted/unwanted waters, i.e., high turbid but safe floodwater will by-pass the lake and drain to the downstream. Only a small part of excessive floodwater is allowed to enter the lake, which can mitigate the flood disaster, and the majority of the water stored in the lake comes from the falling limb in the hydrograph shown in Fig. 2, thus clean and low turbid water is stored. Currently the lake's storage capacity is only 1/10 of the annual incoming water volume, if SPP is used and then the

sediment carried by this part of water will be deposited in the lake, approximately only 1/10 of the total incoming sediment, the remaining sediment will by-pass the lake and go to the downstream. Hence, if the SPP is applied, every year only 173 million tons of sediment will deposit in the lake, therefore the reduction of sedimentation rate will be

$$\frac{173-17.3}{173}=90\%$$

Thus the SPP strategy can reduce significantly the sedimentation rate, and the life span of the lake can be extended.

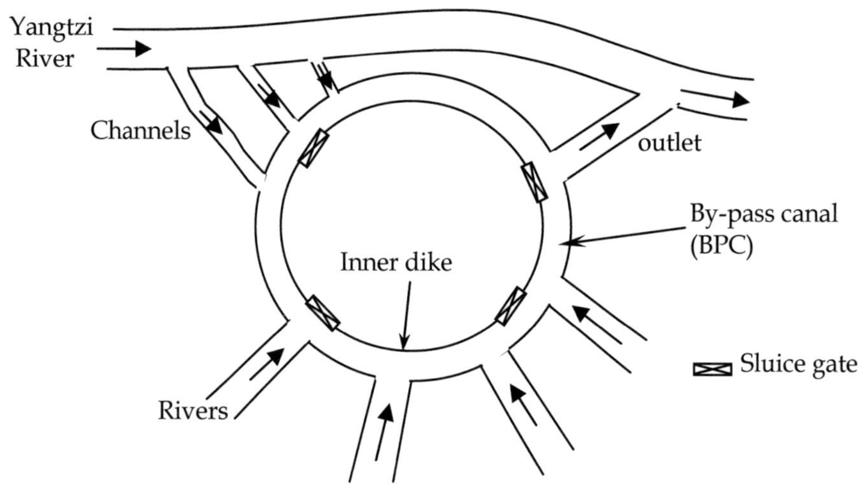

Fig. 6. Simplified river-lake system for the Dong-Ting Lake

2) Flood disaster: it has been mentioned that the cause of flood disaster is that the lake's storage is occupied by the runoff before the arrival of peak discharge, hence the excessive flood water incurs disasters. With the strategy of SPP, the floodwater can be separated as safe water and excessive floodwater. The former refers to that flow in BPC is safe, the incoming flow rate is less than BPC's design capacity and water level is below the BPC's design highest water level. The excessive floodwater is water that overflows the rivers and incurs the flood disaster if no retention area is provided for its temporally storage. All sluice gates are closed before the arrival of the excessive floodwater, thus in the rising limb in Fig. 2, all river flows together with sediment and non-point pollutants will by-pass the lake. Once the excessive peak flow that needs a retention area appears, the sluice gates are opened, at this time the lake is still empty, thus the flood disaster can be mitigated. After that, the water level in BPC will be lowered to the safe level, then all gates can be closed again to wait for the next flood wave, or if the weather prediction shows no more heavy rains in the catchment, the gates can be opened again to store as more clean water as possible for the next drought. Thus, at the end of wet season, the lake is full of good quality water.

3) Water shortage. Every year, after the flood season, this basin has a long dry period, and the basin often suffers the water shortage problem in this period. For example, in 2011, this region was struck by severe drought, affecting residents and agriculture, and more than a million people did not have adequate supplies of water. About 709,000 hectares of farmland in the basin had been hit and no water for irrigation. The central region of the lake had become a vast grassland. On May, 2010, the water surface area was only 780km^2, about 30% of normal water surface area.

But 10 months ago, in July, 2010, this lake suffered the flood disaster with the highest water level reading 32.9m above the sea level, and it had to rely on the upstream reservoir's regulation to mitigate the flood disaster. Thus, it is obvious that, without control, the lake cannot play a dominant role for droughts as the water has lost quickly after the previous flood. 2011 droughts can be avoided if the SPP strategy was applied and the floodwater in 2010 was impounded by the inner levees and gates shown in Fig. 6.

5. Algal blooms control in Tai-Hu Lake, China (too dirty)

Taihu Lake, the third largest freshwater lake in China, is located in the highly developed and densely populated Yangtze River Delta. The lake is shallow with an average depth of 1.89 m, but a large surface area of 2,238 km^2, and a volume of 4.66×10^9 m^3. The annual water input averages 7.66×10^9 m^3 and the residence time of its waters is about 300 days. There are over 219 inflow rivers or tributaries but only three main outflow rivers. The average of rainy days is 132 days/year and the average annual rainfall is about 1,145 mm. Rainfall varies seasonally with wet seasons between June and September, i.e., during the typhoon season or flood period, and the dry seasons from October to next May and constitutes the long dry period.

Taihu Lake plays multifunctional roles including floodwater storage, irrigation and navigation. It also serves as a major water resource for drinking, aquaculture and industrial needs, as well as being a source of entertainment and tourist interest. Its drainage basin extends over 37,000 km^2 and is bounded by the Yangtze River to the north, the East China Sea to the east and mountainous areas to the west. While the basin accounts for 0.4% of the total area of China and 2.9% of the nation's population, it provides more than 14% of China's Gross Domestic Production (GDP). The GDP per capita is 3.5 times as much as the contry's average and its urbanization level ranks the first in China. This basin is vital for eastern China, where the lake water supports more than 60 million people (about 600-900 person/km^2 on average), including the water supply to cities such as Wuxi, Suzhou, and Shanghai, one of the largest cities in the world.

With the tremendous economic growth and increased population in its basin, Taihu Lake has begun to suffer from various environmental stresses, including deterioration of its water quality with increasing nutrient and other chemical inputs. The lake is becoming increasingly eutrophicated and has experienced annual lake-wide cyanobacterial blooms in recent decades; this has affected the drinking water supply of surrounding cities. Taihu Lake now receives annually approximately 30,635,000 kg total nitrogen (TN) and 1,751,000 kg total phosphorus (TP) from a combination of municipal and industrial wastewaters and agricultural soil runoff; chemical oxygen demand on chromium (CODCr) is 131,223,000 kg (Qin et al, 2007).

Consequently, algae blooms have appeared and continued. The lake is often covered by algae blooms in summer, autumn and even spring. In 2007, a severe algae bloom caused a drinking water contamination crisis for 4.43 million people in Wuxi city. An article in Science (Yang et al. 2008) reported that the concentration of dimethyl trisulfide in a water sample collected on 4 June 2007 from the drinking-water intake was 11,399 mg/liter—high enough to yield strong septic and marshy odors.

More than 50% of rainfall in the catchment appears from June to September, so this period can be defined as the wet season, and the remaining 8 months is the dry season. Unlike rainwater, industrial and municipal wastewater releases have little seasonal variation. The hydrograph of rainwater and wastewater is simplified as shown in Fig. 2. Integrating the runoff Qr with respect to time from January to December, one has:

$$\int Q_r dt = V_o = 7.66 \times 10^9 (m^3) \tag{1}$$

where V_o = annual water yield in the basin.

From June to September, the water volume can be estimated as half of the total water yielded from the catchment as its rainfall is half of the annual rainfall, i.e.,

$$\int_{June}^{Sept} Q_r dt \approx \frac{V_o}{2} = 3.83 \times 10^9 (m^3) \tag{2}$$

Similarly, the wastewater yielded in the basin can be determined by

$$Q_w \times 365(d) \times 86400(s/d) = W_o \tag{3}$$

where W_o = yearly total volume of wastewater yielded in the basin and discharged to the waterways.

Currently all wastewater flows into the lake and its average concentration C_o is

$$C_o = \frac{W_o}{V_o} \tag{4}$$

In the wet season from June to September, the concentration C_1 is

$$C_1 = \frac{4W_o/12}{V_o/2} = \frac{2}{3} C_o \tag{5}$$

where $4W_o/12$ is the amount of wastewater yielded in 4 months (wet season), whilst Vo/2 is the amount of clean water yielded in the same period. Eq. 5 shows that in wet seasons, the water in inflow-rivers is relatively clean when compared to the water in dry seasons.

Water Quality in the Lake: Yang and Liu (2010) estimated the amount of wastewater entering the lake if the scheme shown in Fig. 3 is used. They assumed that river water with good quality is 50% of the total water resources, and it will be allowed to enter the lake via the sluice gates with the amount of $3.83 \times 10^9 m^3$, and the lake's storage capacity is the sum of the dead volume and the effective volume, i.e., 4.6×10^9 m^3.

It should be stressed that while the rainwater from June to September flows into the lake via sluice gates, this does not mean that the sluice gates will remain open for the 4-month period. Instead they will always be closed even in the flood period if the river water is not clean enough. Thus, the first flush of each storm event will by-pass the lake in order to prevent the non-point source pollution. In the wet season, there is an average of 46.9 rainy days. The sluice gates will be opened, and only on these days will the floodwater be discharged to the lake. The concentration C_{in} on these days is

$$C_{in} = \frac{46.9W_o / 365}{V_o / 2} \approx \frac{1}{4}C_o \tag{6}$$

It can be seen that with the aid of sluice gates and BPC, the pollutant concentration entering to the lake can be significantly reduced. In other words, only 25% of contaminants yielded by its catchment in a year will be released into the lake to mix with the clean water while 75% of wastewater yielded in a year will by-pass the lake and be discharged to the downstream via the three outlets. While we have only discussed inflowing water concentrations, in principle, the concept can be extended to all other parameters, such as sediment inputs, BOD, TP, TN etc. Our estimation shows that if the SPP strategy is used, in about 3.5 years, the quality of lake water can be restored, the damaged eco-system can be remediated, and the algal blooms will disappear as nutrient levels decline.

Water Quality in BPC: Taihu Lake has a residence time of 300 days and the slow water movements together with high concentrations of nutrients contribute to the problem of algal bloom in lakes. However, if the residence time of water is short, say 0.1 to 1 days, the high velocity of the water in the By-Pass Canal will prevent organic aggregation and transport phytoplankton into low light environments; turbulence will also keep phytoplankton and aggregates dispersed. Thus, there should be no problem of algal blooms in the canal. Higher water velocities do and can improve water quality in Taihu Lake, and this has been found in the lake: East Taihu Bay is a long (27.5 km) and narrow (greatest width is 9.0 km) bay located in east of Dongshan Peninsula; it connects with the West Taihu Lake at a narrow interface. East Taihu has an area of 132 km^2 (5.9% of the total Lake Taihu surface area), with an average depth of only about 1.2 meters. The cross section of East Taihu Bay is much smaller relative to West Taihu Lake, but it is the main channel draining the lake. About 70% ~ 80% of the total outgoing discharge flows from this bay; therefore the flow velocity in this bay is higher than the velocity in the West Taihu Lake as this bay is much shallower and narrower. Similarly, water quality in the East Taihu Bay is better than the quality in the West Taihu Lake even the wastewater discharge received by the former bay is 4 to 5 times of the wastewater discharge received by the latter (Yang, 2004). This observation supports the inference that an increase of flow velocity can improve the water quality.

From the above discussions, it is reasonable to conclude that the proposed scheme shown in Fig. 3 could significantly improve the water quality of Taihu Lake. Improvements are based on clean water being stored and protected by the inner levee while polluted water is retained (and concentrated) in the surrounding canal with algal blooms prevented by high flow speeds. Moreover, it is possible to further improve water quality in the canal by ecological remediation techniques and/or by flushing the wastewater in the canal using the clean water from the lake. High velocity water has strong ecological self-purification capability.

6. Potential application of SPP strategy to other lakes in the world

In this section, some typical lakes in the world will be discussed, and the application of the proposed SPP strategy to these lakes will be assessed.

6.1 Dianchi Lake, China

Lake Dianchi is the sixth largest freshwater lake in China and a capital city of this province with population of 6 million is located adjacent to the lake (see Fig. 7). The lake is divided into two parts: the northern, smaller part is called Caohai, with a surface area of 7.5km^2 and an average depth of 2.5 m; the larger southern and main body of the lake is called Waihai, with a surface area of 292km^2 and an average depth of 4.4 m. There are more than 20 major rivers flowing into the lake from the east, south and north. The lake water has been used to support industrial development, urban drinking water, navigation, tourism and irrigation etc.

Due to the rapid growth of population and economic development in the basin (6 million in 2006, only 1.5 million in 1980), this lake has received more and more wastewater that is yielded from the catchment. In 1995, the wastewater was up to 185 million tons, among them, TP (total phosphorous) 1021 tons, TN 8981 tons, more than half of which was non-point source pollutants from the agricultural fertilizer or pesticide. The algal blooms have been emerged from the late 1980. The lake has been listed in the "Three Important Lakes Restoration Act in China" by the central government. Huge investment has been spent for the remediation, but none of them has been proved effective. The city currently faces a difficult dilemma: whereas on one hand many efforts have been undertaken to improve the local water environment, the pollution problem is still overwhelming. On the other hand, the city is growing and its dependence on the lake, though already severely problematic, is still growing (Huang et al., 2007).

Gray and Li (1999) reported that if Dianchi Lake is to have the high water quality as it had in the 1960s, the TP inflow through surface water should be less than 60 tons per year. Although the government has made many attempts to reduce the TP, Huang et al (2007) drew a pessimistic conclusion: "The TP load reduction envisaged as realistic would only stabilises the lake water quality by about the year 2008; unfortunately, interventions could not return the lake to its former pristine condition."

This is understandable as the government does not use the strategy of SPP. Similar to Taihu lake, once the proposed strategy is used, only the clean water is allowed to enter the lake, when the water crisis is curable. The required inner dike is about 163km for the lake to implement the SPP strategy, and the construction of this levee and its associated sluice gates is only a small fraction of the cost required by other alterative.

6.2 Lake Biwa, Japan

Lake Biwa is located in the central part of Japan, and it is the largest lake in Japan with a surface area of about 681 km^2, water volume about 27.6 km^3, average depth of 44 m and the maximum depth of 104 m. The lake extends north to south for about 65km, and there are about 40 inflow rivers with residence time of 5 years and a single outlet river, located at the southernmost end of the lake. Lake Biwa consists of a larger north basin and a smaller south basin, and it is a valuable water resource for 14 million people living in this area.

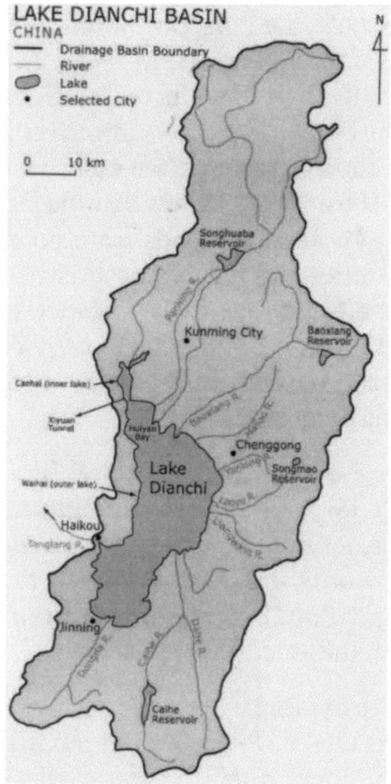

Fig. 7. The Lake Dianchi basin in China

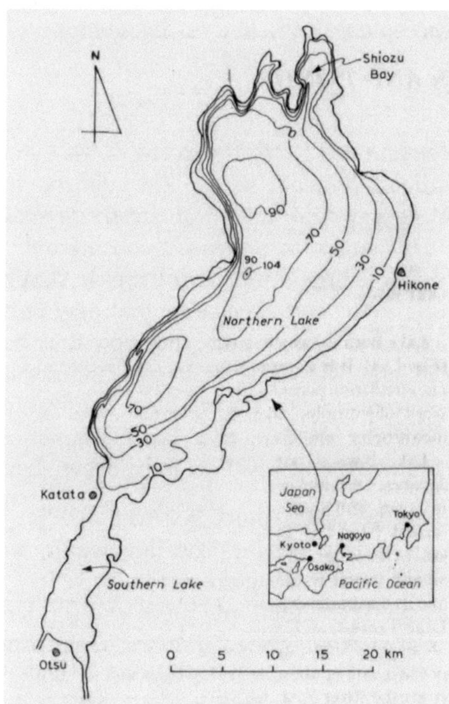

Fig. 8. The Biwa Lake, Japan (after Mori et al. 1984)

The problems that the lake has include: the eutrophication of the lake caused by wastewater, which led to algal blooms in the lake has started in 1980s. After that, even the percentage of domestic wastewaters treated in the basin had increased from 1.7% in 1975 to 78.2% in 2004, and the removal rates of N and P have also been improved, the chemical oxygen demand and the NO_3 and PO_4 concentrations in the surface layer water have not been improved as expected (Nakano, et al. 2008). The analysis shows that the lack of improvement in the water quality of the lake is that the environmental loads from nonpoint sources are unexpectedly greater than those from point sources such as domestic wastewater. Especially, N and P of agricultural origin have a great effect on the eutrophication of the lake. This citation clearly indicates that without SPP strategy, it is impossible to provide very clean water to the users with the control of point-source pollution only, as the agricultural wastewater has been increased steadily to increase the food production.

Application of SPP strategy to the lake: The cause of nutrophication in Biwa lake is very clear: no measured has been taken to separate the agricultural wastewater with the clean rainwater, because this lake has not had the industrial and domestic wastewaters like other lakes in the world. In other words, the water quality of lake Biwa can be improved significantly if N and P from agricultural sources are reduced significantly, but so far no any technology except SPP can reach this goal.

Currently, all N and P yielded from farm lands in the basin are collected by the lake, and eventually it mixes with the clean lake water and the mixing leads to the eutrophication in the lake. However, if the 133km inner dike is constructed along the 10m contour, then all unwanted river water with high concentration of N and P will by-pass the lake, and only the clean river flow is allowed to enter the lake. Thus the N and P level in the lake can be significantly lowered to an acceptable level, and no algal blooms will occur.

7. Conclusions

Meeting increasing water demand has always been one of the main challenges of civilization and could be the most difficult problem in the 21st century. With the rapid growth of population together with agricultural and industrial development, the clean water demanded in the 21st century would be increased considerably. On the other hand, the wastewater produced by human being would also increase significantly. The limited clean water may become scarce due to the climate change that may reduce the rainwater supply, increase the evaporation loss, and wastewater pollution that has extremely deteriorated quality of water, especially in large water bodies. In this circumstance, if there is no effective strategy to manage our water resources, the modern civilization may collapse that had happened in the history. Aiming at this, this book chapter proposed the SPP water management strategy. The following conclusions can be drawn from this study:

1. The analysis shows that the water must be managed for its quality and quantity simultaneously. In other words, the appropriate water management must include particle management or impurity management;
2. Currently the water has been separated into many groups, like wastewater, river water rainwater, floodwater etc. It is suggested that the water can be further separated as wanted and unwanted water based on the purpose of water management. Generally, the unwanted water includes heavily polluted or highly turbid water, and the wanted water could be excessive floodwater or clean river water;

3. The SPP strategy refers to water separation, clean water protection and prevention from external pollution. Currently in the world, all reservoirs and lakes' water are exposed to external wastewater with no countermeasures taken to protect the stored water, and the incoming river water has never been separated. Consequently, all clean lake/reservoir water has been polluted and the storage capacity is quickly lost by sedimentation.
4. In this book chapter, we discussed some typical lakes/reservoirs in China and Japan, like Sanmemxia Reservoir, China's second, third and sixth largest lakes and Lake Biwa in Japan. To solve problems in these lakes, the SPP strategy is very effective. The water quality can be restored to acceptable level in a short period and the life span of these lakes can be extended significantly. All construction needed for SPP is only an inner dike with sluice gates. The longest dike for these lakes is about 240km, and the maximum height of these dike is about 10m.
5. Generally, for a given catchment, there always exist threats like flood disasters (too much), droughts (too little), deterioration of water quality (too dirty) and siltation (too turbid). SPP strategy is effective to all these problems.
6. In this book chapter, we only outlined the application of the SPP to the mentioned lakes, but in principle, the strategy is valid to all other large water bodies. Next, more detailed research like physical and/or numerical models are needed before it can be used in practice. Eventually, it is subject to the decision maker/politician's decision.

8. Acknowledgment

The works are supported, in part, by the open fund provided by State Key Laboratory of Hydraulics and Mountain River Engineering at Sichuan University, China (SKLH-OF-1002); the open fund from Nanjing Institute of Geography and Limnology, China (2010SKL005). National Science Foundation of China (50679046 and 51061130547) and the Ministry of Science and Technology of China (2007CB714150).

9. References

Davis, J.R. and Koop K. (2006). "Eutrophication in Australian rivers, reservoirs and estuaries-a southern hemisphere perspective on the science and its implications. Hydrobiologia, 559, 23-76.

Du, Y., Cai, S., Zhang, X. and Zhao, Y. (2001). Interpretation of the environmental change of Dongting Lake, middle reach of Yangtze River, China, by 210Pb measurement and satellite image analysis. *Geomorphology*, 41, 171-181.

Gleick, P.H. et al . (2004). *The world's water 2004-2005. The Biennial Report on Freshwater resources*. Island Press.

Gray, A.V., Li, W., 1999. Case study on water quality modelling of Dianchi Lake, Yunnan Province, South West China. *Water Science and Technology*, 40(2), 35–43.

Hallegraeff, G.M., 1993. A review of harmful algal blooms and their apparent global increase. *Phycologia*, 32, 79-99.

Heisler, J., Glibert, P. M., et al.. (2008). "Eutrophication and harmful algal blooms: A scientific consensus." *Harmful Algae*, 8(1), 3-13.

Huang, D. B., Bader, H. P., Scheidegger, R., Schertenleib R., and Gujer, W. (2007). Confronting limitations: New solutions required for urban water management in Kunming City, *J. Enviro. Management*, 84, 49–61

Kuhnle, R.A., Simon, A., and Knight, S.S. (2001). "Developing linkages between sediment load and biological impairment for clean sediment TMDLs". Wetlands Engineering and River Restoration Conference, ASCE, Reno, Nevada.

Milliman, J.D. and Meade, R.H. (1983). World-wide delivery of river sediment to the oceans. *J. Geol.*, 91, 1-21;

Milliman. J.D. and Syvitski, J.P.M. (1992). Geomorphic/tectonic control of sediment discharge to the ocean: the importantnce of small mountainous rivers. *J. Geol.*, 100, 525-544.

Mori, S., Saijo, Y. and Mizuno, T. (1984). Limnology of Japanese lakes and ponds, In: *Ecosystems of the world: Lakes and Reservoirs*, Taub, F.B., 303-329. Elsevier, 0-444-42059-2, Tokyo.

Nakano, T., Tayasu, I., Yamada, Y., Hosono, T., Igeta, A., Hyodo, F., Ando, A., Yu S., Tanaka, T., Wada, E., Yachi, S. (2008). "Effect of agriculture on water quality of Lake Biwa tributaries, Japan". *Science of the Total Envir.*, 389, 132-148.

Newcombe, C.P. and Jensen, J.O.T. (1996). "Channel suspended sediment and fisheries: a synthesis for quantitative assessment of risk and impact". North American Journal of Fisheries and Management, 16(4), 693-727.

Qin, B.Q., Wang, X.D., Tang, X.M., Feng, S. and Zhang, Y.L. (2007). "Drinking Water Crisis Caused by Eutrophication and Cyanobacterial Bloom in Lake Taihu: Cause and Measuremen. *Advances in Earth Scienc.* 22(9): 896-906 (in Chinese).

Smith, V.H. and Schindler, D.W. (2009). "Eutrophication science: where do we go from here?". *Trends Ecol Evol.* 24(4): 201-7.

USBR (U.S. Bureau of Reclamation) 2006, *Erosion and sedimentation manual*, U.S. Department of the Interior, Bureau of Reclamation, Technical Service Center, Sedimentation and River Hydraulics Group, Denver, Colorado.

White, R. (2001). Evacuation of sediments from reservoirs. Thomas Telford, 0727729535, London.

Yang M., Yu J.W., Li Z.L. Guo Z.H., Burch M. and Lin T.F. (2008). "Taihu Lake Not to blame for Wuxi's Woes". *Science*, 319, 158.

Yang, S.Q. (2004). The Global and China's water crisis and solutions in the 21st century. Tianjing Univ. Press, 7-5618-2069-0, Tianjing, China (in Chinese).

Yang, S.Q. and Liu P.W. (2010). "Strategy of water pollution prevention in Taihu Lake and its effects analysis". *J. Great Lake Research*, 36(1), 150-158.

Zhang J. (1995), Geochemistry of trace metals from Chinese river/estuary systems: an overview. *Estuar Coast Shelf Sci*; 41:631–58.

11

Managing the Effects of the Climate Change on Water Resources and Watershed Ecology

Ali Erturk
Istanbul Technical University, Department of Environmental Engineering, Maslak, Istanbul, Turkey

1. Introduction

Based on the monitoring data and climate projections, scientists highly agree that freshwater resources are vulnerable and have the potential to be strongly impacted by climate change in the long-run. However, there is no consensus about the degree of impact of human activities on climate change. Using simulation techniques, Intergovernmental Panel for Climate Change (IPCC) estimates the expected changes in the climate on a global scale for different emission scenarios. The results from global estimations are used to drive other simulations that run on regional scaled smaller domains at higher spatial resolution.

Assuming that climate change scenarios will be realized in the future, it is possible to foresee that there will be effects of climate change on watershed ecology and on the water resources. Considering only two of the climate change related variables; temperature and precipitation one can conclude that

- Risks of flooding may increase.
- Droughts may happen more frequently and for longer periods directly affecting the water demand changing the quantity and quality of available water.
- Increase in water demand may result in insufficient capacity of reservoirs and transfer of water from other watersheds might be necessary.
- Changes in water quantity and quality will in turn affect food availability, stability, access and utilization.
- Water quality of surface runoff from urban and rural areas may change.
- Function and operation of existing water infrastructure (including water treatment, hydropower, drainage and irrigation systems) may be affected.

This chapter is devoted to the impacts of climate change on freshwater resources; their availability, quality, quantity, uses and management is evaluated. Impacts on ecology are mentioned. Several management alternatives to reduce the potential adverse effects of climate change are identified; merits and tradeoffs involved are discussed. The discussions on this chapter is about what the ecological impacts of climate change on aquatic ecosystems and water resources will be and what precautions can be taken to sustain watershed ecosystems and water resources together with the demands of our socioeconomic system rather than "how we can prevent the climate change".

1.1 A general summary of climate change

In the last two decades, global climate change has continuously been gaining importance. Scientists from different disciplines agree that our climate is continuously changing. The question open to discussion and debate is the importance and the relative influence of human activities on the climate.

The first study related to climate change was initiated in California University by a climatologist named Charles Keeling. In 1958, he started to monitor the amount of atmospheric carbon dioxide in Mauna Loa observatory located on Hawaii. His studies, which were continued by his son Ralph Keeling after his death, indicated that atmospheric carbon dioxide is continuously increasing. Figure 1, also known as the Keeling Curve illustrates the trend of this increase.

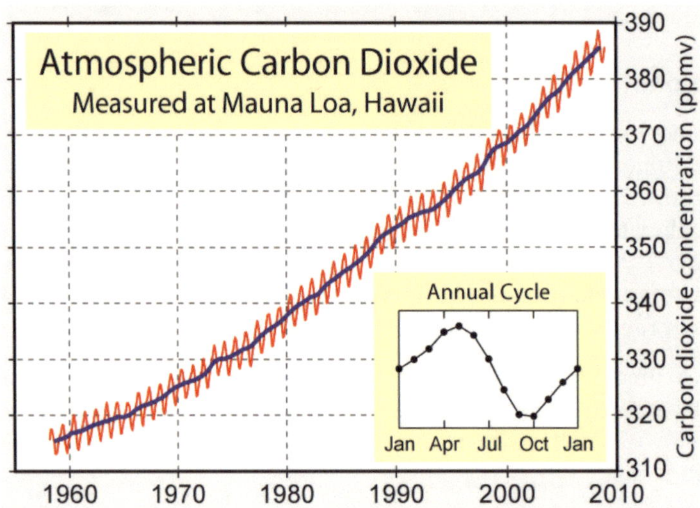

Fig. 1. Keeling curve

The atmospheric carbon dioxide investigation of Charles Keeling is considered to be the first milestone in climate change studies.

In 1970's and 1980's climate change was usually considered as global warming by scientist from various disciplines and interpreted according their knowledge. In late 1980's however it was clear that a more integrated approach is needed to investigate the climate change. Global climate change was considered a complex topic with many aspects and policy and decision makers needed objective information on climate change including

- Its reason
- Environmental and socioeconomic effects
- Possible solutions

To cover these needs, World Meteorological Organization (WMO) and United Nations Environmental Programme (UNEP) founded together the International Panel of Climate Change (IPCC) in the year 1988.

Since early 1990's, IPCC published comprehensive reports in regular intervals the last one being published in 2007 (IPCC, 2007). Simulation results from IPCC modelling studies indicated that without any limitations in human activities related to industrial emissions, 1.8°C of global temperature increase and 6 - 30 cm sea level rise may be expected 2030 taking the beginning of the industrial revolution (second half of the 18th century) as reference.

Intergovernmental Panel on Climate Change (IPCC, 2007) refers in the Fourth Assessment Report, AR4, to the warming of the global climate system and states that "most of the observed increase in globally averaged temperatures since the mid-20th century is very likely [this likelihood statement can be interpreted as probability in excess of 90%; comment added] due to the observed increase in anthropogenic greenhouse gas concentrations" (Szwed et al., 2010). It is also expressed that on the global average, surface temperatures have in¬creased by about 0.74°C between years 1906 and 2005 during which the warming has not been steady and not kept the same both temporally and spatially (IPCC, 2007). According to the recordings taken since 1901 only a few areas have been cooled, among which one of the most notable one is the northern North Atlantic near southern Greenland. However, during this period warming has been experienced more over the continental interiors of Asia and northern North America. As referred by IPCC (2007), the most evident warming signal has occurred in parts of the middle and lower latitudes whereas the duration of the frost-free season has increased in most mid- and high-latitude regions of both hemispheres. Besides, most mountain glaciers and ice caps have been shrinking since 1850s.

Observations so far indicate that over most land areas, cold days and nights have got warmer and fewer, while hot days and nights have got warmer and more frequent. Area affected by drought has been increased. This trend is expected to continue in the future.

The effects of climate change have been highly sensed in sectors like agriculture, energy and water related applications. As stated by Szwed et al., (2010), agriculture in the northern Europe has been temperature-restricted, while in the south it has been water-restricted. Both conditions may have lead to decrease in the crop yields and require the selection of new irrigation techniques, new crop patterns etc. for the sustainability of agricultural production. Water-related studies frequently mention that water budgets may become increasingly stressed. High evapotranspiration and low precipitation in summer leads to depletion of the water storage.

Moreover, researches and model-based studies indicate that weather-related extremes are expected to get more frequent and/or more severe and coping with these events will become more difficult. Countries facing such conditions are attempting to take mitigation measures and develop national and/or regional adaptation strategies.

2. Simple methods to quantify the climate change

2.1 Data

Data related to climate change are on two different temporal scales. The first of these data is the so-called paleoclimatological data that is indirect. Stable isotope data dating back hundreds thousands of years back is used to reconstruct the past atmospheric composition and basic atmospheric conditions such as temperature, precipitation. The second type of data is the historical data from observations dating approximately two centuries back at most. Long term time series of meteorological data can be used to analyse the recent climate dynamics.

Meteorological data can be obtained from state agencies; some of the free data is available on the internet as well. WMO is responsible to organize and manage the meteorological data globally. Meteorological data on territories without meteorological stations can be interpolated using many techniques included models. Results obtained using these techniques are post processed, reanalyzed and published; usually on regional scale such as Europe for example. There are several European community projects that provide such data.

Globally, meteorological data from 1960 to 2000 was used to calibrate and validate global climate change models that will be briefly described in the next section.

2.2 Models related to prediction of climate change

There is not a general term of "climate change prediction model". Several models on different spatial and temporal scales are linked to provide high resolution hydrological forcing data on changed climate, which will help to predict the response of watershed ecosystems. Each of these type models are described before.

2.2.1 General circulation models

Global Circulation Models (GCM) solve the geophysical fluid dynamics of the atmosphere. They have the same general structure as the numerical meteorological models used for weather prediction. The main difference is that weather prediction models are run for several days or a week, while GCMs are run years even centuries. Therefore GCMs have to be developed using energy conserving algorithms. Another difference is the spatial and temporal discretization. The weather prediction models are run on a horizontal resolution of several ten kilometres, whereas GCMs have a horizontal resolution of several degrees of longitude and latitudes.

Climate change related studies are conducted using global Atmosphere-Ocean General Circulation Models (AOGCM). AOGCMs provide results that give general information on global scale and boundary forcing for higher resolution regional climate change models. Seveal well known AOGCMs are briefly described in following paragraphs.

Hadley Centre Coupled Model, version 3 (HadCM3) is a coupled atmosphere-ocean general circulation model (AOGCM) developed at the Hadley Centre in the United Kingdom (Gordon et al., 2000; Pope et al., 2000; Collins et al., 2001). It was one of the major models used in the IPCC Third Assessment Report in 2001. HadCM3 includes two components; the atmospheric model HadAM3 and the ocean model that includes a sea ice model. Simulations often use a 360-day calendar, where each month is 30 days. HadAM3 has a horizontal resolution of 3.75×2.5 degrees in longitude × latitude. This gives 96×73 grid points on the scalar (pressure, temperature and moisture) grid; the vector (wind velocity) grid is offset by 1/2 of a grid box resulting in a resolution of approximately 300 km. The timestep is 30 minutes (with three sub-timesteps per timestep in the dynamics).

The coupled global model ECHAM4/OPYC3 was developed in co-operation between the Max-Planck-Institute for Meteorology (MPI) and Deutsches Klimarechenzentrum (DKRZ) in Hamburg, Germany. The ECHAM model is an atmospheric circulation model. The reference horizontal resolution is 300 km , but the model is set up to use finer and coraser resolutions. The time step at reference horizontal resolution is 40 minutes. ECHAM4 is coupled with ocean circulation model OPYC3.

2.2.2 Regional climate models

Regional climate models (RCM) have a higher spatial and temporal resolution. They provide more detailed information then the GCMs, however they work on a smaller domain. RCMs work by increasing the resolution of the GCM in a small, limited area of interest. An RCM usually cover an area the size of Western Europe or southern Africa. GCMs determine very large scale effects of changing greenhouse gas concentrations, volcanic eruptions etc. on global climate. The climate (temperature, wind etc.) calculated by the GCM is used as input at the open boundaries of the RCM. RCMs can resolve the local impacts given small scale information about orography, land use etc., giving weather and climate information at fine horizontal resolutions such as 50 or 25km. The outputs of RCMs are used to force finer spatial resolution models that are used to predict the response of watershed ecosystems to climate change that are briefly explained in the next section.

2.3 Models related to prediction of response of watershed ecosystems to climate change

Many environmental models that are useful to predict the response of watershed ecosystems to climate change are available. Most of these models are freely available, even open source. Giving detailed information about environmental modelling is beyond the scope of this text. For such information, the reader is referred to other standard texts such as Schnoor (1996), Chapra (2008) or Simonovic (2009).

Basically, there are two general types of models that are used to predict the response of watershed ecosystems to external forcing such as climate change: The watershed models and the aquatic ecosystem models.

2.3.1 Watershed models

Watershed models are derived from hydrological models, usually from a formerly known hydrological model so that they contain all the key hydrological processes. They also contain sediment transport and terrestrial biogeochemical cycle related processes.

Since 1960's many hydrological models were developed and some of them evolved to general purposed watershed models. However, just a minor fraction of them were designed to simulate the hydrology and ecology of entire watershed using the coupled modelling approach and even fewer of them were continuously developed and became widely used, freely available open-source modelling tools. SWAT (Arnold et. al., 1999) and HSPF (Brickner et. al, 2001) are good examples for such modelling tools. WASH123D (Yeh et. al, 2005) is more comprehensive than SWAT and HSPF as a hydrological model but has simpler water ecology related facilities. MIKE-SHE is also capable to simulate the watershed hydrology and ecology; however it is neither free nor open source; and needs additional products such as MIKE-11 to be coupled with. Other models such as SWAP (Van Dam, 2000; Van Dam et al., 2008), PIHM, Hydrogeosphere (Therrien et al, 2010) are general hydrological model taking almost all the compartment of the hydrological and can be easily linked to landscape and water ecology models.

Among all the models discussed SWAT and HSPF are the most widely used ones and generally most applicable ones. The applicability of SWAT was reviewed by Gassmann et al

(2007). There is also a literature database on SWAT website, which indicates that SWAT and its variants were applied 816 times in studies published by peer reviewed journal articles reporting hundreds of applications on different watersheds all over the world. HSPF on the other hand is widely used as well. The bibliography provided by developers contains more than 300 entries. The performance of SWAT and HSPF were compared by several authors (Im et al., 2003; Nasr et al., 2004; Saleh and Du, 2004; Sigh, et al., 2004), where both models were applied to the same watersheds. In these studies, both models produced comparable results; however HSPF produced slightly more accurate results in river discharges, whereas SWAT was better in reproducing the nutrient loads.

2.3.2 Aquatic ecosystem models

Aquatic ecosystem models are the successors of water quality models; however there is not a standard definition of a "water quality model" and "a water ecology model" or a very strict border between them. Many well known aquatic ecosystem models or their predecessor water quality models were developed in late 1970s. There are many well written texts related to water quality and aquatic ecosystem modelling, so the reader is referred to those texts. Information related how to obtain them can be reached by simple internet queries. Following paragraphs will give brief information on some well known models that may be useful for aquatic ecosystem modelling especially on estimating the possible impacts of climate change on aquatic ecosystems.

The Water Quality Analysis Simulation Program (WASP) is a water quality model that was developed in early 1980s by United States Environmental Protection Agency. It is a good model for initial studies. The latest version of WASP (Version 7.5) includes an advanced eutrophication module that can simulate the nutrient cycle and primary production up to three phytoplankton groups as well as the detritus cycle. Unfortunately higher trophic levels of the aquatic food web are not covered by the advanced eutrophication module. WASP can be driven by external hydrodynamic simulation models.

CE-QUAL-W2 is a hydrodynamic and water quality model in 2D (longitudinal-vertical). It is applicable to large watershed/water resource systems that contain these types of water bodies such as lakes, rivers and reservoirs. The current model release enhancements have been developed under research contracts between the Corps and Portland State University. The model can simulate basic eutrophication processes such as temperature-nutrient-algae (multi groups)-dissolved oxygen-organic matter and sediment relationships. Additionally, zooplankton (muti groups) can also be simulated.

CE-QUAL-R1 is a one dimensional (vertical) reservoir model developed by Hydrologic Engineering Center (United States Army Corps of Engineers). It can simulate nutrients and phytoplankton (three groups) and zooplankton like CE-QUAL-W2. Additionally, a simplified simulation of fish can be conducted. The model is designed to simulate anaerobic processes and dynamics of reduction processes as well.

AQUATOX is originally developed to assess the fate and effect of chemicals in experimental containers. With the improvements of former versions of AQUATOX, the model has reached the 3rd release, which has the capability of risk assessment combined with the fate and effect of pollutant and toxic chemicals in the aquatic environments. The way AQUATOX characterizes the aquatic system is different from many other models do. Mostly the ecosystem models represent the individuals by the changes in their numbers; hence, they called as population models. However, AQUATOX simulates the ecosystem by changing the concentrations of all components such as chemicals, sediments, and even organisms including the ones on the higher trophic levels of food web. The model is intended to assess dynamic effects of various stressors such as temperature, toxic chemicals, nutrients, sediment; which is applied to aquatic environments from experimental tanks to lake systems.

3. Effects of climate change on water resources and watershed ecology

Climate change may have short and long term effects on watershed ecosystems resources. Short term effects take place because of the extreme events that are related to climate change. Floods are good examples for such extreme events. During a flood shock loading of sediments, organic matter and nutrients can be transported into lentic freshwater ecosystems such as lakes and reservoirs. Aquatic ecosystems respond to such sudden forcing by instantaneous changes in water quality. Recovery of the system that may take from a couple of weeks up to a couple of years depends on following factors:

- the intensity of the effect
- internal structure of the system
- operating schedule (in case of engineered systems)

Long term effects on water resources occur due to climatic trends and extended periods of droughts.

The relation between the components of the historical water balance and climatic variables may be needed as reference in order to quantify the effect of climate change on the water balance of a watershed. This task is straightforward if historical data on both; the climatic variables and the water balance components exist. If one of them is missing the other one can be reconstructed using simulation techniques. Kavvas et al., (2009) used a regional hydro climatic model (RegHCM-TE) for Tigris-Euphrates watershed located in the Middle-East for reconstructing the historical precipitation data to perform water balance computations for infiltration, soil water storage, actual evapotranspiration and direct runoff.

3.1 Change of water quantity reaching the water resources

Climate change may result in average temperature and total precipitation increase. However the temporal and spatial heterogeneity of meteorological parameters may increase as well resulting in prolonged dry season and increased in flood frequencies in wet season. Average temperature in the warm season may increase and average temperature in the cold season may decrease as well.

Changes in precipitation and temperature do not only change the total amount of runoff to freshwater systems from their catchments but also the temporal distribution of water inputs. Generally, intensification of the global hydrological cycle is expected as a result of temperature increase. However, if the land surface hydrology is dominated by the winter snow accumulation and spring melt, temperature increase is likely to cause a change in the outflow hydrographs of the watersheds where time of peak flow will be shifted towards winter. Detailed information related to this phenomenon is provided by Barnett et al., (2005) in great detail. Forbes et al., (2011) analyzed the water cycle in a small snow dominated

Canadian catchment (Beaver Creek, Alberta) using a hydrological simulation model (ACRU agro-hydrological modeling system) and concluded that regions with snowmelt-dependent water supply may experience severe changes to the hydrological regime due to temperature increase. The consequences were reported by Forbes et al., (2011) as less available soil water with potential negative impacts on agriculture, and also increased stresses for the natural vegetation, lower streamflows in late summer and fall with potentially adverse impacts on the aquatic ecosystem and anyone who withdraws water from the river.

Furthermore, as temperatures rise the winter precipitation may shift from snow to rain and the timing of peak streamflows in many continental and mountainous regions will change. The spring snowmelt peak flow may shift to earlier days of the year or even get eliminated entirely and winter flows increase (Kundzewicz et al., 2008).

Changes in frequencies and intensities of extreme events such as floods and droughts are projected as well. According to IPCC (2007), the proportion of total rainfall from heavy precipitation events will increase and tropical and high latitude areas will experience increases in both the frequency and intensity of heavy precipitation events.

Döll & Flörke (2005) stated that many of the current water-stressed areas will suffer from decreasing amount of water since both the river flows and the groundwater recharges are expected to decline. In addition, Kundzewicz et al., (2008) reported that drought frequency is projected to increase in many regions, in particular, in those areas where reduction of precipitation is projected.

3.2 Capacity shortage in river/reservoirs systems because of the increased water demand and water transfer among watersheds

Temperature increase may increase evaporation from surface waters and evapotranspiration and thus water loss from plants and soil will result in increased irrigation water demand. However, Barnett et al., (2005) states that there is little agreement on the direction and the magnitude of historical and/or predicted evapotranspiration trends. Temperature increase alone is expected to enhance evaporation and eventually evapotranspiration. On the other hand, temperature increase also affects other variables such as wind speed, humidity, cloudiness that have their amplifying/dampening effects on the evaporation and evapotranspiration as well. Therefore, the magnitude and the direction of the total response of evapotranspiration to temperature increase should be considered as spatially and temporally variable. This should be considered when deciding on the operational schedule of reservoir systems and especially on those that have the purpose of irrigation water storage and supply.

Temperature increase may also stimulate water consumption. Increased water consumption may result in future shortage of reservoir capacity that is sufficient today. In this case two options are available:

1. Promoting decreased water consumption
   - change in way of life in urban areas
   - change in crop patterns/irrigation methods
   - shifting to water saving processes in the industry
   - application of ecological sanitation in rural areas

2. Transfer of water from another watershed. Water transfer from other watersheds should be planned carefully and managers should not only consider the quantity but also take into account the ecological effects on both watersheds. More information on this topic is given in the mamagement section.

According to Mirza et al., (2003) the benefits of expected annual runoff in several regions such as South-Eastern Asia will be tempered by negative impacts of increased variability and seasonal runoff shifts on water supplies. Flood risk will increase especially in low-lying river deltas. Furthermore, additional precipitation during the wet season in those regions may not solve the water stress problem occurring in dry season if the extra water cannot be stored because of the shortage of reservoir capacity. Similarly; Barnett et al., (2005) states that changes in precipitation patterns will not offset the problems as associated with warming.

3.3 Change of water quality in runoff

Another response of ecosystems to climate change is the change in the quality of surface runoff from agricultural land, forests and urban areas.

Changing meteorological conditions may necessitate changes in crop patterns and thus manure/fertilizer/pesticide applications and irrigation schedules may change. Some areas may loose the ability of any agriculture whereas other frozen wastelands may become appropriate for agriculture. Hence, water quality of surface runoff from agricultural areas is expected to be affected in the future due to the direct and indirect impacts of climate change.

Forests, depending on their ecological characteristics emit nutrients and organic matter that are transported into aquatic ecosystems sooner or later. Forest ecology is complex and more inertial compared to aquatic ecology. In other words, their response to external forcing is slower and less predictable making it much harder to estimate the short term effects of climate change on surface runoff quality from the forests. Annual and seasonal average temperature increase generally eases the photosynthesis rate and plant yield changing vegetation and forests. Increase in temperature and changes is other meteorological variables may cause forests to succeed in higher elevations. However, extreme increase in temperature may result in higher plant respiration rates and shift the photosynthesis-respiration balance towards respiration. Droughts have an adverse effect on forests favouring succession of steppes and shrubs. Soil organisms will be affected by climate change as well, thus the biogeochemical cycles are likely to be shifted to different equilibria.

Change in both natural vegetation and soil biology will cause different water quality and quantity from forest runoff.

Increase in storm event intensities and frequencies will result in more wastewater containing storm water release to receiving water bodies in case of combined sewer systems. Also sudden events related to precipitation and temperature may also affect the performance of wastewater treatment systems. In case of droughts, accumulation of contaminants on land can be expected as there will not be storm event for extended periods. Hence, a storm event following an extended dry period will have an increased shock loading effect on water resources.

3.4 Deteoration of water quality in aquatic ecosystems

Climate change may affect the ecological processes in lentic ecosystems which in turn will affect the water quality. Increase in average annual water temperature affects the primary producers following ways:

- Temperature changes will affect both; the photosynthesis and the respiration rates. Initially, increases in temperature will promote higher photosynthesis and respiration rates. However, for each group of primary producers, there is an optimum temperature range. If the water temperature exceeds the upper limit for optimum conditions, temperature stress will decrease photosynthesis rates and increase respiration rates. This mechanism will accelerate the nutrient recycle and making nutrients available for primary producers adapted to higher temperatures causing a shift in dominant phytoplankton group. More increase of temperature may completely suppress some phytoplankton groups and/or cause sudden breakdown of their blooms eventually leading to decreased water quality.
- Phytoplankton groups that can adapt to higher temperatures for example cyanobacteria will be favoured. Cyanobacteria that are generally better adapted to higher temperatures may dominate the algal community. Genera such as *Anabaena* and *Aphonizomenon* produce algal toxins, taste and odour problems. Some species of cyanobacteria are capable of nitrogen fixation and hence increase nitrogen in aquatic ecosystems through internal loading.
- More days with suitable light conditions for algal blooms or longer photoperiods during a day may occur if the cloud cover changes due to the climate change. Those conditions may extend the vegetation period as well as earlier blooms may be possible. A large portion of the nutrient inputs to lotic ecosystems generally occur in late winter and early spring related to rain events and snowmelt. In this period although the nutrient concentrations increase in water, lower water temperatures limit phytoplankton growth. However, if the water temperatures increase, two factors needed for phytoplankton growth, more suitable temperature and high nutrient concentrations, will synergistically favour phytoplankton growth. If these conditions are followed by better light availability, phytoplankton blooms will be stronger and more frequent and adversely affect the water quality.

Increase in water temperature will increase the biological activity in aquatic ecosystems, hence increase the oxygen demand. Ironically, higher temperature decreases the saturation concentration of dissolved oxygen in water as well. Combining these effects with accelerated primary production and more internal detritus loads as its conclusion, it is possible to foresee that oxygen scarcity, hypoxia or anoxia events may increase especially in deep lakes. These conditions cause stress for aquatic organisms increasing their mortality, which also means even more detritus. Those changes in internal dynamics of aquatic ecosystems are likely to decrease their capacity to assimilate external organic matter loads.

4. Effects of climate change on water resources systems

Climate change may impact the components on man-made water resources systems as well.

Deteoriation of reservoir water quality may cause operational problems related to equipment. Those problems are:

- Anaerobic corrosion: Anaerobic bacteria will reduce certain substances, such as sulphate, and consequently corrosion may come about. The oxidation of iron atoms into ions and ferrous sulphate ions reduced sulphide ions act as a catalyst.
- Increased suspended sediment load and sedimentation may decrease reservoir capacity.
- Increased primary production and phytoplankton biomass may cause clogging in filter systems.

Increase of organic matter in raw water may increase energy and chemical consumption in water treatment systems. Possible problems are:

- Increased odour
- Increased colour and turbidity
- Increased algal toxins because of cyanobacteria growth

As stated previously, climate change is likely to increase the water demand that will increase water abstraction through the water distribution network. This situation increases the operational load on water distribution and may cause problems such as pressure drop in water distribution networks followed by urban water shortage risk and related problems listed below:

- shortage of water storage volumes
- pumping stations
- increased costs
- public health problems

5. Management

It is stated by Rosenzweig et al., (2007) that some climate change impacts on hydrological processes have already been observed and further changes are projected. Thus, mitigation measures are needed to be taken as well as adaptation to climate change is necessary. Below common adaptation measures are referred.

5.1 Efficient and effective use of water

When water demand increases and water availability decreases one of the most widely used solution towards decreasing water more consumption is using the available water effectively and more efficiently. Water demand management considers measures to improve efficiency of water use.

Among sectors, agriculture is the leading sector in terms of water consumption. Climate change is expected to directly and indirectly increase demand for agricultural irrigation. Adaptation measures to climate change in the agricultural sector include changes of agrotechnical practices (e.g., use of crop rotation, advancing sowing dates) and introduction of new cultivars (heat-wave- and drought-tolerant crops). Soil moisture should also be conserved (e.g. through mulching). Besides, timing and frequency of irrigation need to be optimized considering the crop requirements. This is important for reducing irrigation return flows which in turn deteriorates the quality of the receiving water.

Industrial water consumption may also be reduced by developing less water using technologies as well as in-plant control measures. Clean technologies should be preferred due to their optimized water consumption.

Domestic uses may be decreased by encouraging public to use water-saving home appliances, through water pricing, legal sanctions and raising public awareness. In the big cities in developing countries, water loss through leakages in the water distribution lines constitutes a significant amount. Thus, it must be aimed to decrease water losses below 10% by renewing the old pipelines.

5.2 Alternative water resources

In cases of severe water scarcity, reducing water consumption may not be a remedy and thus searching for alternative water resources may become crucial.

Desalination of seawater or brackish water is considered as an important option of producing freshwater. Recent technologies and advances in the sector allow producing freshwater at affordable costs when higher amounts are intended. However, water withdrawals for desalination purposes may alter the well-being the related ecosystem. Thus, it is necessary to take into account the environmental impacts that might occur due to the planned water withdrawals. Also brine that is produced in desalination process should be properly disposed.

Another alternative source is reuse of treated wastewater. It is known that treated wastewater may be used for irrigating green land, parks and gardens in big cities. It can also be used for irrigating agricultural land if the national standards are satisfied in terms of irrigation water quality. Industries can also utilize treated wastewater in their processes providing that the quality of the goods manufactured remain unchanged (Asano et al., 2007).

Aquifers can be thought as storages where water loss through evaporation is relatively low. Thus, recharge of groundwater aquifers with treated wastewater is applied in different countries such as Israel and Spain (Esteban & Miguel, 2008; Salgot, 2008). However, it should be underlined that advanced treatment is necessary to protect the aquifers from pollution.

Another option is ecological sanitation (ECOSAN) practices. By such applications generated wastewater is separated into three streams at the source (yellow water, grey water and black water) that may be recycled after applying simpler treatment techniques. For example treated grey water may be used for irrigation and for recharge of aquifers. However, in most of the cases existing and usually old fashioned infrastructure is not compatible with ECOSAN. Reuse and/or disposal of each wastewater stream should be carefully planned. For example, yellow water could be used instead of fertilizer but if not desalinated salinity in human urine can harm the crops and the soil (Beler-Baykal et al., 2011).

5.3 Inter-basin water transfer

Szwed et al., (2010) states that water transfer from an area of relative abundance to an area of scarcity may smooth the spatial water variability. It is applied in many arid and semi-arid regions. Three points are important in water transfer: Feasibility regarding engineering works, hydrological conditions and ecological conditions of the basins. Pre-screening in terms of engineering works focus on costs of the work and on the length of water transmission lines. Besides, head loss/energy consumption of the pumps, natural and artificial barriers along the pipeline and its vicinity are also important factors to be considered.

Inter-basin water transfer depends on the availability of excess water from where the water is withdrawn. Especially the climatic conditions of both basins gain importance. If both basins face drought conditions in the same years, water transfer among them should not be considered as a feasible option. Both basins must be surveyed prior to realization of water transfer regarding their hydrological characteristics. During these surveys, long-term hydrological data must be analyzed. Watershed ecology is equally important. Socio-cultural conditions and economical characteristics should also be taken into consideration and sustainability should be kept in mind during water withdrawals. There are still contradicting opinions on inter-basin water transfer. They argue that inter-basin water transfer may no longer be viable in a future with climate change, as climate change stresses almost every source of freshwater. Also taking more water from the natural system has biological, ethical, and increasingly legal limitations (Karakaya and Gonenc, 2005; Hall et al., 2008). Consequently, it is advised to consider inter-basin water transfer to be considered as the last solution to water scarcity.

5.4 Maintaining the sustainability of watershed ecosystems

Natural aquatic ecosystems are among the important water resources supporting life. It is very important to maintain the ecological flows of these systems. Ecological flows are usually determined by some practical statistical approaches, assumptions and methods supported by scientific research conducted at site. During these studies it must be considered that aquatic ecosystems are in interaction with terrestrial ecosystems. Thus, any change in aquatic or terrestrial ecosystem will have an effect on the other one. For example, the decrease in surface water levels will affect the groundwater levels and dependent ecosystems. Evapotranspitation increase due to climate change has also effect on the decrease of groundwater levels. As this condition may lead to change in the vegetation cover which in turn lead to habitat change regulation of groundwater use becomes more important. As renewal of groundwater lasts long, planning must be done prior to facing water scarcity.

5.5 Revision of infrastructure

Changes in water quality in water resources will necessitate revision of existing water-related infrastructure. New components of the infrastructure should be designed according to possible extremes that would occur. Resilience of the infrastructure should also be enhanced.

Water treatment systems must be designed and operated according to drinking water standards under raw water inflow with varying water quality. On the other hand, different wastewater treatment options that seem not feasible today may be available in a world with higher annual average temperature. One example is the upflow anaerobic sludge blanket (UASB) process that is used to treat municipal wastewater in warmer countries such as India currently. Such technologies that are more cost-efficient could be applied in higher latitudes once further meteorological conditions change due to climate change.

6. References

Arnold, J.G., Srinivasan, R., Muttiah, R.S. and Williams, J.R., 1998. Large area hydrologic modeling and assessment part I : Model development. *J. American Water Resour. Assoc.*, 34(1), 73.

Asano, T., Burton, F.L., Leverenz, H.L., Tsuchihasti, R. and Tchobanoglous, G. 2007. Water reuse: Issues, technologies, and application. Metcalf &Eddy/AECOM, McGraw Hill, USA.

Barnett, T.P., Adam, J.C., Lettenmaier, D.P. 2005. Potential impacts of a warming climate on water availability in snow-dominated regions, *Nature*, 438, 303–309.

Beler-Baykal, B., Allar, A.D., Bayram, S. 2011. Nitrogen recovery from source separated human urine using clinoptilolite and preliminary results of its use as fertilizer, Water Science and Technology, 63(4), 811-817.

Bicknell, B.R., J.C. Imhoff, J.L. Kittle, Jr., Jobes, T.H. and A.S. Donigian, Jr., 2001. Hydrologic Simulation Program - FORTRAN (HSPF), user's manual for version 12.0, USEPA, Athens, GA, 30605.

Chapra, S., 2008. Surface Water Quality Modeling, Waveland Press.

Collins, M.; Tett, S.F.B., and Cooper, C. 2001. The internal climate variability of HadCM3, a version of the Hadley Centre coupled model without flux adjustments. *Climate Dynamics* 17: 61–81

Döll, P., Flörke, M. 2005. Global-Scale Estimation of Diffuse Groundwater Recharge. Frankfurt Hydrology Paper 03, Institute of Physical Geography, Frankfurt University, Frankfurt am Main, Germany.

Esteban, I.R. and Miguel, E.O. 2008. Present and future wastewater reuse in Spain. *Desalination*, 218, 105-119.

Forbes, K.A., Kienzle, S.W., Coburn, C.A., Byrne, J.M., Rasmussen, J. 2011. Simulating the hydrological response to predicted climate change on a watershed in southern Alberta, Canada, *Climatic Change*, 105, 555–576.

Gassman, P.W., Reyes M.R., Gren C.H. and Arnold, J.G. 2007. The Soil and Water Assessment Tool: Historical Development, Applications and Future Research Directions. *Transactions of ASABE*, 50(4), 121.

Gordon, C.; Cooper, C., Senior, C.A., Banks, H., Gregory, J.M., Johns, T.C., Mitchell, J.F.B., and Wood, R.A. 2000. The simulation of SST, sea ice extents and ocean heat transports in a version of the Hadley Centre coupled model without flux adjustments. *Climate Dynamics* 16 (2–3): 147–168.

Hall, N.D., Stuntz, B.B., and Abrams, R.H. 2008. Climate Change and Freshwater Resources, *Natural Resources & Environment*, 22(3), 32-35.

Im, S., Brannan, K., Mostaghimi, S. and Cho. A. J. (2003). Comparison of SWAT and HSPF Models for Simulating Hydrologic and Water Quality Responses from an Urbanizing Watershed, *Proceedings of the 2003 ASAE Annual International Meeting*, Nevada, USA, Paper No: 032175

IPCC, 2007. Climate Change 2007: The Physical Science Basis. Contribution of Working Group I to the Fourth Assessment Report of the Intergovernmental Panel on Climate Change [Solomon, S., D. Qin, M. Manning, Z. Chen, M. Marquis, K.B. Averyt, M.Tignor and H.L. Miller (eds.)]. Cambridge University Press, Cambridge, United Kingdom and New York, NY, USA.

Karakaya, N. and Gonenc, I.E. 2005. Interbasin water transfer, Proceedings of 2nd National Water Resources Engineering Symposium, Izmir, Turkey (in Turkish).

Kavvas, M.L., Chen, Z.Q., Anderson, M.L., Ohara, N., Yoon, J.Y. 2009. A Hydroclimate Model of the Tigris-Euphrates Watershed for the Study of Water Balances, in Proceedings of Conference on Transboundary Waters and Turkey, Editors: Mehmet Karpuzcu, Mirat D. Gürol, Senem Bayar, Istanbul.

Kundzewicz, Z.W., Mata, L.J., Arnell, N.W., Döll, P., Jimenez, B., Miller, K., Oki, T., Şen, Z. and Shiklomanov, I. 2008. The implications of projected climate change for freshwater resources and their management, *Hydrological Sciences Journal*, 53, 1, 3-10.

Mirza, M.M.Q., Warrick, R.A. and Ericksen, N.J. 2003. The implications of climate change on floods of the Ganges, Brahmaputra and Meghna rivers in Bangladesh. *Climatic Change*, 57, 287–318.

Nasr, A., M. Bruen, P. Jordan, R.Moles, G. Kiely, P. Byrne and B. O'Regan, 2004. Physically-based, distributed, catchment modelling for estimating sediment and phosphorus loads to rivers and lakes: issues of model complexity, spatial and temporal scales and data requirements. *National Hydrology Seminar-2004*, Ireland

Pope, V.D.; Gallani, M.L., Rowntree, P.R., and Stratton, R.A., 2000. The impact of new physical parameterizations in the Hadley Centre climate model — HadAM3. *Climate Dynamics* 16 (2–3): 123–146.

Rosenzweig, C., G. Casassa, D.J. Karoly, A. Imeson, C. Liu, A. Menzel, S. Rawlins, T.L. Root, B. Seguin, P. Tryjanowski, 2007. Assessment of observed changes and responses in natural and managed systems. Climate Change 2007: Impacts, Adaptation and Vulnerability. Contribution of Working Group II to the Fourth Assessment Report of the Intergovernmental Panel on Climate Change, M.L. Parry, O.F. Canziani, J.P. Palutikof, P.J. van der Linden and C.E. Hanson, Eds., Cambridge University Press, Cambridge, UK, 79-131.

Saleh, A., Du., B. 2004. Evaluation of SWAT and HSPF within basins program for the upper north bosque river watershed in central Texas. *Transactions of ASAE*, 47, 1039

Salgot, M. 2008. Water reclamation, recycling and reuse: Implementation issues. *Desalination*, 218, 190-197.

Schnoor, 1996. Environmental Modeling, Wiley & Sons

Singh, J., Knapp, V.H, and Demissie, M. (2004). *Hydrologic Modeling of the Iroquois River Watershed Using HSPF and SWAT*. Illinois State Water Survey Contract Report 2004-08.

Simonovich, S. 2009 Managing Water Resources Unesco Publishing

Szwed, M., Karg, G., Pińskwar, I., Radziejewski, M., Graczyk, D., Kędziora, A., Kundzewicz, Z.W., 2010. Climate change and its effect on agriculture, water resources and human health sectors in Poland, *Nat. Hazards Earth Syst. Sci.*, 10, 1725–1737.

Therrien, R., McLaren, R.G., Sudicky, E.A., Panday, S.M. (2010). *HydroGeoSphere A Three-dimensional Numerical Model Describing Fully-integrated Subsurface and Surface Flow and Solute Transport*. Groundwater Simulations Group

Van Dam, J.C., P. Groenendijk, R.F.A. Hendriks and J.G. Kroes, 2008. Advances of modeling water flow in variably saturated soils with SWAP. *Vadose Zone J.*, 7(2), p. 640.